Biological Diversity

Biological Diversity

Origin, Evolution and Conservation

Editors

Arun Kumar Sharma
Debal Ray
Soumyendra Nath Ghosh

West Bengal Biodiversity Board
Kolkata

Viva Books

New Delhi | Mumbai | Chennai | Kolkata | Bengaluru | Hyderabad | Kochi | Guwahati

Publisher's note

First Published 2012

Viva Books Private Limited

- 4737/23, Ansari Road, Daryaganj, New Delhi 110 002
  Tel. 42242200, 23258325, 23283121, Email: vivadelhi@vivagroupindia.net
- 76, Service Industries, Shirvane, Sector 1, Nerul, Navi Mumbai 400 706
  Tel. 27721273, 27721274, Email: vivamumbai@vivagroupindia.net
- Jamals Fazal Chambers, 26 Greams Road, Chennai 600 006
  Tel. 28294241, 28290304, Email: vivachennai@vivagroupindia.net
- B-103, Jindal Towers, 21/1A/3 Darga Road, Kolkata 700 017
  Tel. 22816713, Email: vivakolkata@vivagroupindia.net
- 7, GF, Sovereign Park Aptts., 56-58, K. R. Road, Basavanagudi, Bengaluru 560 004
  Tel. 26607409, Email: vivabangalore@vivagroupindia.net
- 101-102, Moghal Marc Apartments, 3-4-637 to 641, Narayanguda,
  Hyderabad 500 029, Tel. 27564481, Email: vivahyderabad@vivagroupindia.net
- First Floor, Beevi Towers, SRM Road, Kaloor, Kochi 682 018
  Tel. 0484-2403055, 2403056, Email: vivakochi@vivagroupindia.net
- 232, GNB Road, Beside UCO Bank, Silpukhuri, Guwahati 781 003
  Tel. 0361-2666386, Email: vivaguwahati@vivagroupindia.net

www.vivagroupindia.com

ISBN: 978-81-309-2113-6

Published by Vinod Vasishtha for Viva Books Private Limited, 4737/23 Ansari Road, Daryaganj, New Delhi 110 002.

Printed & bound by Syndicate Binders, New Delhi.

G0000/G0300

Contents

Section – III
Indian Scenario

Section – IV
Conservation Initiatives

Preface

The myriads of life forms, from unicellular microbes to complex eukaryotes, form components of the biosphere in which all living beings live in mutual interdependence. This vast assemblage in this planet is the outcome of successive evolution over a period of four billion years. In the evolutionary history of this earth, human beings though comparatively recent in origin, have led to manipulation and unjudicious utilization of this pool of life to meet their own needs and satisfaction. This diversity of life, as no two individuals are alike, is the essence of survival of human beings in this planet. It helps in solving the unresolved questions on evolution, throws light on the life process before and after the great extinctions, provides clue to the understanding of the selection process in nature, besides being a tool to improve the quality of our life.

Being a party to the Convention on Biological Diversity (CBD), India has taken several novel and innovative steps to conserve biological diversity and consolidate its sovereign right over them. Out of these, the most important is to enact the Biological Diversity Act in the year 2002 and the Biological Diversity Rules in the year 2004 aimed at achieving three major objectives, such as (i) conservation of biodiversity, (ii) sustainable use of biodiversity and (iii) equitable sharing of benefits arising out of the use of biodiversity and associated traditional knowledge. For implementation of the act, National Biodiversity Authority and State Biodiversity Boards have come into existence. The proactive steps towards conservation of biodiversity taken by India have been acknowledged worldwide.

This book, in a major departure from similar books, deals with origin, principles and processes on biodiversity. Thus, this book will help the readers to understand the basic issues and the emerging areas of research. Although this is a multi-authored book, the chapters have been so chosen, so as to maintain continuity of discussion. The first section of the book deals with the origin and evolution of biodiversity. The second one deals with the drivers of loss of biodiversity, along with an article on mass extinctions. The third section deals with the unique richness of biodiversity in India, from microbes to man. The fourth and last section highlights a few conservation initiatives taken by India, including one on global initiative.

The authors of all the articles are well-known for their eminence in the respective fields. The editors gratefully acknowledge their contributions in this volume.

It is very unfortunate that Dr. L.K. Banerjee, the celebrated expert of mangroves, passed away shortly after completing the manuscript.

The editors convey their thanks to the employees of the West Bengal Biodiversity Board for their assistance in this endeavour. The editors also thank Mr. S. Sreetilak, General Manager Publishing and the Editorial Board of Viva Books for their interest.

Section – I

Origin and Evolution

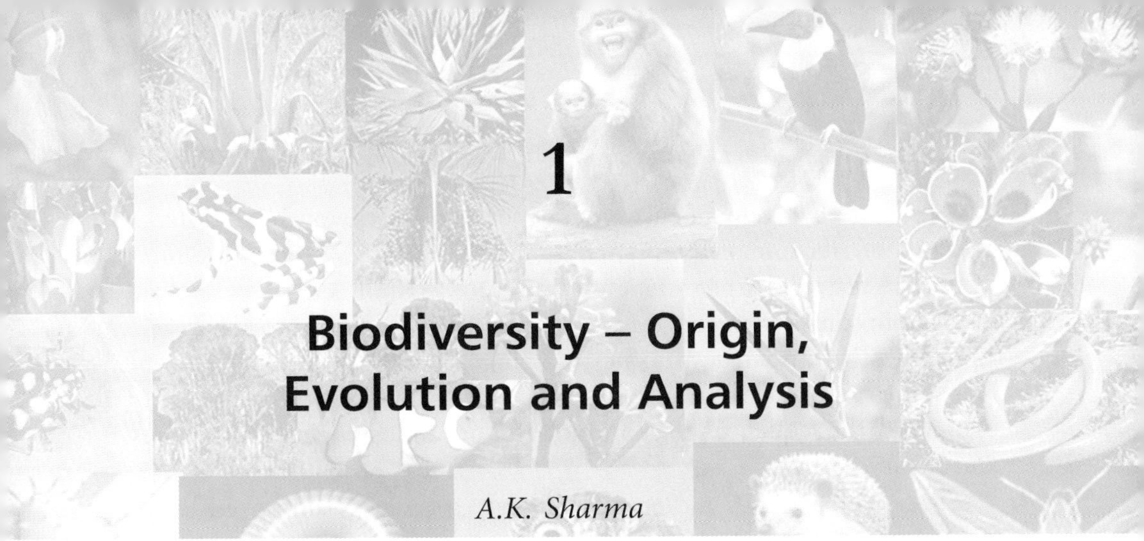

1

Biodiversity – Origin, Evolution and Analysis

A.K. Sharma

BASICS OF BIODIVERSITY

The International Year of Biodiversity is being observed throughout the world with due solemnity and seriousness it deserves. Biodiversity signifies the sum total of all the genomes in this planet and is unique in representing life in this entire universe. It would, therefore, be befitting to recapitulate the path it has traversed since the beginning of life at the successive phases of evolution. Depending on the zone of occurrence, as well as the status and nature of manifestation, the terms *ecological* or *species diversity* are often applied. But in reality, true and stable biodiversity must entail genetic difference and that comes under the category of *genetic diversity*. The basic changes underlying stable diversity are the changes in genetic architecture. In simple terms, it involves genic changes or chromosomal alterations – numerical or structural. The progress of evolution from simplicity to complexity is initially triggered at the gene level leading to a series of changes reflected phenotypically as new types, varieties, or species. The steps in evolution of complexity in biodiversity are dealt with hereafter, followed by the different parameters of the measure of plant biodiversity with specific examples.

ORIGIN OF PRIMITIVE GENOME – AND INITIATION OF DIVERSITY

It is recorded (Hoffman, 1992) that the unicellular organisms which can grow in extreme heat and metabolize sulphur – the *eocytes* – are the closest surviving relatives of eukaryotes, higher organisms – the origin of which could be traced to more than 2 billion years ago.

Study of phylogeny of eukaryota depends primarily on small or large subunits of ribosomal RNA (rRNA), although 5s rRNA and a number of protein sequences provide additional proof (Knoll 1992, Sogin et al. 1987). Ribosomes from eocytes are structurally more similar to eukaryotic ribosomes in ultrastructural pattern than they are to other bacterial ribosomes. Moreover, in one of the proteins EF-Tu, which is highly conserved, the sequences of 11 amino acids are almost identical in eocytes and eukaryotes.

The ancient RNA molecule (Gilbert, 1986) was supposed to be stable, symmetrical, 50–100 residues long, rich in guanine and cytosine residues (Dorit et al. 1998 vide review by Sharma, 1995). It possibly had a homogeneous codon pattern with a comma-free read-off. The primitive translational machinery considered likely was a tRNA-like structure, though not so complex. The transfer RNA has thus been assigned a unique role in the pre-protein RNA world (Robertson, 1994) while biodiversity was yet to be perceived.

It is also claimed that protamine is a genetic fossil (Mori, 1993) of the primordial gene products in the ancient state of diversity. It is present in the nuclei of the sperm of vertebrates and is a highly DNA-binding basic protein, rich in arginine. The basic unit might ultimately have been a repetition of CGCAGG hexanucleotides, which served as the building block of the primitive gene in the biosphere while duplication and mutation resulted in a sequence similar to the protamine gene. It is suggested that the high GC content of the 120 nucleotide sequences of the protamine, makes it chemically stable necessary for survival in the primitive state with high fidelity for replication.

GENES AND BIODIVERSITY

Evidences indicate that genes, the basic material responsible for diversity, were constructed not initially from DNA but from a surprisingly small number of genetic building blocks of RNA which existed even 3–3.5 billion years ago in ancient strata (Cavalier Smith, 1991). Much later, the DNA appeared in evolution of life. Several thousands of these blocks, the exons, underwent shuffling and recombination in new ways during evolution and it is estimated that up to 7000 exons might have been necessary to construct all the proteins in the biosphere.

The process was at least at certain early stages of evolution, possibly facilitated by *introns*, the so-called non-essential sequences, the role of which has been much debated. The two contrary views (Cavalier Smith, 1985) can be summed up by stating that *introns* were either derived from transposable or parasitic elements inserted between unsplit genes or they are ancient, as old as the pre-cellular life, and the genes were split from the beginning.

The antiquity of introns is indicated in the topographical evidence that introns are found in the identical location on the genomes of distantly-related organisms such as corn, chicken and fungus (Palmer and Logsdon, 1991). This fact is an index that such introns in diverse organisms must have been inherited from a common ancestor in evolution of diversity.

High degrees of repeats (Crick, 1971) are also found in intron sequences. Such sequences have the property of amplification and mobility (Lambourtz and Belfort, 1993). Their presence in identical sequences in different loci suggests the property of dispersion as well. These sequences with the property of movement contribute greatly to mutation and diversity and termed "Dynamic" as well (Sharma, 1983; Sharma and Sharma, 1984).

These repeated sequences, present in introns, often called non-essential, and 'selfish' (Orgel and Crick, 1980), are present universally in major segments of chromosomes and are highly conserved in a wide diversity of species across the biosphere. They occupy a significant portion of the genes, are present in multiple copies and are capable of insertion at different sites. Introns have been found in all major groups of organisms. Though considered as non-essential, these sequences provide valuable clues to the study of evolution of biodiversity.

With the progress in evolution, (vide de Duve, 2007; Rivera and Lake, 2004) in addition to the development of nuclear membrane, nucleus and prokaryotic origin of organelles through endosymbiosis – such as plastids and mitochondria, certain events often appear which are rather striking and stand as landmark in the hierarchy of evolution. Such a rare event occurred at the transition from Precambrian to Cambrian, when there was a sudden increase in the diverse flora and fauna, approximately between 1.2 to 2 billion years back. In other words, it was a *spurt* in biodiversity. This era was marked by three important biological processes, namely *acquisition of chloroplastids with capacity to utilize so far energy for photosynthesis*, *multicellularity* and *sexual reproduction*. The sequences, or more precisely, regulatory systems, which initially triggered multicellularity and sexual reproduction are still not precisely known. From a biological standpoint, such events leading to sudden origin of widely diverse forms across the globe, surpassed in significance, plate movements, drifts in the continents, or even periodic ice ages.

DIVERSIFICATION OF LAND PLANTS

The next step in the evolution of biodiversity is the origin of land plant with progressive complexity of diverse forms. It is the widely accepted concept that land plants originated from aquatic ancestors which were the inhabitants of intertidal marks, where sexual reproduction occurred mostly during high tide. It has also been stated that certain forms, in order to escape competition, established themselves on land, where sexual reproduction was possible only during rains or copious dews. Plants, therefore, had to develop the alternative method of increase of individuals by the production of spores, through sporophytes by elongation of a post zygotic phase. Later, the evidences indicated that green algal group, the Charophyceae, might be the progenitors of land plants. In any case, the origin of land plants through algal ancestors, though accepted, suffered from lack of evidence, since the exact connecting link between the aquatic and land plants could not be found. Lately, intron sequences from chloroplastids of algae like *Coleochaete* and *Chara* (Manhert and Palmer, 1990) and of *Lamprothamnion* have been homologized with those of *Marchantia, the Bryophyte*. In fact, introns were initially found in the chloroplasts of land plants. The later discovery of such introns in green algae brings the two different groups together and confirms the lineage. The long-awaited gap in the evolution of biodiversity from algae to land plants and origin of land habit could be bridged through molecular evidences.

DIVERSIFICATION OF FLOWERING PLANTS

The succeeding step in plant evolution is the gradual origin of terrestrial flowering plants. Different ideas have been proposed so far for origin of angiosperms ranging from Benettlitales of Gymnosperms to *Magnoliaceous* forms. Intron sequences of *Magnoliaceous* chloroplastids as also the megagametophytes provided clue to their ancestry. Even the human ancestry has been deduced from mitochondrial intron sequences. However, in recent years, clear embryological and organellar genomic evidences have been obtained to indicate that within the 15 million years of Angiosperms, three major lineages of diversity developed in flowering plants. All analyses suggest that the monotypic New Caledonian taxon *Amborella trichopoda*, of Clusiaceae, Austrabaileyaceae, and Nymphaeales are the most ancient lineages of flowering plants (Friedman, 2006). This complex, sister to other angiosperms hold key to the earliest evolutionary history of biodiversity of flowering plants.

CHROMOSOMES, ECOLOGY AND BIODIVERSITY

In plants, gene mutation, chromosome rearrangements, polyploidy and aneuploidy have undoubtedly played an important role in evolution and origin of biodiversity. There are cases such as of *Crepis*, *Anemone*, etc. where the different species can be identified on the basis of their karyotype and chromosome number. In *Lilium*, on the other hand, almost all species are characterized by a common chromosome number of $2n$=24, but the species differ only in minute details of the chromosomes. Thus biodiversity is associated with numerical differences of chromosomes in certain genera whereas by structural differences in certain others.

A very good example of ecological adaptation of certain diverse chromosome races is provided by the family Liliaceae. In the genus *Smilax* both diploid and polyploid species occur in the low altitude, whereas the alpine species has a constant number of $2n$=32 chromosomes (Sen, 1975). In *Chlorophytum* too, polyploidy ($2n$=42, 56), is common in Himalayan representatives such as *C. arundinacea* Baker and *C. nepalense* Baker (Sharma and Raju, 1967). The Eastern Himalayan population of *Kniphofia aloides* Meench ($2n$=14) of the lower altitude is distinct in their karyotype from the Western Himalayan counterparts of the same altitude – a clear reflection of the climatic influence on biodiversity.

In the Himalayas, there are certain families or genera where a correlation could be established between biodiversity, ecological adaptation and genetic architecture. A distinct example of this correlation is manifested in the species of *Rhododendron* – where species differ in different attitudes in the Eastern Himalayas along with the different degrees of polyploidy. A rare example of species growing strictly in the alpine zones showing correlation with adaptation has been recorded in the species *Clintonia alpina* Kunth of Liliaceae which is distributed throughout the Himalayas, ranging from Western to Eastern border. This species with strictly alpine distribution is not found below 10,000ft. All the populations have a constant karyotype and a common chromosome number of $2n$=28. Such a wide distribution of the species in the arctic conditions of the Himalayas, coupled with the constancy in karyotype, suggests that a rigid selection pressure does not allow the survival of any species variant, thus maintaining stability of identity. Parallel behaviour has been recorded as well as in a few species of orchids grown in the alpine zone, as in the genus *Polygonatum* of the tribe Polygonateae of Liliaceae.

In Araceae, there are three species of *Arisaema* which grow in the alpine zone between 9–12,000ft. namely *A. nepenthoides* Mart., *A. sikkimense* Stapf. and *A. wallichianum* Hock. All of them contain B chromosomes in the complement which have not been recorded in any of the Himalayan species of this genus. The B chromosomes have also been shown to have a high content of repeated DNA exerting certain control over crossing over and other vital functions. Their very presence in the arctic species indicates certain selective advantages for their survival – a reflection of the influence of B chromosomes in the stability of diversity at certain ecological regime.

DNA SEQUENCES AND BIODIVERSITY

Remarkable impact of the study of DNA sequence is recorded in the area of biodiversity.

Polymorphism in DNA sequences contributing to evolution and biodiversity is well documented. In later years, the role of single nucleotide polymorphism (SNP) has been repeatedly reiterated in diversification.

The analysis of *DNA fingerprinting* in documentation of biodiversity is well established. It is based on the premise that each individual has a unique DNA pattern. Polymorphism in genome is detected by variable number of repeat sequences arranged in tandem. Leaving aside mere genetic diversity, molecular hybridization detected through synthetic oligonucleotide probes complementary to short tandem repeats, may lead to trace the pathways and trends in evolution.

Short Tandem Repeats and Transposable Sequences

In plants, short tandem repeats are recorded to be common in the genome, of which bi- and tetranucleotides are most abundant. For example, the species specific nature is indicated by one in every 11 Kb in *Petunia*, one in every 25 Kb in *Brassica*, whereas it is one in every 42, 58 and 156 Kb in *Arabidopsis*, *Zea* and *Hordeum* respectively. In rice, there is a variety of STRs, whereas only one type is found in *Hordeum*. In dicots, STRs are three times that in monocots. In algae, all STRs contain GC base pairs whereas it is 50% in monocots and 15% in dicots. The nature of the sequence in distinct assemblages, their frequencies in broader taxonomic groups, and differences in different plant groups indicate their role as landmarks in evolution of biodiversity and determination of phylogenetic status.

The study of genomics vis-à-vis DNA sequences has further revealed the remarkable role of transposable sequences in evolution of biodiversity. Since the discovery of mobile sequences long back by Barbara McClintock and their regular occurrence in several plants as revealed in recent years by different authors, their major influence in evolution is becoming evident. Even if they cannot be regarded as a part of the programmed gene action, their susceptibility to stress makes them candidates for species evolution. The magnitude of its role in evolution is demonstrated in the classical example of the origin of recessive wrinkled seeded character in pea. It has been established by the John Innes group that round seeded character in pea is due to the presence of formation of insoluble branched starch by the enzyme SEB. In wrinkled recessive, this enzyme is absent and the starch is soluble and loses water on drying, leading to wrinkled nature of seed on maturity. Later it was noted that the coding sequence for SEB is 3.4 Kb long whereas in wrinkled seed, the sequences are 4.2 Kb long. This absence of SEB activity is entirely due to a 0.8 Kb insert in SEB sequence making it non-functional leading to recessive character. Thus the classical recessive gene in pea is nothing but due to the activity of a mobile sequence insertion. This reported discovery has opened up the possibility that the transposable sequence might have played a much greater role in evolution and origin of diversity than could be visualized earlier.

DNA Barcode and Species Diversity

Lately, an international program has been launched by the major Natural History Museums, Herbaria, Gardens and Institutions in the West (vide Phil. Trans. Roy. Soc. Lond. : B 360: 1807–1980, 2005) for the preparation of "Barcode of Life Initiative" with the aim of producing an encyclopaedia of life covering approx 10 million species on earth (Savolainen et al. 2005). DNA barcoding is a diagnostic strategy through which individual species can be identified through short DNA fragments which provides a strict identity to the genome.

Barcoding is based on the premise that some short pieces of DNA sequences can be found which show little variation within a species as compared to variation or differences between species. One such DNA fragment, the mitochondrial cytochrome oxidase gene "*cox-1*" subunit and "*rbcl*" in plant gene are extensively used in barcoding studies specially in animals and plants respectively. In addition, nuclear ribosomal genes too have been used in some animals as well.

The criteria for the choice of DNA sequences as barcode markers are principally, sufficient variability to provide identity to the taxon on the one hand, and linkage with allied species on the other. In addition, it must have minimum variability within a taxon as well as PCR stability. It is treated as a shortened version of the entire genome.

However, eukaryotic groups in general and land plants in particular, pose certain problems to barcoding in view of the polyploidization and hybridization extensively occurring during evolution (vide Pennisi, 2007). However, some of the plastid markers as well ITS of large nuclear ribosomes have been found to be useful barcodes. Barcode of life data system (BOLD) has been proposed and applied in animals but in order to be universal, it is to include multiple DNA regimes.

In order to meet the complexity of land plants with intense hybridization and polyploidy, the barcode has to be multilocus (MBC), as single locus may not be able to detect introgression and hybridization. More or less, one gene and two spacers have been found to be of much use. However, a portion of the gene "*inatK*" has been used successfully to demarcate African species of Podostemaceae (cf. Laura Kelly – Kew. Org. – 2010). Multilocus barcodes may serve as a tool not only for species identification but would also aid in determination of phylogeny and trend of evolution.

Barcoding technology, has a strong bearing on the control of the access to and benefit sharing of our bioresources (Einavarzala, 2010) both for commercial and non-commercial purposes. It serves as an aid to the owners of specific bioresources to monitor, conserve and enrich the resource.

EPILOGUE

The above resume is just a glimpse, though in no way complete, on the different facets of biodiversity – the only living component in this planet, if not the universe. However, one of the important aspects, and possibly the most crucial in the present scenario is the effect of climate change and green house effect on our bioresources, looming large over the horizon. This aspect has been dealt with in a different article in this volume by Prof. V. Meher-Homji. In fact, an analysis of the impact of global climate change on the entire biota and the measures of resistance to meet the challenge is an urgent task needing strong collaboration between scientists dealing with life sciences on the one hand and earth scientists and climatologists on the other. It is a serious challenge for this decade and decades to come.

References

Crick, F. (1971). General model for the chromosomes of higher organisms. *Nature*, 234: 25–27.

Cavalier Smith, T. (1991a). Intron phylogeny: A new hypothesis. *Trends in Genetics*, 7: 145–148.

Cavalier Smith, T. (1985). Selfish DNA and the origin of introns. *Nature*, 351: 110.

de Duve, Christian. (2007). The origin of eukaryotes: A reappraisal. *Nature Review Genetics,* 3: 395–403.

Dorit, R.L., Schoenbach and Gilbert, W. (1990). How big is the universe of exons? *Science*, 250: 1377–1382.

Einavarzala, H. (2010). DNA barcoding: Access to biodiversity and benefit-sharing policy issues in the Indian context. *Curr. Sci.*, 99: 594–600.

Friedman, W.E. (2006). Embryological evidence for developmental lability during early angiosperm evolution. *Nature*, 441: 337–340.

Gilbert, W. (1986). Origin of life: The RNA world. *Nature*, 319: 618.

Hebert, P.D.N., Gywinska, A., Ball, S.L. and de Waard, J.R. (2003). Biological identifications through DNA barcodes. *Proc. Roy. Soc.* B, 270: 313–321.

Hoffman, M. (1992). Researchers find organism they can really relate to. *Science*, 257 : 32 [doi: 10.1126/science.1621091].

Knoll, A.H. (1992). The early evolution of eukaryotes: A geological perspective. *Science*, 256: 622–627.

Kumar, V. (1960). Cytological studies in the Eastern Himalaya species of *Smilacina. Proc. 47th Ind. Sci. Congr.*, 360–361.

Lambouritz, A.M. and Belfort, M. (1993). Introns as mobile genetic elements. *Ann. Eur. Biochem.*, 62: 587–622.

Manhart, J.E. and Palmer, J.D. (1990). The gain of two chloroplast tRNA introns marks the green algal ancestors of land plants. *Nature*, 345: 268–270.

Marshall, E. (2005). Will DNA bar codes breathe life into classification? *Science*, 307: 1037.

Moore, G. (1995). Cereal genome evolution: Pastoral pursuits with 'Lego' genomes. *Curr. Opin. Genet. Development*, 5: 717–724.

Mori, N. (1993). A genetic fossil: Protamine gene as a primordial gene. *Naturwiss.*, 80: 222–224.

Orgel, L. and Crick, F.H.C. (1980). Selfish DNA: The ultimate parasite. *Nature*, 204: 604–607.

Palmer, D. and Logsdon, J.(Jr.) (1991). The recent origins of introns. *Curr. Opin. Genet. Dev.*, 1: 470.

Pennisi, E. (2007). Wanted: A barcode for plants. *Science*, 318: 189–191.

Rivera, M.C. and Lake, J.A. (2004). The ring of life provides evidence for a genome fusion origin of eukaryotes. *Nature,* 431: 152–155.

Robertson, H.D. (1994). Life before DNA. *Science*, 264: 1479–1480.

Savolainen, V., Cowan, R.S., Rodocick, A.K. and Lane, R. (2005). Towards writing the encyclopaedia of life: An introduction to DNA barcoding. *Proc. Trans. Roy. Soc.* B, 360: 1805–1811.

Sen, S. (1975). Cytotaxonomy of liliales. *Feddes Report.*, 86: 255–305.

Sharma, A.K. (1983). Additional genetics materials in chromosomes. In: *Chromosome Conference II* (eds. Brandham PE and Bennett MD). George Allen and Unwin, London, 35–42.

Sharma, A.K. (1995). Antiquity of the plant kingdom and molecular world. *Curr. Sci.*, 68: 801–806.

Sharma, A.K. and Raju, T.D. (1967). Cytological analysis of six species of Chlorophylum. *Bull. Bot. Soc. Bengal,* 21(1): 37–46.

Sharma, A.K. and Sharma, A. (1984). Trends in chromosome evolution in the plant kingdom. In: *Chromosome in Evolution of Eukaryotic Groups* (eds. AKS and AS). CRC Press, Florida, 2: 227–239.

Sharma, A.K. and Sharma, A. (2006). Genomic sequences and species diversity. *Perspectives in Cytology and Genetics, XII.* (*Proc. Ind. Cytol. Genet.*). Das, R.K., Chatterjee, S. and Sadhukhan, G.C. (Eds.). pp. 1–4.

Sogin, M.L., Gunderson, J.H., Elwood, H.J., Alonso, R.A. and Peatlic, D.A. (1989). Phylogenetic meaning of the kingdom concept: An unusual ribosomal RNA from *Giardia lamblia. Science*, 243: 75–77.

2

Polyploidy in Relation to Plant Evolution and Speciation[1]

Prem P. Jauhar

"To create a little flower is the labour of ages *except by allopolyploidy.*"

Haldane 1958

INTRODUCTION

Polyploidy is recognized as a major force in plant evolution and speciation. Among the polyploids, allopolyploids comprising divergent genomes are preponderant in nature. Interspecific or intergeneric hybridization accompanied by chromosome doubling through functioning of unreduced gametes results in the formation of amphidiploids or allopolyploids (Harlan and de Wet, 1975). The allopolyploids have several sets of chromosomes from related species, which are capable of pairing among themselves. Therefore, establishment of a genetic control of chromosome pairing was essential to confer meiotic and reproductive stability on the allopolyploids. Thus, the allopolyploids enjoy the benefits of both stable polyploidy and hybridity and are adaptable to adverse environmental conditions. Many of our crop plants are of allopolyploid origin. Tetraploid durum wheat (*Triticum turgidum* L., $2n = 4x = 28$; AABB genomes) and hexaploid bread wheat (*Triticum aestivum* L., $2n = 6x = 42$; AABBDD genomes) present excellent models of evolution through allopolyploidy and various steps in their origin are represented in Figure 4.

To gain an insight into the mechanism of evolution of durum wheat, the present author produced its haploids (polyhaploids) ($2n = 2x = 14$, BA

[1] Mention of trade names or commercial products in this publication is solely to provide specific information and does not imply recommendation or endorsement by the US Department of Agriculture.

genomes) and studied them meiotically. The polyhaploids produced unreduced gametes through meiotic restitution. They also produced seed obviously through functioning of unreduced gametes. By reproducing these evolutionary steps we essentially repeated what happened in nature thousands of years ago. Clearly, the study of various aspects of chromosomes helps unravel several aspects of polyploidy. Various aspects of polyploidy in relation to evolution and the role of genetic control of chromosome pairing in stabilizing polyploidy and facilitating speciation will be reviewed and discussed in this article.

DIPLOIDY

A diploid species has only two sets of chromosomes, one derived from each of its parents. Figure 1a shows a mitotic metaphase cell with 14 somatic chromosomes. Each chromosome has its own homologous partner to pair with, resulting in the formation of 7 bivalents at metaphase during meiosis (Fig. 1b, 1c). Such diploid organisms show regular meiosis and

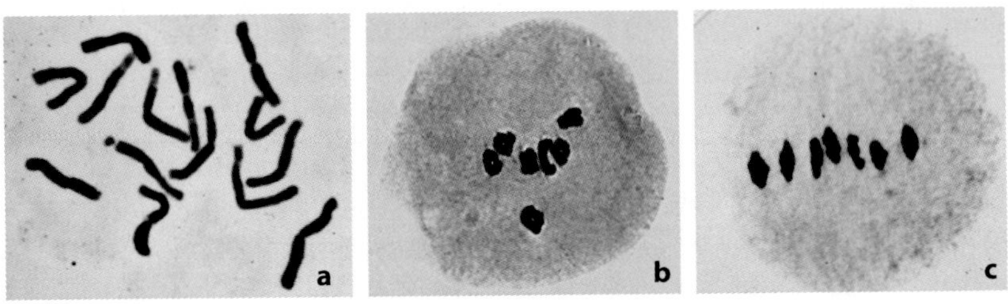

Fig. 1. Somatic and meiotic chromosomes of a diploid species ($2n = 14$). **(a)** 14 somatic chromosomes; **(b)** and **(c)** Pollen mother cells (pmcs) showing regular meiosis with 7 bivalents; note both ring and rod bivalents.

are reproductively stable and fertile. *Ornithogalum virens* Lindl. (family Liliaceae) with $2n = 6$ chromosomes is a favourable plant for cytological studies. With only three pairs of chromosomes, this plant forms three bivalents during various stages of meiosis I, viz., zygotene-pachytene (Fig. 2a), diplotene (Fig. 2b), diakinesis (Fig. 2c) and metaphase I (Fig. 2d and 2e), which separate as 3:3 at meiotic anaphase I (Fig. 2f) and then show equational division (Fig. 2g, 2h and 2i). Thus, this Liliaceous plant is very useful for teaching cytology even to beginners (Jauhar and Storey, 1982).

Several crop plants are supposedly diploid. However, genetic mapping studies have shown that several crop species traditionally considered diploid are in fact ancient polyploids and have a redundancy of chromosomes.

Some of these seemingly diploid species like the maize (*Zea mays* L.) could well have arisen as an "ancient polyploid" that can tolerate chromosomal deficiency (see Jauhar, 2006; Vasal et al. 2006). Thus, several diploid species could have had polyploid ancestry.

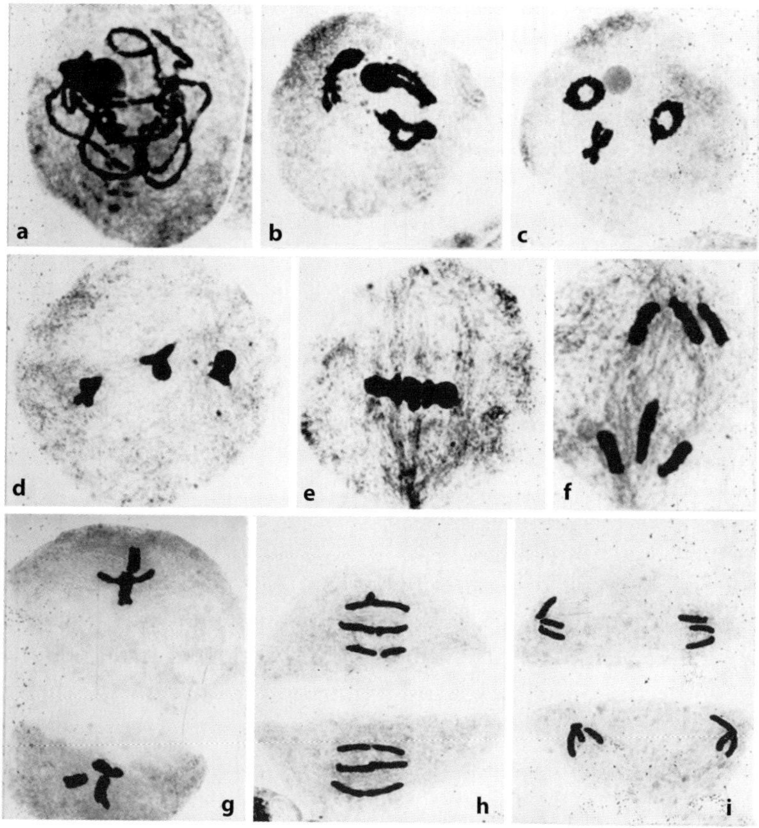

Fig. 2. Stages of meiosis in *Ornithogalum virense* ($2n = 6$), an ideal organism for teaching cytology. Two successive divisions, meiosis I (reductional division) and meiosis II (equational division), result in four cells with three chromosomes each. **(a)** zygotene-pachytene showing pairing of 6 chromosomes. **(b)** Diplotene showing 3 pairs or bivalents (3 II), one associated with the nucleolus; **(c)** Diakinesis with three condensed bivalents, one associated with the nucleolus; note terminalization of chiasmata. **(d)** Metaphase I with three bivalents; **(e)** Metaphase I with three bivalents arranged on the metaphase plate; note spindle fibers. **(f)** Late anaphase I showing separation of three chromosomes each to the poles culminating the completion of reductional division. **(g)** Meiotic metaphase II showing equational division of three chromosomes; note two chromatids of each chromosome. **(h)** Chromatids separating at anaphase II. **(i)** Telophase II showing the 3:3 distributions of chromatids to microspores comprising a tetrad that completes the equational division. (Source: Jauhar and Storey, 1982)

POLYPLOIDY: AN IMPORTANT EVOLUTIONARY FORCE

Polyploidy – a condition with multiple sets of chromosomes – is recognized as a major force in evolution and plant speciation. Polyploidy has played a major role in the evolution of all angiosperms (Soltis and Burleigh, 2009; Proost et al. 2011). Plants with more than two sets of chromosomes are called polyploids. A set of chromosomes is generally called a *genome* and, thus, polyploids have more than two genomes. Plants with more than two copies of the same genome are called *autopolyploids*, e.g., AAAA, BBBB, etc. The potato (*Solanum tuberosum* L.) and alfalfa (*Medicago sativa* L.) are good examples of successful autopolyploids. On the other hand, polyploids with more than two copies of different genomes from divergent sources are referred to as *allopolyploids*, e.g., AABB, AABBDD, etc., and are highly successful in nature. Several of our crop plants, e.g., durum wheat (AABB) and bread wheat (AABBDD) are very good examples of successful allopolyploids (see Jauhar, 2006).

Preponderance of Allopolyploids in Nature: An Important Evolutionary Force

Allopolyploidy is an important mechanism of speciation and is wide spread in flowering plants (Jenczewski and Alix, 2004; Wood et al. 2009; Chen, 2010; Doyle and Egan, 2010; Proost et al. 2011; Yang et al. 2011). Many of the crop plants are allopolyploids that have arisen through interspecific or intergeneric hybridization in conjunction with chromosome doubling, resulting from the production of unreduced gametes with the sporophytic chromosome number (see Jauhar, 2003; Rapp et al. 2009). Durum wheat, bread wheat, and oats (*Avena sativa* L., $2n = 6x = 42$; AACCDD genomes) are some of the most important allopolyploid crop plants (see Jauhar, 2006, 2007).

Mechanism of Allopolyploidization: Steps of Evolution of Polyploid Wheats

The mode of origin of some allopolyploids is fairly well understood. We will take the example of hexaploid bread wheat (AABBDD genomes), the most important cereal crop that sustains humankind. Tetraploid durum wheat evolved about half a million years ago and is the forerunner of bread wheat. Three wild grasses, *Triticum urartu* Tumanian ex Gandilyan (Fig. 3a) – the donor of A genome (Nishikawa, 1983; Dvořák et al. 1993), *Aegilops speltoides* Tausch (Fig. 3b) – the donor of B genome (Sarkar and Stebbins, 1956; Wang

Fig. 3. Wild progenitors of durum wheat (AABB) and bread wheat (AABBDD): **(a)** *Triticum urartu* (the donor of genome A); **(b)** *Aegilops speltoides* (the donor of genome B); **(c)** *Triticum tauschii* (the donor of genome D). The first two species crossed in nature to give rise to tetraploid durum wheat (Fig. 3d), and incorporation of genome D from *T. tauschii* resulted in hexaploid bread wheat (Fig. 3e). See Figure 4 for details of steps of evolution.

et al. 1997), and *Triticum tauschii* Coss. (Fig. 3c) – the donor of D genome (Kihara, 1944; McFadden and Sears, 1946), are the progenitors of bread wheat. Different steps of evolution of durum wheat (Fig. 3d) and bread wheat (Fig. 3e) are shown schematically as step 1 and step 2 in Figure 4.

Meiotic restitution plays an important role in the formation of unreduced gametes that result in sexual polyploidization. Normal meiosis is a well orchestrated and specialized cell division with one round of DNA replication but with two successive nuclear divisions, referred to as meiosis I and meiosis II, that result in normally reduced, haploid gametes. Meiosis I involves pairing of homologous chromosomes at zygotene-pachytene and their subsequent segregation to the poles leading to reduction in chromosome number and is therefore called reductional division. The ensuing meiosis II is simply a mitotic division or equational division that separates sister chromatids and forms four haploid daughter cells with half the parental chromosome number. However, there are some exceptions to this rule. Haploids of diploid species have only one dose of each chromosome and therefore they rarely show any pairing at stages equivalent to meiosis I and hence there is no reductional division.

Haploids of polyploid species, like the polyploid wheats, are called polyhaploids, or amphihaploids, to be more precise. Again, such amphihaploids show very little or no pairing firstly because they have only one dose of each chromosome, but more importantly because of the presence of *Ph1* that restricts pairing to homologous partners (see Jauhar et al. 2009a). Because chromosomes in amphihaploids generally remain as univalents, meiosis may not be reductional, but simply equational, and may result in the formation of unreduced gametes (see Fig. 5). Such unreductional meiotic cell division is also called meiotic restitution and has major evolutionary consequences (see, for example, Jauhar, 2007). Meiotic restitution occurs in intergeneric hybrids (that are essentially amphihaploids) and plays a remarkable role in producing amphidiploids or allopolyploids and may be a major mechanism for the widespread occurrence of polyploidy in nature (Fukuda and Sakamoto, 1992; Jauhar et al. 2000; Jauhar, 2003, 2006, 2007; Ramanna and Jacobsen, 2003; Cai et al. 2010). See also Figure 4.

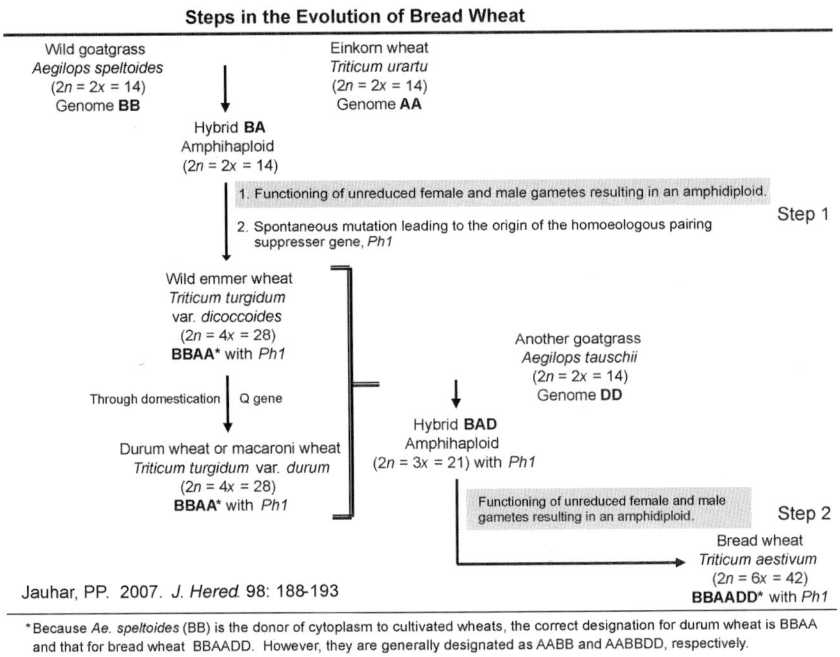

Fig. 4. Diagrammatic representation of different steps of evolution of durum wheat (Step 1) and bread wheat (Step 2). Events shown in step 1 lead to the evolution of durum wheat (AABB with *Ph1*) and step 2 shows the evolution of bread wheat (AABBDD with *Ph1*). However, because the genome of the female parent is generally shown first, the hybrid between *Aegilops speltoides* (the donor of genome B) and *Triticum urartu* (the donor of genome A) is shown to have the genomic constitution BBAA. The same holds for bread wheat (BBAADD) as shown in step 2. (Source: Jauhar, 2007)

RETRACING THE STEPS OF EVOLUTION

To verify the occurrence of certain steps in the course of evolution (Fig. 4), one can go back on the evolutionary ladder and try to repeat some of the steps. Thus, starting from durum wheat AABB genomes (or BBAA genomes because the genomic contribution of the female parent is shown first) with *Ph1* we produced its haploids or amphihaploids ($2n = 2x = 14$; BA genomes) by crossing with maize (Almouslem et al. 1998) and studied their meiosis. The amphihaploids showed restitution (Fig. 5a-d) in 5–10% of the pollen mother cells (pmcs) resulting in the formation of unreduced gametes with 14 chromosomes (Jauhar et al. 2000). The so called meiosis I was not reductional, but simply equational (Fig. 5b, c) that resulted in the formation of unreduced gametes (Fig. 5d). The fusion of unreduced male and female gametes resulted in partially fertile disomic durum wheat that set seed. Fertility of the derived disomic durum and the presence of two doses of the 4A·7B translocation corroborated the precise duplication of all chromosomes (Jauhar et al. 2000).

Thus, starting with disomic durum wheat, we obtained amphihaploids and then reconstituted the parental disomic durum. Similar studies were done with hexaploid bread wheat in which the induced polyhaploids or amphihaploids ($2n = 2x = 21$; BAD genomes) produced unreduced gametes through meiotic restitution and thus produced hexaploid wheat (Jauhar, 2007). Reconstitution of both durum and bread wheat through meiotic restitution under experimental conditions is a simulation of the evolutionary steps that occurred in nature at the time of origin of polyploid wheats and could perhaps be described as "evolution in the test tube." Meiotic restitution resulting in unreduced gametes is certainly a potent force in producing allopolyploids in nature.

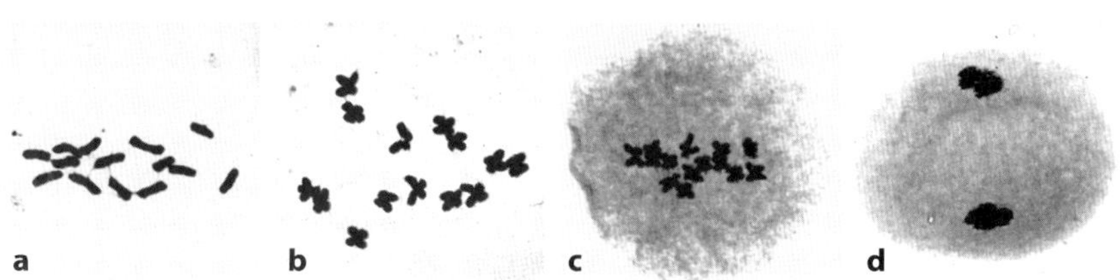

Fig. 5. Meiotic stages in durum wheat haploids or amphihaploids ($2n = 2x = 14$; AB) showing restitution and formation of unreduced gametes with 14 chromosomes. **(a)** Haploid pollen mother cell (pmc) with 14 univalents; there is no pairing among them because of the presence of *Ph1* and hence no reduction division. **(b)** A pmc at meta-anaphase showing all 14 univalents, each split into two chromatids. **(c)** A pmc at meta-anaphase showing all 14 univalents (each with two chromatids) arranged on the metaphase plate. **(d)** The sister chromatids move to the two poles in an equational division, resulting in unreduced gametes with 14 chromosomes. (Source: Jauhar et al. 2000)

Genetic Control of Chromosome Pairing and Disciplined Meiosis: An Essential Step in Evolution

As discussed above, the allopolyploids have several sets of chromosomes, e.g., durum wheat (AABB) and bread wheat (AABBDD). The corresponding (homoeologous) chromosomes of the constituent genomes in these allopolyploid species are related to the extent that they are capable of pairing with one another. Therefore, to restrict pairing to homologous partners and maintain diploid-like, disciplined meiosis, a regulatory mechanism would be needed. Thus, in both durum and bread wheat, the *Ph1* gene in the long arm of chromosome 5B allows pairing between homologous partners, while suppressing homoeologous pairing. Such a controlled pairing ensures diploid-like meiosis and disomic inheritance (Riley and Chapman, 1958; Sears and Okamoto, 1958; Jauhar and Joppa, 1996; Jauhar, 2006). The *Ph1*-regulation was acquired at the initial stage of the origin of durum wheat and was and is obviously essential for the evolution and survival of allopolyploid wheats (see Fig. 4).

Elucidation of Allopolyploid Nature of a Polyploid Species

Durum wheat with AABB genomes is an allotetraploid that shows diploid-like paring resulting in the formation of 14 bivalents (Fig. 6). Such a precise pairing is enforced by the presence of *Ph1*, the A-genome chromosomes as well as the B-genome chromosomes forming precisely seven bivalents. However, the bivalents formed by the two genomes cannot be distinguished by conventional staining (Fig. 6a). We developed a multicolour fluorescent (fl) genomic in situ hybridization (GISH) technique (Jauhar and Peterson,

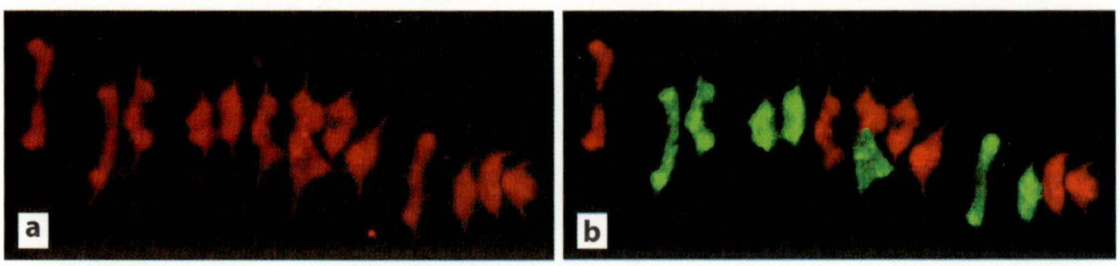

Fig. 6. Meiotic chromosomes of allotetraploid durum wheat ($2n = 4x = 28$; AABB genomes) showing 14 bivalents. **(a)** Chromosomes counter-stained with propidium iodide (PI) are all red in colour; thus, the chromosomes of the A and B genomes cannot be differentiated in this photograph. **(b)** The same cell as in (a) analyzed with multicolour fluorescent genomic in situ hybridization (fl-GISH) and visualized using fluorescein isothiocyanate (FITC). The seven bivalents of the A genome are shown in yellow-green and the seven bivalents of the B genome are shown in red. Using this sophisticated staining technique, the genomic constitution of durum wheat can be revealed. (Source: Jauhar, 2010, *Nucleus* 53: 3–12)

2006) that helps distinguish bivalents formed by the two genomes. Figure 6a shows 14 bivalents counterstained with propidium iodide (PI). Fl-GISH analysis of the same cell in Figure 6b shows 7 A-genome bivalents (fluorescing green) and 7 B-genome bivalents (red). This cytological technique is an elegant means of demonstrating the allopolyploid nature of durum wheat. Multicolour fl-GISH can help distinguish intergenomic chromosomes and their pairing or segmental interchange in interspecific and intergeneric hybrids (Jauhar and Peterson, 2006; Molnár et al. 2009; Qi et al. 2010).

Genetic Buffering in Allopolyploids

In contrast to diploid plant species that have two sets of chromosomes, allopolyploids have several sets of chromosomes and hence duplication of genetic material. Polyploidy leads to a certain degree of redundancy in chromosome material. Therefore, some of the chromatin may be dispensable as evidenced by the survival of plants deficient in some chromosomes.

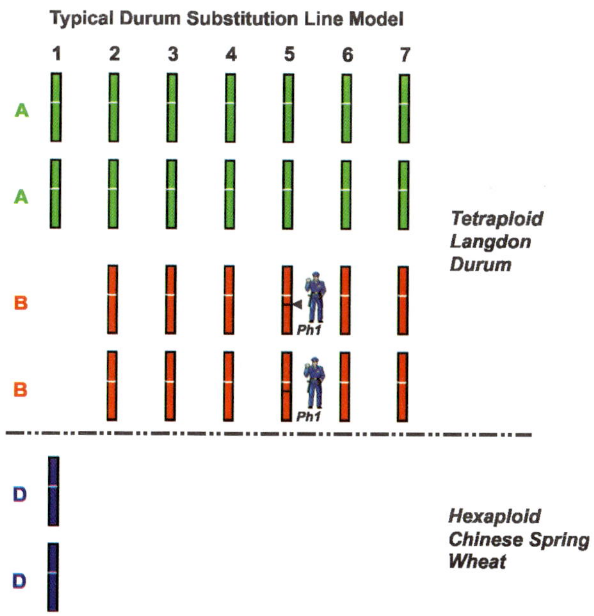

Fig. 7. A typical substitution line in durum wheat showing genetic buffering. It shows all the seven pairs of chromosomes of the A genome (green colour), and only six pairs of chromosomes of the B genome. The missing chromosome 1B pair is replaced by chromosome 1D pair from the D genome of hexaploid wheat variety Chinese Spring; 1D performs the function of the missing chromosome 1B and therefore the plant is by and large normal.

Durum wheat, for example, can survive even when one of its chromosome pairs is substituted by its homoeologue from bread wheat (Fig. 7). In such a case, the added chromosome can effectively take over the function of the missing chromosome and the plant is by and large normal. This sort of genetic buffering in polyploid wheats lends them to cytogenetic manipulation in producing alien substitution or alien addition lines (see for example, Jauhar et al. 2009b) or any other form of genetic engineering (Ceoloni and Jauhar, 2006).

POLYPLOIDY, GENETIC CONTROL OF PAIRING, AND UNUSUAL REPRODUCTIVE ISOLATION

As stated above, allopolyploids are preponderant in nature and they show diploid-like pairing, disomic inheritance, and reproductive stability because of the presence of a regulatory mechanism that controls chromosome pairing. A comparison of the regulatory mechanisms in polyploid species of the family Poaceae is shown in Table 1. In some allopolyploid species, like the durum wheat, bread wheat, and hexaploid oats, the genetic control of chromosome pairing is hemizygous effective, i.e., one dose of the gene(s) can effectively control pairing (references in Table 1). Thus, there is no, or very little, pairing in the haploids (polyhaploids) of these species (Jauhar, 1977).

However, hexaploid tall fescue (*Festuca arundinacea* Schreb.) has a genetic control that is hemizygous-ineffective, i.e., two doses of the regulator are needed to enforce diploid-like pairing (Jauhar, 1975a,b). Such a genetic control has major evolutionary implications. Tall fescue is a cross-pollinated, self-incompatible species with several ecotypes that cross very easily to produce fertile hybrids. Its ecotypes from various parts of the world show regular, diploid-like meiosis forming 21 bivalents (Figs. 8a-d, 9a,b). Hybrids between different ecotypes also show regular meiosis with 21 bivalents, and are fertile like the parental ecotypes. However, certain hybrid combinations

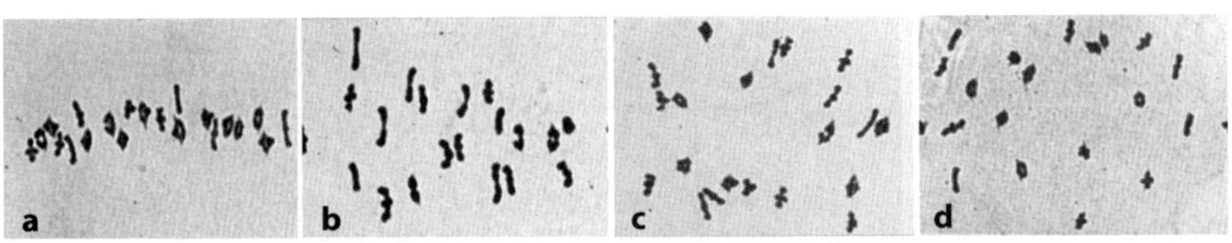

Fig. 8. Regular chromosome pairing in different ecotypes of allohexaploid tall fescue ($2n = 6x = 42$). **(a-d)** All ecotypes show 21 bivalents at meiotic metaphase I; there is some variation in chiasma frequency. (Source: Jauhar, 1975c, 1993)

Table 1: Comparison of the Chromosome-Pairing Control Mechanisms Discovered in Polyploid Species of the Family Poaceae

Features	*5B-Control*	*Reference*	*A-Control*	*Reference*	*C-Control*	*Reference*
Species in which chromosome pairing is regulated	Tetraploid and hexaploid wheats	Okamoto, 1957; Sears and Okamoto, 1958; Riley and Chapman, 1958	Tetraploid and hexaploid oat species of *Avena*	Rajhathy and Thomas, 1972, 1974	*Festuca arundinacea*, *F. rubra* and other polyploid fescues	Jauhar, 1975a, b
Location	$5B^L$	Sears and Okamoto, 1958; Riley, 1960	A genome?	Jauhar, 1975c	C genome?	Jauhar, 1975c, 1977
Effectiveness	Hemizygous-effective	Riley and Chapman, 1958; Riley, 1960	Hemizygous-effective	Nishiyama and Tabata, 1964; Rajhathy and Thomas, 1972	Hemizygous-ineffective	Jauhar, 1975a, b
Dosage effect	+	Feldman, 1966, 1968; Martinez et al., 2001	+	Jauhar, 1975c; Ladizinsky, 1973	+	Jauhar, 1975c
Genetically repressible	+	Riley, 1960; Dvořák, 1972	+	Rajhathy and Thomas, 1972, 1974	+	Jauhar, 1975b, c, 1977, 2006
Species/ genotype suppressing control	*Aegilops speltoides*	Riley et al., 1961	*Avena longiglumis* Accession CW57	Rajhathy and Thomas, 1972, 1974	Diverse ecotypes of tall fescue	Jauhar, 1975c, 1991, 2006

e.g., between North African and North European ecotypes, show irregular meiosis because of multivalent formation (Fig. 9c). In some cases, hybrids between geographically-diverse ecotypes form both multivalents and univalents, resulting in high or complete sterility. Jauhar (1975b, 1993, p. 64) conducted extensive studies on tall fescue accessions from all over the world and inferred that a high multivalent formation in hybrids cannot be explained on the basis of differentiation of parental chromosomes, as claimed by several European workers (Jauhar, 1975c). If such a chromosome differentiation were present, it would result in lack of pairing rather than extensive pairing in the inter-ecotype hybrids.

The crosses shown in Figure 10 present an enigmatic situation. Central Russian, Tunisian, and Israeli ecotypes are meiotically regular (Fig. 10a) and fertile. As expected, hybrids between these ecotypes are also meiotically regular and fertile (Fig. 10a). However, a different situation is presented in hybrids among some ecotypes. The hybrids between the Algerian ecotype (A) and the Israeli ecotype (B) are meiotically regular and fertile; and hybrids between the Israeli ecotype (B) and Tunisian ecotype (C) are also meiotically regular and fertile. But, interestingly, the hybrids between the

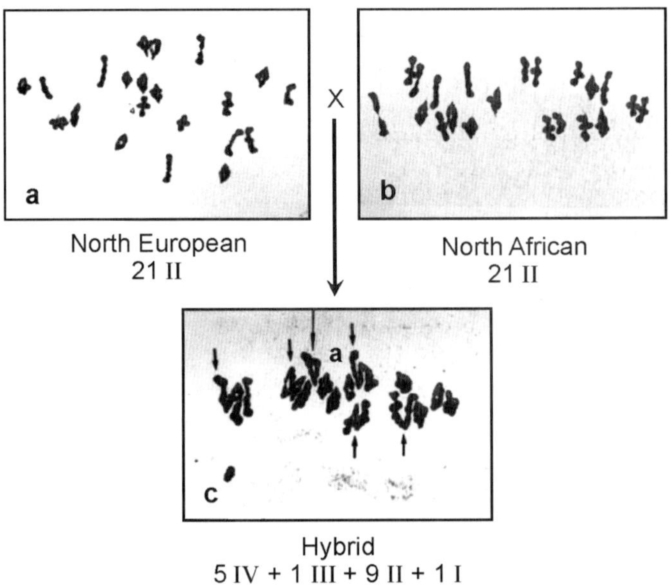

Fig. 9. Chromosome pairing at meiotic metaphase I in a North European ecotype **(a)** and North African ecotype **(b)** of tall fescue both showing regular pairing with 21 II. The hybrid **(c)** between the two ecotypes shows highly irregular meiosis with 5 IV + 1 III + 9 II + 1 I and is sterile. (Source: Jauhar, 2010, *Nucleus* 53: 3–12)

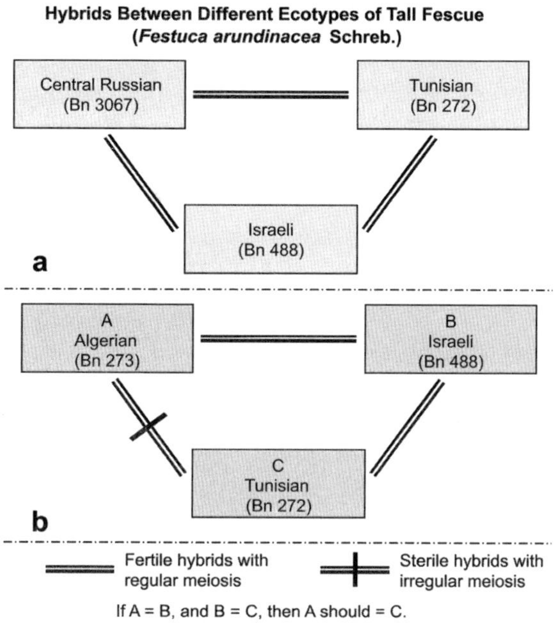

Fig. 10. Crossability among different ecotypes of tall fescue, and meiosis and fertility in their hybrids. The Central Russian, Tunisian, and Israeli ecotypes show regular meiosis and are inter-fertile producing fertile hybrids with regular meiosis. Note that some crosses (Algerian × Tunisian ecotypes) produce sterile hybrids with abnormal meiosis. This is a genetic block (as shown in one arm of the triangle) causing meiotic irregularity and sterility. (Source: Jauhar, 2010, *Nucleus* 53: 3–12)

Tunisian ecotype (C) and the Algerian ecotype (A) are meiotically irregular and sterile (Fig. 10b). Meiotic regularity and sterility in A × C hybrids is hard to explain, because they would be expected to be meiotically regular and fertile. Logically, if A = B, and B = C, then C should = A. In other words, if the chromosomes of ecotypes A and B are meiotically similar enough to be able to pair, and the chromosomes of ecotype B are also meiotically congruent to the chromosomes of the ecotype C, then the chromosomes of ecotype C would be expected to be meiotically compatible with those of the ecotype A, but they are not. This meiotic incongruity between C and A, resulting in irregular meiosis and complete sterility, can be satisfactorily explained on the basis of genetic regulation of chromosome pairing. Both the Algerian and Tunisian ecotypes show regular meiotic pairing, but meiosis in their hybrids breaks down (Fig. 10b). It seems that the Algerian genotype interferes with the regulatory mechanism in its hybrids with the Tunisian ecotype, or interaction between the Algerian and Tunisian genotypes interferes with the regulatory mechanism in their hybrids, leading to extensive homoeologous pairing and sterility (Jauhar, 1975c, 1991, 1993).

The gene-repressing effect of some genotypes causing homoeologous pairing in hybrids between ecotypes is clearly of evolutionary significance. This is a novel means of creating reproductive isolation among geographically isolated ecotypes of the same species for further speciation and evolution (Jauhar, 1975b,c, 1993). Such an evolution may well be occurring in nature.

BRIEF DISCUSSION AND CONCLUSION

Eukaryotic evolution is accompanied by duplication of chromosomes or genetic material. Duplications of whole genomes, chromosomes, chromosome segments, or genes have played an important role in the evolution of eukaryotic genomes (Goffeau, 2004). Polyploidy could well be a recurrent process leading to multiple rounds of polyploidization and continues to be a force in plant evolution (Adams and Wendel, 2005). Extensive silencing, elimination of duplicate genes, and rapid genomic alterations may occur with the onset of polyploidy (see Adams and Wendel, 2005; Patterson et al. 2010).

Allopolyploids are preponderant in nature and may have resulted from intergeneric or interspecific hybridization coupled with genome doubling through the fusion of unreduced gametes formed in the parental species or in the resultant hybrids. Sexual polyploidization resulting from functioning of unreduced gametes is a major source of allopolyploids in nature. Many of the crop species that feed the world are of allopolyploid origin and likely arose via sexual polyploidization in one or two steps that may be described as abrupt or cataclysmic evolution. The evolution of sex would be extremely important because of the invaluable evolutionary benefits of meiosis and consequent genetic recombination linked with sexual reproduction. Sexually reproducing organisms are adaptable under ever-changing environmental conditions; and among the sexual polyploids, allopolyploids are clearly much more successful. Autopolyploidy, on the other hand, does not add any new genes and confers little adaptive advantage.

Normally, evolution is a slow process and perhaps that led Blake (1792) to make a statement: "To create a little flower is the labour of ages." Here, a "little flower" means a new species and Blake meant that evolution was a notoriously slow process. However, it has been known for some time that evolution can in fact be rapid or even cataclysmic as evidenced by the production of durum wheat in one step and of bread wheat in two steps (Fig. 4). This fact led Haldane (1958) to change Blake's famous dictum to read: "To create a little flower is the labour of ages *except by allopolyploidy*." The italicized portion was added by Haldane. Triticale, a man-made cereal that was produced as a hybrid between *Triticum* species and *Secale cereale* L. provides a remarkable example of accelerated evolution through allopolyploidy. We have since known that the successful establishment of an allopolyploid depends on meiotic regularity and reproductive stability that can be effectively achieved by a precise genetic control on chromosome pairing as described earlier in polyploid wheats (Riley and Chapman, 1958; Sears and Okamoto, 1958) and hexaploid oats (Rajhathy and Thomas, 1972, 1974; Jauhar, 1977).

In view of the importance of genetic control of chromosome pairing, Jauhar (1975c) stated that "polyploidy and sexuality cannot coexist in nature without such a regulatory mechanism." Natural polyploids without a genetic control of chromosome pairing are apomictic and hence blind alleys from the evolutionary standpoint. Most, if not all, disomic polyploids in nature have some form of genetic regulation of pairing to achieve regular meiosis and disomic inheritance. These facts of allopolyploidy and evolution led Jauhar (1975d) to further modify the Blake-Haldane statement to read "To create a little flower is the labour of ages except by allopolyploidy *coupled with genetic control of chromosome pairing*." The italicized portion in the modified statement was added by Jauhar (1975d). Jauhar (2003) further emphasized that sexual reproduction and polyploidy may not coexist without a precise control of chromosome pairing. Research on chromosome pairing is useful in elucidating multifarious evolutionary phenomena including polyploidy and speciation.

It may be concluded that a combination of sexuality, allopolyploidy, and genetic regulation of chromosome pairing provides an ideal recipe for evolution of successful crop species. That the allopolyploids have remarkable benefits of polyploidy and hybridity is made possible by the genetic control of chromosome pairing. Among others, hexaploid bread wheat and its tetraploid forerunner, durum wheat, are stabilized natural hybrids of progenitor wild diploid species. These are some of the marvels of evolution.

ACKNOWLEDGEMENTS

I am grateful to Mr. Terrance Peterson and Ms. Anna Hagemeyer for their invaluable help in the preparation of this article.

References

Adams KL, Wendel JF (2005). Polyploidy and genome evolution in plants. *Curr. Opinions Plant Biol* 8: 135-141.

Almouslem AB, Jauhar PP, Peterson TS, Bommineni VR, Rao MB (1998). Haploid durum wheat production via hybridization with maize. *Crop Sci.* 38: 1080-1087.

Blake W (1792). The Marriage of Heaven and Hell. (Privately printed)

Cai X, Xu SS, Zhu X (2010). Mechanism of haploidy-dependent unreductional meiotic cell division in polyploid wheat. *Chromosoma* 119: 275–285.

Ceoloni C, Jauhar PP (2006). Chromosome engineering of the durum wheat genome: Strategies and applications of potential breeding value. In: *Genetic Resources, Chromosome Engineering, and Crop Improvement, Volume 2: Cereals.* Eds. Singh, RJ, and Jauhar, PP, CRC Press, Boca Raton, FL, USA. pp. 27-59.

Chen ZJ (2010). Molecular mechanisms of polyploidy and hybrid vigor. *Trends Plant Sci.* 15: 57-71.

Doyle JJ, Egan AN (2010). Dating the origins of polyploidy events. *New Phytologist* 186: 73-85.

Dvořák J (1972). Genetic suppression of homoeologous chromosome pairing in hexaploid wheat. *Can. J. Genet. Cytol.* 14: 39-42.

Dvořák J, Diterlizzi P, Zhang HB, Resta P (1993). The evolution of polyploid wheats-identification of the A-genome donor species. *Genome* 36: 21-31.

Feldman M (1966) The effect of chromosomes 5B, 5D, and 5A on chromosome pairing in *Triticum aestivum*. *Proc. Natl. Acad. Sci.* USA; 55: 1447-1453.

Feldman M (1968). Regulation of somatic association and meiotic pairing in common wheat. In: *Proceedings of the 3rd International Wheat Genetics Symposium*: eds. Finlay DW, Shepherd KW. Australian Academy of Science, Canberra (1968); pp. 169-178.

Fukuda K, Sakamoto S (1992). Studies on the factors controlling the formation of unreduced gametes in hybrids between tetraploid emmer wheats and *Aegilops squarrosa* L. *Jap. J. Breed.* 42: 747-760.

Goffeau A (2004). Evolutionary genomics: Seeing double. *Nature* 430: 25-26.

Haldane JBS (1958). The theory of evolution, before and after Bateson. *J. Genet.* 56: 1-17.

Harlan JR, de Wet JMJ (1975). On Ö. Winge and a prayer: The origins of polyploidy. *Bot. Rev.* 41: 361-390.

Jauhar PP (1975a). Genetic control of diploid-like meiosis in hexaploid tall fescue. *Nature* 254: 595-597.

Jauhar PP (1975b). Genetic regulation of diploid-like chromosome pairing in the hexaploid species, *Festuca arundinacea* Schreb. and *F. rubra* L. (Gramineae). *Chromosoma* 52: 363-382.

Jauhar PP (1975c). Genetic control of chromosome pairing in polyploid fescues: Its phylogenetic and breeding implications. In: *Rep. Welsh Plant Breeding Station*, Aberystwyth, Wales, U.K., pp. 114-127.

Jauhar PP (1975d). Polyploidy, genetic control of chromosome pairing and evolution in the *Festuca-Lolium* complex. Proc. 178th Meeting of British Genetical Society, John Innes Institute, Norwich, England, April 1975. *Heredity* 35: 430.

Jauhar PP (1977). Genetic regulation of diploid-like chromosome pairing in *Avena. Theor. Appl. Genet.* 49: 287-295.

Jauhar PP (1991). Recent cytogenetic studies of the *Festuca-Lolium* complex. In: *Chromosome Engineering in Plants: Genetics, Breeding, Evolution, Vol. 2B.* Eds. Tsuchiya T, and Gupta PK, Elsevier Science Publishers, Amsterdam, The Netherlands; p. 325-362.

Jauhar PP (1993). *Cytogenetics of the Festuca-Lolium Complex: Relevance to Breeding.* Springer-Verlag, Berlin, Heidelberg, New York, Paris, Tokyo. 260 pages. ISBN 0-387-52113-5.

Jauhar PP (2003). Formation of 2*n* gametes in durum wheat haploids: Sexual polyploidization. *Euphytica* 133: 81-94.

Jauhar PP (2006). Cytogenetic architecture of cereal crops and their manipulation to fit human needs: Opportunities and challenges. In: *Genetic Resources, Chromosome Engineering and Crop Improvement, Volume 2: Cereals*. Eds. Singh RJ, and Jauhar PP, CRC Press, Boca Raton, FL, USA; pp. 1-25.

Jauhar PP (2007). Meiotic restitution in wheat polyhaploids (amphihaploids): A potent evolutionary force. *J. Hered.* 98: 188-193.

Jauhar PP (2010). Genetic control of chromosome behaviour: Implications in evolution, crop improvement, and human biology. *Nucleus* 53: 3-12.

Jauhar PP, Joppa LR (1996). Chromosome pairing as a tool in genome analysis: Merits and limitations. In: *Methods of Genome Analysis in Plants*. ed. Jauhar PP. Boca Raton, London, Tokyo : CRC Press. pp. 9-37.

Jauhar PP, Peterson TS (2006). Cytological analyses of hybrids and derivatives of hybrids between durum wheat and *Thinopyrum bessarabicum*, using multicolour fluorescent GISH. *Plant Breeding* 125: 19-26.

Jauhar PP, Storey WB (1982). *Ornithogalum virens* ($2n$ = 6): A favorable organism for teaching cytology. *J. Hered.* 73: 243-244.

Jauhar PP, Doğramaci-Altuntepe M, Peterson TS, Almouslem AB (2000). Seedset on synthetic haploids of durum wheat: Cytological and molecular investigations. *Crop Sci.* 40: 1742-1749.

Jauhar PP, Xu SS, Baenziger PS (2009a). Haploidy in cultivated wheats: induction and utility in basic and applied research. *Crop Sci.* 49: 737-755.

Jauhar PP, Peterson TS, Xu SS (2009b). Cytogenetic and molecular characterization of a durum alien disomic addition line with enhanced tolerance to Fusarium head blight. *Genome* 52: 467-483.

Jenczewski E, Alix K (2004). From diploids to allopolyploids: The emergence of efficient pairing control genes in plants. *Critical Rev. Plant Sci.* 23: 21-45.

Kihara H (1944). Discovery of the DD-analyser, one of the ancestors of vulgare wheat. *Ag. Hort.* (Tokyo) 19: 889-890.

Ladizinsky G (1973). Genetic control of bivalent pairing in the *Avena strigosa* polyploid complex. *Chromosoma* 42: 105-110.

Martinez M, Naranjo T, Cuadrado D, Romero C (2001). The synaptic behaviour of *Triticum turgidum* with variable doses of the *Ph1* locus. *Theor. Appl. Genet.* 102: 751-759.

McFadden ES, Sears ER (1946). The origin of *Triticum speltoides* and its free-threshing hexaploid relatives. *J. Hered.* 37: 81-89.

Molnár I, Benavente E, Molnár-Láng M (2009). Detection of intergenomic chromosome rearrangements in irradiated *Triticum aestivum – Aegilops biuncialis* amphiploids by multicolour genomic in situ hybridization. *Genome* 52: 156-165.

Nishikawa K (1983). Species relationship of wheat and its putative ancestors as viewed from isozyme variation. In: *Proceedings of the 6th International Wheat Genetics Symposium*, Kyoto, Japan, pp. 59-63.

Nishiyama I, Tabata M (1964). Cytogenetic studies in *Avena*. XII. Meiotic chromosome behaviour in a haploid cultivated oat. *Jap. J. Genet.* 38: 311-316.

Okamoto M (1957). Asynaptic effect of chromosome V. *Wheat Inform.* Serv. 5: 6.

Patterson AH, Freeling M, Tang H, Wang X (2010). Insights from the comparison of plant genome sequences. *Ann. Rev. Plant Biol.* 61: 349-372.

Proost S, Pattyn P, Gerat T, van der Peer Y (2011). Journey through the past: 150 years of plant evolution *Plant J.* 66: 58–65.

Qi Z, Du P, Qian B, Zhuang L, Chen H, Chen T, Shen J, Guo J, Feng Y, Ziyou P (2010). Characterization of a wheat-*Thinopyrum bessarabicum* (T2JS-2BS·2BL) translocation line. *Theor. Appl. Genet.* 121: 589-597.

Rajhathy T, Thomas H (1972). Genetic control of chromosome pairing in hexaploid oats. *Nat. New Biol.* 239:217-219.

Rajhathy T, Thomas H (1974). Cytogenetics of oats (*Avena* L.). *Miscellaneous Publications of the Genetics Society of Canada* No. 2. p. 90.

Ramanna MS, Jacobsen E (2003). Relevance of sexual polyploidization for crop improvement—a review. *Euphytica* 133: 3-18.

Rapp RA, Udall JA, Wendel JF (2009). Genomic expression of dominance in allopolyploids. *BMC Biology* 7: 18-27.

Riley R (1960). The diploidisation of polyploid wheat. *Heredity* 15: 407-429.

Riley R, Chapman V (1958). Generic control of the cytologically diploid behaviour of hexaploid wheat. *Nature* 182: 713-715.

Riley R, Kimber G, Chapman V (1961). Origin of genetic control of diploid like behaviour of polyploid wheat. *J. Hered.* 52: 22-25.

Sarkar P, Stebbins GL (1956). Morphological evidence concerning the origin of the B genome in wheat. *Amer. J. Bot.* 43: 297-304.

Sears ER, Okamoto M (1958). Intergenomic chromosome relationships in hexaploid wheat. *Proc. X Intern. Cong. Genet.* 2: 258-259.

Soltis DE, Burleigh JG (2009). Surviving the K-T mass extinction: new perspectives of polyploidization in angiosperm. *Proc.Natl.Acad. Sci.* USA, 106: 5455-5456.

Vasal SK, Riera- Lizarazu O, Jauhar PP (2006). Genetic enhancement of maize by cytogenetic manipulation, and breeding for yield, stress tolerance, and high protein quality. In: *Genetic Resources, Chromosome Engineering, and Crop Improvement, Volume 2: Cereals.* eds. Singh RJ, and Jauhar PP, CRC Press, Boca Raton, FL, USA. pp. 159-197.

Wang GZ, Miyashita NT, Tsunewaki K (1997). Plasmon analyses of *Triticum* (wheat) and *Aegilops*: PCR single-strand conformational polymorphism (PRC-SSCP) analyses of organellar DNAs. *Proc. Natl. Acad. Sci.* USA 94: 14570-14577.

Wood TE, Takebayashi N, Barker MS, Mayrose I, Greenspoon PB, Rieseberg LH (2009). The frequency of polyploid speciation in vascular plants *Proc. Natl. Acad. Sci.* USA 106: 13875-13879.

Yang X, Ye C-Y, Cheng Z-M, Tschaplinski TJ, Wullschleger SD, Yin W, Xia X, Tuskan GA (2011). Genomic aspects of research involving polyploid plants. *Plant Cell Tiss. Organ Cult.* 104: 387-397.

3

Can Dynamic Conservation of Endemics be Delinked from the Biodiversity in its Totality?

P.S. Ramakrishnan

INTRODUCTION

Endemics represent unique biodiversity – taxa that are restricted to a particular geographical space or confined to a given ecological unit (ecosystem/landscape unit); UNEP (1995) whilst making a more recent global biodiversity assessment implies that it is somewhat problematic to precisely define endemism, and therefore suggesting that definitions are tailored to particular needs. Endemism may be due to a complex set of factors – a combination of historical factors, ecological process, and/or inherent biological properties of the species concerned. Recent estimates suggest that out of approximately 10 million species, only a very small fraction of about 1.55 million has been validly named (Raven, 1998), a vast majority of them largely located in the tropics remains yet to be discovered and named! Much of what is known and the un-known are confined to the developing tropics and there again what is often-times referred to as 'hot-spots' of biodiversity. What are referred to as endemics are largely restricted to these 'hot-spots', where in traditionally speaking, the forest dwellers (referred to as 'traditional societies') living close to nature and natural resources, depending upon biodiversity, both natural and human-managed. How then can we look at endemics as distinct from the rest of the biodiversity in the context of any effective conservation strategy? It is in this context, '*endemism*' which is unique biodiversity [species that have restricted distribution, being confined to restricted sites, ecosystems or geographical regions UNEP (1995)] has to be looked at from its conservation viewpoint. Here again, whilst, we have a relatively better grasp over larger species that are distributed at the aboveground level, what is available at the belowground level, particularly the microbial diversity is anybody's guess!

Coming to the 'endemics' *per se*, our grasp over them obviously remains is far from complete: (i) although major centres of endemism is known, more work is needed to get into the specifics related to their uniqueness, area of distribution, and population structure and size; (ii) in the absence of incomplete basic biodiversity inventory it is difficult to get a feel of the magnitude of the problem - which species are endemic and to which region; (iii) it is assumed that many of the endemics are highly vulnerable to anthropogenic disturbances, and yet we know so little about perturbation linked tolerance levels of endemics under diverse ecological situations. Whilst a comprehensive understanding of all these aspects is desirable, and realizing that developing predictive models of the biotic and abiotic correlates if endemism still remains elusive and at best can be seen as only a distant objective (UNEP, 1995).

In any case, from conservation viewpoint, effective conservation of endemics has to be seen as an integral component of the total biodiversity. Therefore effective conservation policies and the linked governmental/non-governmental policies to be effective, an integrated natural cultural landscape conservation and management strategy alone could be effective. Further, to be effective, community participation is critical; this demands taking an integrative socio-ecological system viewpoint in our approaches. This is what the emerging trans-disciplinary areas of study – ecological economics or sustainability science is all about (Ramakrishnan, 2009a).

TRADITIONAL SOCIETIES AS CUSTODIANS OF BIODIVERSITY

Traditional societies (those living close to nature and natural resources around) have always managed biodiversity (endemics included, they being an integral component of biodiversity) and linked natural resources sustainably, until humans from the industrializing/industrialized world stepped in to exploit the natural resources. For traditional societies nature and natural resources not only form the basis for their food security and livelihood needs (Ramakrishnan, 2008a), but they form the basis for their very cultural identity – intangible values (Ramakrishnan, 2008b). In spite of their deeply rooted connectivity with nature and natural resources around them, traditional societies oftentimes said to be 'primitive' have all along been blamed for destroying biodiversity around them

It is in this context, our detailed studies on shifting agriculture (jhum) in north-east India which was initiated in the early 1970s, a community-participatory research that was spread across a few decades now

(Ramakrishnan, et al.1992a, 2006a) becomes significant. In spite of the fact that the jhum farmer clears only just about 2-3 ha of forest land at a given time, they have been held responsible for deforestation and land degradation in the north-eastern hill regions of India. In an ecological sense, this indeed is a small perturbation which contributes towards sustaining and indeed enhancing biodiversity, a basic principle associated with perturbation levels and linked ecosystem function (Fig. 1); indeed, it is not surprising that we were not able to find any evidence about the jhum farmer being responsible for inflicting damage on forest ecosystems in the region. On the contrary, since this land use practice of the tribal farmers left to themselves, would cause only small perturbations, indeed, contributes towards sustainable

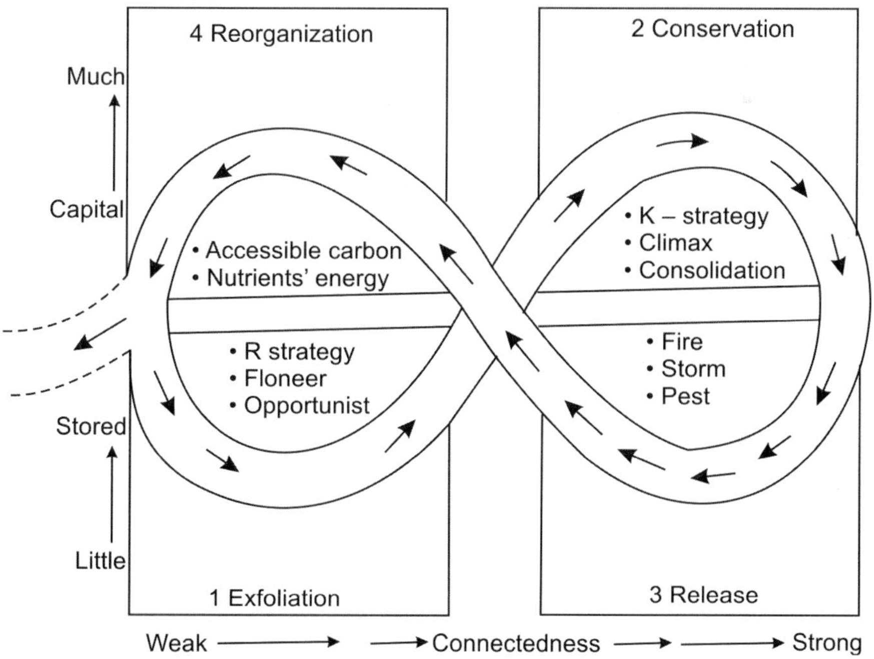

Fig. 1. Interconnectedness of the four major phases and the flow of events of ecosystem organization and function, in the context of small perturbations built into ecosystem functioning; they contribute towards biodiversity conservation/management within the system both in time and space. The arrows indicate the speed at which recovery of the system occurs after a perturbation, starting with the 'release' phase of flow; closeness of the arrows indicate rapid change. Y-axis is the amount accumulated capital of carbon and nutrients; X-axis shows the degree of connectedness between variables within the system. The exit room the cycle is indicated on the left, where a flip is likely into a less or more productive and organized system (from: Holling, 1995).

management of both crop and the associated biodiversity in the landscape, whilst ensuring sustainable livelihood of these societies (Fig. 2).

That exploitative external pressures arising from policy formulations of the governmental agencies and market pressures (national and international) - in other words, pressures external to the systems coming from outside - are responsible for large-scale deforestation linked land degradation came out vividly also from a more broad-based tri-Academy research initiative

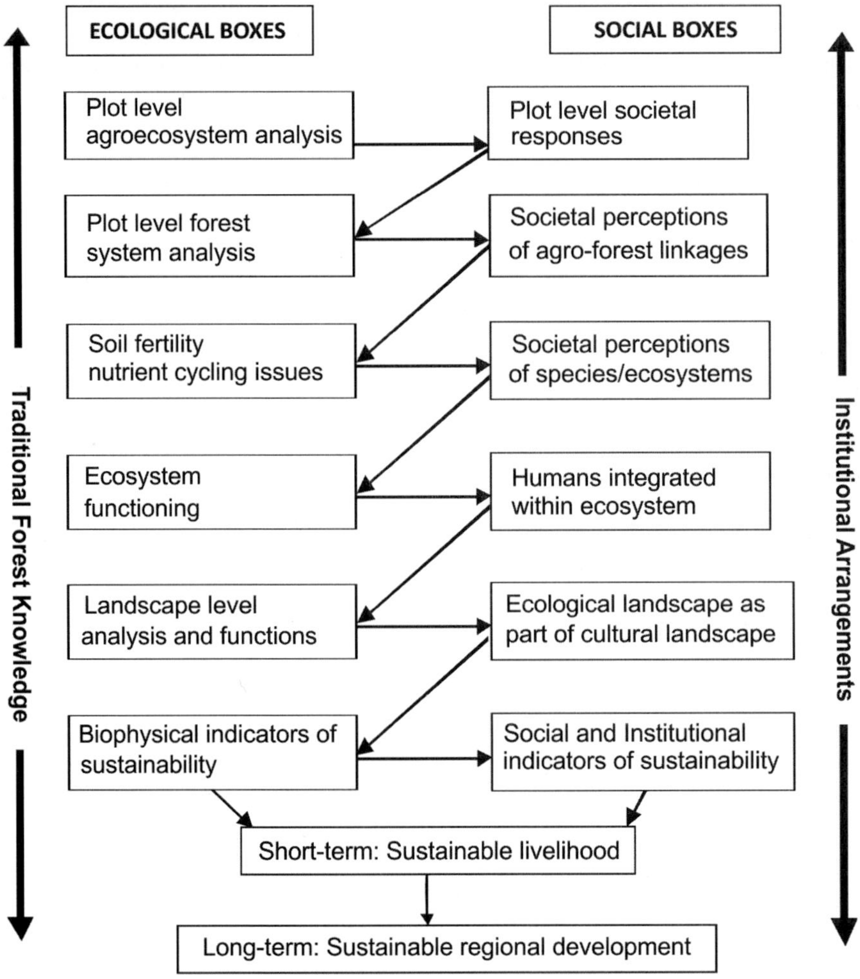

Fig. 2. An iterative process for building linkages between ecological and social sciences and analyses of land-use dynamics from plot level to landscape level, leading to sustainable livelihood/development of traditional societies living in biodiversity-rich areas (from Ramakrishnan, 2006a).

involving India, China and USA also (Indian National Science Academy et al. 2001). This tri-Academy initiative also triggered a global literature review in this subject area, under the auspices of the International Geosphere Biosphere Programme – IGBP (Lambin et al. 2001). What we know now is that that a myth got propagated all along in the scientific literature to suggest that poverty and population pressures are responsible for deforestation and land degradation, but without any rigorous scientific evidence.

From a biodiversity (inclusive of endemics) conservation viewpoint what comes out is that local communities as custodians of biodiversity have only been contributing to enhancing this resource base through smaller perturbations that they are responsible for; in other words, biodiversity depletion is caused by forces far removed from the site, arising from the greed of the industrialized/industrializing societies. This realization, therefore, has implications for *in situ* dynamic conservation strategies that one would like to follow, which obviously has to be local community-participatory in nature.

Knowledge systems: The basis for a sustainable dynamic conservation strategy

Traditional societies, having a vested interest in conserving biodiversity for reasons linked with their sustainable livelihood concerns, should therefore have to be viewed as the real custodians of biodiversity, including endemics. Much of the rich traditional knowledge and wisdom that these societies have are, therefore, closely interwoven with biodiversity which they manipulate in a variety of different ways (Ramakrishnan, 1992a,b; 2006a,b), for meeting with their sustainable livelihood/development needs. It is in this context that the efforts made through routine governmental policies initiated from outside, which obviously is not in agreement with the value systems that traditional societies cherish, have failed all along in the past. Indeed, a sense of frustration and cynicism has set in arising from developmental impositions coming from outside, which has had its adverse impacts on biodiversity conservation efforts too. This is the context in which the efforts to resurrect traditional ecological knowledge (TEK) generated by human living in biodiversity rich areas becomes significant.

Recognizing that we are dealing with biodiversity still as a 'black-box', any meaningful efforts made to conserve the same has to follow a more generalized landscape level conservation approach closely linked with sustainable livelihood/development that is community-participatory that will avoid social disruptions (Ramakrishnan, 1992a, 2006a,b). In other words, text-book biophysically oriented formal ecological knowledge (FEK) focused around ecosystem dynamics (cf. Fig. 1), biomass, productivity, soil fertility, nutrient cycling and soil hydrological process dynamics, no double is important in the contemporary context large-scale biodiversity depletion and land degradation. But TEK linked with societal values has an equally important role in ensuring community participation (cf. Fig. 2).

BIODIVERSITY LINKED TEK: ETHNO-BIOLOGICAL

Lesser-known species of traditional food value

Throughout history, humans have used around 3000 plant species as food collected from the wild, of which at least around 150 being commercially cultivated to some extent (National Academy of Sciences, 1975). Subsequently, emphasis was laid on lesser-known tropical legumes were picked up as resources for the future, for obvious reasons linked with their value as a source of protein, particularly for the weaker sections of the society, obviously for two reasons: (i) that they are an important source of protein for the ever-expanding global human population; (ii) that they are, at the same time, an important source of nitrogen fertilization through the bacteria that live in the nodules of the roots, having the ability to convert gaseous nitrogen to nitrogenous fertilizer, enabling humans to make a drastic departure from excessive chemical fertilizer use. Whilst doing these two global analyses, this global survey undertaken by the U.S. National Academy of Sciences had identified for themselves a range of objectives (Box 1).

Box 1. Under-exploited and lesser-known species of economic value: The agenda for the future (from: National Academy of Sciences, 1975, 1979)

- To identify neglected but seemingly useful tropical plants, both wild and domesticated , that have economic potential
- To select the plants that showed the most promise for wider exploitation throughout the tropics
- To indicate requirements and avenues for research to ensure selected plants reach their full potential
- To identify lesser-known and under-utilized leguminous plants that could meet with the protein needs of the increasing population, particularly the under-nourished of the developing world, by selecting those that have the greatest inherent value
- To further explore strategies and indicate what needs to be done to capitalize on these species so as to develop them to their full potential

This list of the lesser-known species, by no means is exhaustive, since the evaluation, and selection of species for inclusion is based on collective consensus arrived at from among a few hundred species nominations received by them, in response to a global enquiry. Further, this review of what is available as a rich resource for future exploitation as food plants does not get into socio-ecological specific situations from where they originate. Such detailed studies of seemingly simpler systems, that indeed are very complex indeed, becomes critical as we are losing many of these traditional practices, because of pressures that work towards homogenizing land use practices, in the name of modernization of agriculture. From a conservation viewpoint, we are still far away from a complete cataloguing of these lesser-known species, let alone the sub-specific variability available within. From the viewpoint of this discussion on endemics, one is at a loss about endemism embedded within these lesser-known food crops, at the species and sub-specific levels.

Lesser-known species of medicinal value

At another level, alternate medicare systems currently form the basis of primary health care for the majority of the world's population. Recognizing the role of these indigenous practices in addressing the health care needs of developing countries, integration of this alternate systems into the national planning process, and investments for capacity building through research, and to preserve and perpetuate this traditional knowledge base, are increasingly felt at all levels (Bodeker, 1994). India, for e.g., has one of the most ancient medical systems in *Ayurveda* from the Vedic (Hindu religious texts of antiquity) times (circa 1500 BC). With vegetable products dominating this ancient medical system, diet and drugs derived from plants were considered to be part of a holistic approach to treatment of diseases. Apart from *Ayurveda*, , there are a whole variety of traditional and well-recognized medical practices such as *Unani* system largely restricted to the northern part of India and the *Siddha* system in the southern region. Further, with about 2800 culturally distinct communities living in India, and with over 50 million 'tribals' belonging to over 600 ethnic communities that form equally diverse linguistic groups (Dube, 1998), the richness in ethnobiological knowledge available within the country is mind-boggling . About 76 species of animals - 16 invertebrates and about 60 vertebrates, are known to be part of the tribal medicine in India alone. Over 9500 wild plants are recognized to be of ethnobotanical value of which over 7500 as medicines,

3900 as edibles, 500 for fibre, 400 as fodder, 300 as pesticides; many of them have the potential for commercial exploitation (Ministry of Environment & Forests, 1994). With such a rich tradition in India alone and equally diverse systems elsewhere in the tropical world, the opportunities to conserve this biodiversity, the involved endemics and indeed, making use of this resource for human welfare is very vast, particularly when one realizes that these cheap herbal remedies are part of a thriving alternate system of medicine, with minimal side effects in clinical experience (Valiathan, 1998).

To conclude this discussion on lesser-known plants of food and medicinal value, one is still dealing with a 'back-box', in spite of the recent interests that have emerged in this area of study. With limited knowledge on these rich resources of value for the future, dynamically conserving this rich biodiversity in nature alone is the viable option to conserve what is little-known at the species level, let alone the sub-specific level diversity. A calculated guess here again is that endemism is likely to be very high here since much of what is used as food or as medicinal species are human-selected; with rich cultural diversity that one is dealing with here, the linked biological diversity is likely to be equally rich and diverse (Ramakrishnan, 2008b).

BIODIVERSITY LINKED TEK: INFLUENCING ECOSYSTEM/ LANDSCAPE DIVERSITY

Socially valued species

Societal perceptions of TEK that impacts upon ecological processes operate at scalar dimensions, species, ecosystems and landscapes. Of these, socially valued species invariably have an ecological keystone value as they act as triggers towards ecosystem restoration, ensuring biodiversity regeneration that characterizes the given ecological situation. The beginnings of this broader generalization emerged whilst working with traditional societies of diverse ethnic background living the in north-east Indian hill region (Ramakrishnan, 1992a):

- Nepalese Alder (*Alnus nepalensis*) is a species commonly found in many parts of the north-east India, as part of the fallow phase of jhum in the region. During the slash and burn phase of jhum, irrespective of the ethnic background of the pople involved, short stumps of this rapidly pollarding species trees may be left after removing the bole for timber, and smaller branches for household use. Even full-grown trees may often be left behind in the plot, without being disturbed during the slash and burn operation. In many situations, where one may have a rotational fallow system of agriculture, where the burning of the slash has been done away with and where the system is working under very short fallow phase of less than 5 years, or even where sedentary agriculture has come in as in Nagaland, for e.g., the leaves of this species may be used as an organic manure. Ecologically speaking, Nepalese Alder is a non-leguminous nitrogen-fixing tree species, contributing as much as 125 kg nitrogen ha^{-1} yr^{1}, during its peak growth. In other words, this socially valued species has an ecological keystone value, this species in the given systems enabling recovery all most all of the 600 kg nitrogen lost during one cropping season, during about 5-year fallow period; without Nepalese Alder, nitrogen recovery within the system may take up to 10–15 years. In other words, in a situation where 10–15 year fallow cycle is not feasible because of land degradation and attendant population pressure on the land, socially selected species like Nepalese Alder could be seen to stabilize jhum under a 5 year cycle.

- On the basis of interaction with local communities involved with jhum, we looked at a set of randomly selected culturally valued bamboo species picked up at random, during one of our research initiatives in the region. Our research studies on three randomly picked up three culturally valued and commonly preferred bamboo species (*Dendrocalamus hamiltonii, Neohouzeau dulloa, Bambusa tulda* and *B. khasiana*), interestingly enough, showed that these three species contributed towards conservation of three key nutrient elements, namely, nitrogen, phosphorus and/or potassium within the soil profile through leaf litter recycling. Indeed, during fallow plot selection for jhum, the farmers invariably end up selectively picking up a bamboo plot, suggesting thereby that the choice of species is based on experience that a bamboo plot will ensure good crop production.
- Whilst these two above mentioned set of species are preferentially selected slash and burn during the fallow phase, when the fallow phase itself becomes a very short less than 5 years, with little regrowth available during the fallow phase, a lesser-known leguminous food crop species, *Flemingia vestita* is emphasized in the mixed cropping system; often times *F. vestita* may also be

raised as a pure crop. This socially valued crop species common at higher elevations of Meghalaya fixes up to 250 kg nitrogen ha^{-1} yr^{-1}; in other words, this may be viewed as a keystone crop species not only from a nutrition viewpoint but also from the point of sustainability of the cropping system itself

This principle that species of social value for traditional societies invariably have an ecological keystone value was further strengthened not only through a detailed analysis of the complex set of socio-ecological systems found in the north-eastern hill region, but this was followed up through global initiative on conserving the 'sacred' for biodiversity management (Ramakrishnan et al. 1998). Further, this study also suggested that closely related socially valued taxa could be independently selected by different communities far apart from one another, and used by them for their ecological keystone value. To cite such an example, *Quercus* spp. Thus this species is a socially valued one right across the Himalayan region, from the north-west to the north-east, going right up to Arunachal Pradesh in the Indian context, and beyond on the other side of the border. For the Celts, Slavs and Germans, Oak is a divine tree, and now remains 'sacred' confined within the context of Christian monasteries in the Mediterranean Europe; sacred Oak groves was venerated in Asia Minor too by Galatians and Oak groves existed dedicated to Gods and Goddesses of crops, war-God, etc. (Hughes and Chandran, 1998). In the central Himalayan Garhwal region it is socio-culturally valued and has much folk-music exalting the eco-cultural value of the Oaks species (locally called '*Banjh*') for the local communities (UNESCO & CEEE, 2005). In view all what is stated here on the eco-cultural value attached to Oaks by humans living far apart from one another, spread across Asia and Europte too, it is not surprising, to find a carefully pruned dwarf tree of Oak being treasured and conserved as a cultural heritage in the heart of the city centre at Florence, Italy (Plate 1).

In a detailed analysis done on conserving the sacred in the central Himalayan context (Ramakrishnan et al. 1998, 2000), two broad generalizations emerged: (i) the concept of Oak (*Quercus* spp.) form a set of socially valued species in the central Himalayan Garhwal region, with folk-music woven around 'Banjh' : (ii) apart from the fact that Oaks form an important source for fodder and fuelwood of the local people; the leaf litter of these species contribute towards sustainable maintenance of soil fertility through efficient nutrient cycling, with the dense and extensive fine root biomass in the soil profile helps in conserving moisture and releasing the

Plate 1. The socially valued species of Oak (*Quercus* sp.) , occupies a key position in the city centre in Florence, Italy.

Plate 2. The drastically reduced socially valued species of Oak (*Quercus* spp.) in the Garhwal mountain landscape, being lopped by local communities as a valuable fodder for the animals and also as a organic resource for sustainable soil fertility management in traditional agricultural systems.

same in small pulses throughout the dry season. These two properties has all along contributed towards sustainable agriculture, with food security for the mountain people (Ramakrishnan et al. 2000a). In other words, these species play a keystone role, contributing towards the integrity of both natural forested ecosystems as well as in sustaining the traditional rotational agriculture of the mountain people. As a keystone species, Oaks also promote a rich associated biodiversity too.

No wonder, the loss of Oaks formed the real basis for '*chipko*' (hugging the tree) movement, when the timber extractors went to fell trees from the forest in Garhwal. The people in Garhwal mountains already are highly stressed because of rapidly depleting Oak forests, and therefore they are forced to do extensive lopping of the trees for fodder and leaf litter for sustaining soil fertility (Plate 2). What emerges from this analysis too is that socially valued species like *Quercus* has a major role to play in ensuring the food security of the people in the central Himalayan region. Arising from all this, Oak species in the Central Himalayan region could be seen as a powerful tool to trigger for community participation, because of the cascading impacts they have on ecosystem/landscape level processes and functions, which in turn have implications for sustainable livelihood/development concerns of mountain societies in the Central Himalayan region of India (Fig. 3).

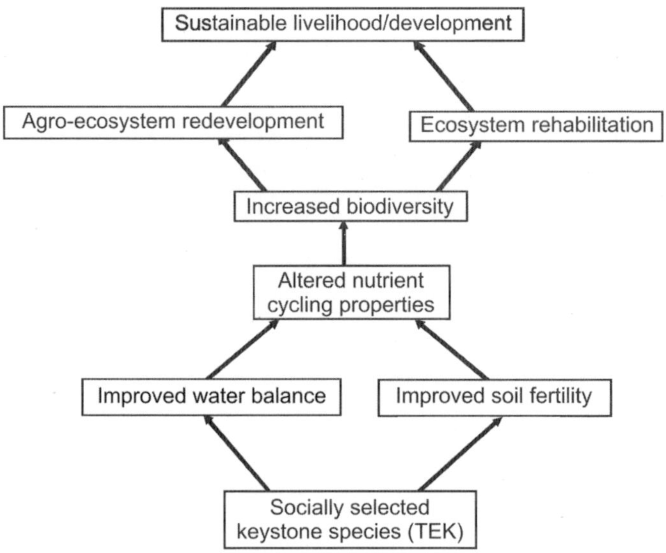

Fig. 3. TEK centred around the socially selected keystone species, *Quercus* spp., acting as a trigger for rehabilitation of the mountain landscape in the Central Himalaya (from: Ramakrishnan, 2000).

Sacred groves and cultural landscapes

If one were to take a broader viewpoint of biodiversity and linked knowledge systems, socially valued ecosystems (sacred groves) that have been conserved all along by local communities oftentimes, is a level in biodiversity at the ecosystem level. In India alone there are a range of estimates on number of surviving sacred groves , going up to about 5000 groves covering about 39,100ha. one extreme; in any case, many of these groves are under varied levels of degradation arising from rapidly changing societal values. In any case, these groves have an important role to play under large-scale degradation that has set in even in biodiversity hot-spot region; one could find very many examples of this situation in the northeastern hill region which is a 'hotspot' of biodiversity. In the northeastern hill region, for e.g., with large-scale deforestation done arising from external pressures on forest resources, oftentimes, sacred groves stand alone as an island of biodiversity in an otherwise degraded and desertified landscape, a glimpse of which is in the foreground (Plate 3).

Plate 3. A well-conserved sacred grove which represents the climax biodiversity-rich forest vegetation with a closed canopy cover of the region, a representative same of what existed in the north-east Indian hill region; in the foreground is a desertified landscape which is now typical of what exists now on a large-scale.

In situations, the 'hotspots' of biodiversity, sacred groves can be viewed as a rich resource for biodiversity, oftentimes with endemics within – a resource of great value in any biodiversity restoration initiative.

At the highest level, is the landscape level biodiversity – landscape units carved out by diverse ethnic groups are socio-ecological system specific, and the concept of community-specific cultural landscapes can appropriately be viewed as representing the highest level in biodiversity, with tangible and intangible values linked to them, which goes down to the lower scalar dimensions of the biodiversity (ecosystem, species, and sub-specific level), a conceptual framework which was seen as very useful for conservation linked sustainable livelihood/development of human societies (Ramakrishnan, 2008a,b).

DYNAMIC CONSERVATION OF BIODIVERSITY ALONG WITH ENDEMICS

Dynamic conservation demands that the strategy has to: (i) enable natural evolutionary processes to function within the given landscape systems without least hindrance, as endemics are an integral part of the biodiversity as a whole in the context of a natural cultural landscape value from which traditional societies seek both tangible and intangible benefits (Plate 4);

Plate 4. Plate Pristine rainforest in the Yoksum (Demojong) region of Sikkim Himalaya, a 'hot spot' of biodiversity.

(ii) ensure community-participation in the process wherein humans remain as effectively integrated as feasible within the given landscape system; (iii) ensure enough flexibility that the pathway enables an improved quality of life for the local traditional societies, appropriately designed with a short-term strategy (the next few decades in view); and at the same time also provides a window for long-term planning for biodiversity conservation; (iv) in the ultimate analysis, an improved quality of life for the people living in the biodiversity 'hotspots' with equal concerns for conflict resolution and human security. The concern for avoiding conflicting situations which is indeed a threat for peace and humans security at the local, regional and global level is a concern which is becoming more and more evident in different parts of the world; we can see these rapidly increasing problems not only in the Indian context where one has to deal with a sizable section of what we often term as 'tribal' population (traditional societies) (Ramakrishnan, 2009b), which indeed is a much wider global problem too (Brauch et al. 2009b). Since we are dealing with a highly marginalized section of the society living in biodiversity rich regions, conservation issues cannot be delinked from livelihood/developmental concerns of these marginalized sections (Ramakrishnan, 2008a,b). This is the context in which the way we have dealt with conservation linked sustainable livelihood/development of traditional societies have to be viewed. This is the context in which a dynamic approach towards *in-situ* conservation of biodiversity linked with a sustainable livelihood strategy as a short-term solution, and sustainable development of these areas from a longer term perspective becomes significant.

The context and approach

During the colonial phase of Indian history, whilst exploiting the rich natural resources including biodiversity located in situations where traditional societies live, though they were of anthropological interest, they were left alone with little or no efforts made to give them a better quality of life. Further, during colonial phase, and subsequently in the post-colonial phase, developmental pathways designed in the area of land use have invariably been alien to them and in a socio-ecological sense damaging. Still firmly rooted in their traditional cultural values, their expectation has always been to have a developmental pathway with which they could relate and therefore participate in the process. In other words, ecological issues are closely linked with economic well being of traditional societies. It is in this context the traditional wisdom of the communities living in biodiversity rich areas expressed through the TEK that they have refined over time and space has to be seen relevant for community participatory approaches for sustainable conservation linked human well being.

As evident from the preceding discussions, traditional societies have always used the rich biodiversity heritage that they have, with twin objectives: (i) biodiversity in all its scalar dimensions (species to landscapes) as the basis for conserving their cultural values; (ii) simultaneously ensuring for themselves a quality of life that they cherished, until the modern and that they have effectively used has to be the basis for a short-term strategy for development; obviously the TEK linked to that biodiversity has to be emphasized in this effort, bringing in textbook based formal knowledge base technologies only to the minimal extend required. In other words, for effective conservation linked development, linking knowledge systems, rather building upon TEK that they have step-by-step and bringing in formal technologies only to the minimal extend that the system demands, to create 'hybrid' technologies will ensure the following:

In this effort, we have to deal with traditional societies living in biodiversity rich areas from two contrasting situations: (i) one where they still are very 'traditional', in their strongly linked with biodiversity and still are maintaining or have the desire to maintain their traditional values. In other words, they aim to maintain their traditional ways without making drastic departures from them, e.g., the traditional Indian societies living in the northeast, central India, remote areas of the Western Ghats, etc. On the other extreme are the less traditional societies who have had some exposure to modern ways of life, and yet suffer from poverty – placed in mountain situations, e.g., those living in the more accessible regions of the Western Ghat and the Himalayan where more intensified plantation economy such as that of rubber, coffee, tea, cardamom, etc. are coming up at the expense of the marginalized labour-class poor. These two sets of marginalized daily wage labour class sections of the society remain largely neglected and wish to move out of the poverty situation, and yet do not have any diversified options to do so.

These two contrasting situations an important basis for this author to get involved with developmental initiative in the north-eastern jhum-affected landscape situations of Nagaland, on the one extreme, and the tea garden linked landscape in the Kodagu region of Karnataka:

(i) Since one was dealing with very traditional diverse and yet having eco-cultural traditions that are comparable, the broad principle for sustainable land use development of the very short-rotation jhum impacted areas, forest regeneration linked with sustainable jhum became the major concern. The socio-cultural attributes of these societies being not ready for a drastic land use change, the Nagaland situation demanded a gradual build-up on TEK step by step, in an incremental mode – the *incremental pathway* (Swift et al. 1996); this approach formed the basis for a developmental initiative that will not cause any major social disruptions, by taking them away from their traditional tribal values; Formal knowledge based agroforestry technologies were brought in only to the minimal extent required, as one is dealing with a landscape which is already altered for reasons beyond the control of the local communities.

(ii) On the other hand, the mountain societies living in the Kodagu region of Karnataka Western Ghats, working as daily wage labour for the tea garden industry; were looking for an improved quality of life for themselves, rather than being just the wage earners; this apart, the tea industry was already in crisis arising from excessive use chemicals to sustain their energy-intensive modern tea gardens. Further, unlike the very traditional Naga people, the daily wage migrant labour class working for the industry are relatively less traditional compared to the people in Nagaland, being therefore more amenable to accept technologies with relatively greater dosage of formal knowledge based inputs. In other words, technologies were designed to fit into the given socio-ecological context in the mountain landscape – what is referred to as the *contour pathway*, which in essence suggests reconstruction of the agroecosystem in response to the pressure of agricultural development, but to do so using effective use of biodiversity in the given landscape context (Swift et al. 1996). In other words, both these somewhat contrasting pathways has the same twin objectives: (i) conserve biodiversity, both above and belowground too; and (ii) pull these societies out of their marginalized existences by providing them with an improved livelihood option/s.

Incremental pathway for a redeveloped jhum landscape:

Linking knowledge systems (TEK with the formal) formed the basis for the *incremental pathway* for jhum landscape development in the north-eastern hill region. As already discussed, the emphasis was on TEK, bringing in formal knowledge only to the extent required to have 'hybrid technologies' that would form the basis for community participation, redevelopment initiative taken up later for the state of Nagaland in this region, the emphasis being on sustainable jhum fallow management based on agroforestry principles, which ensured against drastic departures from their traditional jhum practices (Box 2).

Box 2. Shifting agriculture (jhum) landscape and sustainable development for north-eastern India (Ramakrishnan, 1992a).

For improving the system of land use and resource management in north-eastern India, the following strategies suggested by this author, and his co-workers are based on a multidisciplinary analysis. Many of these proposals have already been put into practice.

- With wide variations in cropping and yield patterns under jhum practiced by over a hundred tribes under diverse ecological situations, where transfer of technology from one tribe/area to another alone could improve the jhum, valley land and home garden ecosystems. Thus, for example, emphasis on potato at higher elevations compared to rice at lower elevations has led to a manifold increase in economic yield despite low fertility of the more acid soils at higher elevations.
- Maintain a jhum cycle of minimum 10 years (this cycle length was found critical for sustainability), when jhum was evaluated using money, energy, soil fertility biomass productivity, biodiversity, and water quality, as currencies) by greater emphasis on other land use systems such as the traditional valley cultivation or home gardens.
- Where jhum cycle length cannot be increased beyond the five-year period that is prevalent in the region, redesign and strengthen this agroforestry system incorporating ecological insights on tree architecture (e.g., the canopy form of tree should be compatible with crop species at ground level so as to permit sufficient light penetration and provide fast recycling of nutrients through fast leaf turnover rates). Local perceptions are extremely important in tree selection, for introduction into the cropping and fallow phases of jhum, as is being done in a major initiative in the state of Nagaland in north-east India.
- Improve the nitrogen economy of jhum at the cropping and fallow phases by introduction of nitrogen-fixing legumes and non-legumes. A species such as the Nepalese alder (*Alnus nepalensis*) is readily taken in because it is based on the principle of adaptation of traditional knowledge to meet modern needs. Another such example is the lesser known food crop legume *Flemingia vestita*, traditionally used by tribals as an important species when jhum cycles decline below 5 years.
- Some of the important bamboo species, highly valued by the tribals, can concentrate and conserve important nutrient elements such as N, P and K. They could also be used as windbreaks to check wind-blow loss of ash and nutrient losses in water.

- Speed up fallow regeneration after jhum by introducing fast growing native shrubs and trees – socially valued species with ecological keystone value, as noted above can easily be identified through community interaction, as we have shown through our research analyses.
- Condense the time-span of forest succession and accelerate restoration of degraded lands based on an understanding of tree growth strategies and architecture from the TEK angle at the same time taking on board social values attached to species as noted above, which could then be the basis for having the right species mix, over space and time.
- Improve animal husbandry through improved breeds of swine and poultry.
- Redevelop village ecosystems through the introduction of appropriate technology to relieve drudgery and improve energy efficiency (cooking stoves, agricultural implements, biogas generation, small hydroelectric projects, etc.). Promote crafts such as smithying and products based on leather, bamboo and other woods.
- Strengthen conservation measures based upon the traditional knowledge and value system with which the tribal communities could identify, e.g., the revival of the sacred grove concept based on cultural tradition which enabled each village to have a protected forest once upon a time although few are now left.
- In the ultimate analysis, have an integrated approach for land use development in a given ecological/cultural landscape; base short-term sustainable livelihood strategy on building upon traditional knowledge and technology; long-term sustainable development plan should based on larger ecologic/economic considerations should gradually built up to avoid social disruptions.

Now well-known as the NEPED (Nagaland Environmental Protection and Economic Development initiative and implemented by through the state of Nagaland in north-east India, sustainable fallow management after jhum through human intervention formed the basis for this developmental initiative. Based on agroforestry management principles, the NEPED project was steered through the India-Canada Environmental Facility, by this author, and was aimed at a redeveloped jhum, based on the value systems cherished by the people of the region. What this project aimed at was to build upon the available TEK and bring in modern science only to the minimal extent that the situations demanded. Obviously, TEK triggered the process of sustainable natural resource management. It is significant to note here that rejuvenating traditional ways of institution building at the village level along with modern elective processes was also put in place to have 'hybrid' village-level institutions – (Village Development Boards (VDBs), which further triggered community involvement in recent times. Such a short-term strategy for development was aimed to empower the Naga society, so that they can gradually make a move towards developing their own strategies, through such an empowered society, for a long-term land use development plan at an appropriate time.

Using socially valued species of ecological keystone value discussed earlier formed the basis for a decentralized village development plan in one of the north-eastern hill States of India. All the villages organized as VDBs, with the specific purpose of rural development in mind, participated in this massive effort over a 10 yr period. It is worth noting here that though the VDBs were established taking into consideration the traditional village organization followed by the given cultural group; all of the VDBs had the same function, namely participatory rural development. Using this institutional mechanism, the highly distorted shifting agricultural systems, which indeed is basically an agroforestry system, but now operating at subsistence or below subsistence level, was redeveloped, by strengthening the tree component that has been weakened due to extreme deforestation in the region (NEPED and IRRR, 1999). Over a dozen tree species identified in consultation with local communities formed the basis for building the agroforestry models linked to the jhum system.

Reliance being placed on participatory testing rather than being transplanted into the field site by the extension agents, about a dozen tree species are being tested in over 200 test plots. Currently it is estimated that the agroforestry technology is being tested in 5500 ha of replicated test plots. Farmers have adopted this, for local-based testing in 870 villages, covering a total area of 33,000 ha (38 ha per villages x 870 villages); in these plots, local adaptations and innovations for activities such as soil and water management are emphasized.

To summarize the range of objectives that were in the back of the mind of diverse stakeholders which were achieved to a great extend: (i) sustainable traditional agricultural development with improved economic returns per unit area; (ii) improved forest cover in the landscape; (iii) contribute towards carbon sequestration to deal with climate change, an aspect that is now being emphasized system; (iv) dynamic conservation TEK available in the region for use by posterity; (v) sustainable soil fertility and soil water management; (vi) though aimed at sustainable agriculture to start with, it ended up with sustainable landscape management in a holistic sense; and (vi) ensure dynamic conservation of the cultural identity of over 3 dozen ethnic groups living in the region.

The 'auto-route' for sustainable plantation landscape system in Kodagu, Western Ghat region:

On the other extreme, are modern societies involved in mono-culture 'green revolution' agriculture and/or plantation systems in a landscape largely devoid of biodiversity, sustaining production through excessive fertilizer application. Obviously what one is seeking for is to '*buffer and improve soil health*' through sustainable organic residue management, with concerns for enhanced biodiversity at both above- and belowground levels. Herein, the revived and adapted TEK-based inputs is restricted to the minimal extent that is necessary to ensure sustainability of the system, by reducing external energy subsidies to the extent desirable, at the same time ensuring increased quality production. This apart we are also dealing with a large number of plantation workers as routine daily wage labour in the industry. To add on to this, additional livelihood activities could enhance the quality of life of these poor workers. This is the context in which *in situ* sustainable management of soil fertility through increasing tree cover in the landscape has to be viewed.

The 'FBO' (Fertilization Bio-organique Dans les Plantations de the) or 'Bio-organic Fertilization for Plantations') *in situ* soil fertility management technology standardized by a group of Indian and French scientists by a group of Indian scientists working with a set of French scientists, as part of an ongoing international initiative on tropical soil biology and fertility (TSBF), looked at land degradation issues, working in the tea garden plantations of the Karnataka Western Ghats of southern India (Box 3). This technology, patented between the Indian and French scientists involving the tea garden managers, has now gone up to the Chinese shores. Here, one could visualize the situation where socially valued species with ecological keystone value for nutrient conservation and release through organic leaf litter residues forms the basis for organic residue management for soil fertility build-up, and enhanced soil biodiversity, without even knowing what is in that diversity since it is difficult to catalogue the same, which then is monitored through using earthworm species of ecological keystone value within the soil sub-system; in other words, appropriately identified organic residue management for sustainable soil fertility build-up makes it possible to have a substantial reduction in inorganic fertilizer application anywhere between 30–50% (Senapati et al. 2002).

Box 3. The FBO technology for buffering intensively managed modern agricultural systems (from: Senapati et al. 2002)

- As part of a systems analysis, a variety of cultured earthworms along with a mixture of high and low quality organic materials in specified proportions, and application procedures are determined for a given situation. The speciality of the capsule is autoregulation of the synchrony of plant demand and soil supply of nutrients through regulated mobilization and immobilization. The capsular implementation is carried out only inside the designed 'fertilization units' to minimize physical disturbance and for high efficiency. *In situ* use of resources within the prevailing cropping system allows flexibility, to decide upon the components, combinations, proportion, placement and application schedule compatable with site conditions.
- Earthworm technology that incorporate organic residues into the soil to stimulate the activities of existing or introduced soil inhabiting earthworms is 'in-soil' technology. FBO prefers this for long-term system sustainability and effectiveness. Success of this technology depends upon the choice of suitable species, the provision of appropriate organic supplies to feed the worms and the maintenance of a minimum diversity in the whole invertebrate communities. Replenishment of the lost eathworm diversity under intensive land use, demands careful selection of existing and eliminated species for rebuilding the community suitable to takeover their role from a threshold level.
- Large-scale production units are laid out, for efficient use of space and to meet the demand for direct 'in-soil' earthworm management. Depending on variable culture conditions it has been quantified that the inoculated population may multiply 10 to 30 times within a span of about three months, at a minimal cost of about about ₹ 200 for a kilogram of live worms, for use in adjoining fields.
- Locally available organic residues are selected for soil amendment along with earthworm species from the reference native systems, in case their population is below a threshold level. Composition of species and their functional proportion depend on the location specificity and soil status. The emphasis is on effective use of local resources and thus minimize labour, expense and time.
- 'FBO' technology package is not equivalent to any single process like vermi-composting, mulching, chemical fertilization, soil replacement, hormone spray etc. nor even combination of some of them.

- Experimental plot level analysis show tea growth, upto 75% to 260% enhancement in intermediate profit over the conventional system within a period of 6 months to one year; this is apart from improved soil physical-chemical and biological qualities for which economic cost of evaluation is not available at present. The overall biological status improved by more than 200% along with about 60% improvement in physico-chemical soil characteristics, over the control the control plots under conventional management.
- Maintaining a minimal functional diversity, this technology does not favour an exaggerated development of a single species that accumulate large biogenic structures and result in severe problems of soil compaction. Application of too much and too little of organic matter without monitoring of quality, quantity and placement is linked to the functional integrity of the system. Inoculation of alien species of earthworms should be weighed against the future implications for biotic community interactions within the soil, and the consequent issues of system sustainability.

The end results achieved arising from this developmental initiative could be seen to have had the following dimensions: (i) increased tree cover in the landscape context, with emphasis on traditionally valued tree species coming into the plantation landscape as a whole; (ii) improved yield of tea with better quality for the leaves arising from increased use of organic residues for improved soil fertility; (iii) overall improvement in aboveground biodiversity with implications for improved belowground biodiversity too and of course (iv) overall sustainability of the landscape system as a whole. What we are trying to achieve is to compensate for the ill-effects that would otherwise be caused arising from intensified land use/land conversion that would impact upon biodiversity, both natural and human-managed, both of which can impact upon endemism at the species/sub-specific level (Fig. 4).

What is to be recognized in the ultimate analysis from this case study is that there are multiple benefits accruing through this initiative: (i) sustainable soil fertility management without compromising on productivity; (ii) improved quality output – in this case better flavoured and more organic-oriented tea leaves; (iii) organic fertilizer production initiative using tree species as part of agroforestry practices through socially valued species providing added income opportunities for the marginalized sections of the society which otherwise have to contend with only daily wage as labourers in the garden; and (iv) a much more diversified natural landscape in which the agroforestry tree component as an integral component of the tea garden landscape, contributes much to enhance a rich associated biodiversity, both aboveground and belowground.

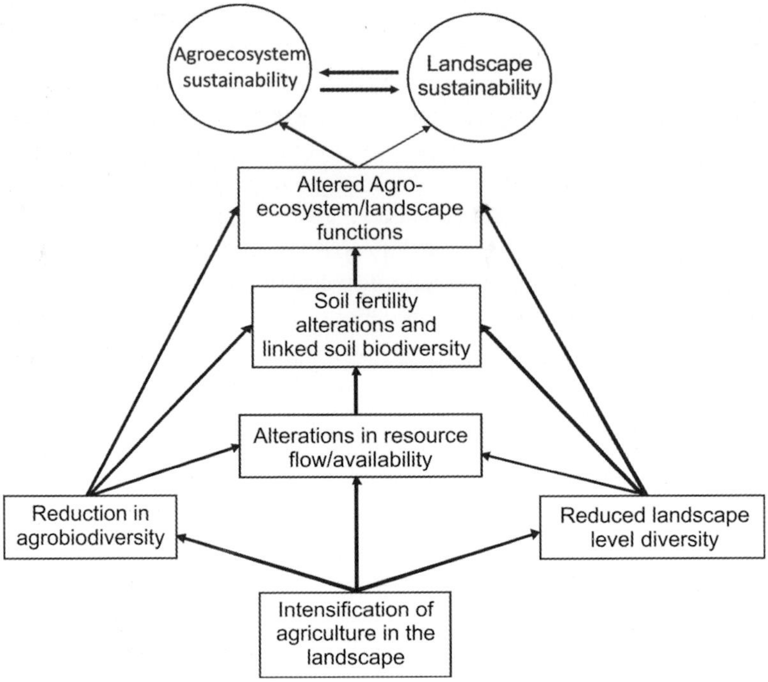

Fig. 4. Intensification of agriculture, biodiversity changes, changing resource flow, aboveground biodiversity impacting upon belowground biodiversity and linked soil fertility alterations in soil fertility levels, altered agroecosystem/ landscape functions with sustainability implications (adapted from Swift and Anderson, 1993).

CONCLUSIONS

In an interesting article entitled 'Ecological economics in post-autistic world', Costanza (2003) implies that that "the main mainstream economics had remained "autistic" - in its deficits in communication and social interaction with other disciplines, preoccupation with mathematical fantasy, language impairment in its limited and specialized vocabulary, and excessive attachment to certain objects (assumptions and models)". As Costanza puts it, it is only in recent times, in the context of an unpredictable future where one is faced with very many real world problems often unpredictable (Senapati et al. 2002); this is the context in which conservation of biodiversity in all its scalar dimensions has a key role to play. 'Ecological economics' or what has come to be known as 'sustainability science' deals with our ability to move beyond and come up with an integrative metaparadigm in research linked applications in the field (Box 4).

Box 4. Ecological economics: An integrative metaparadigm (Costanza, 2003)

- Ecological economics is a transdisciplinary effort to link the natural and social sciences broadly, and especially ecology and economics (Costanza, 1991). The goal is to develop a deeper scientific understanding of the complex linkages between human and natural systems, and to use that understanding to develop effective policies that will lead to a world which is ecologically sustainable, with fair distribution of resources (both between groups and generations of humans and between humans and other species), efficiently allocating scarce resources including "natural" and "social" capital. This requires new approaches that are comprehensive, adaptive, integrative, multiscale, pluralistic, evolutionary and which acknowledge the huge uncertainties involved.
- One of the basic organizing principles of ecological economics is to focus on complex interrelationships existing between ecological sustainability (including system carrying capacity and resilience), social sustainability (including distribution of wealth and rights, social capital, and coevolving preferences) and economic sustainability.
- Ecological economics is based on shared assumptions and theory – a *metaparadigm*. Rather than espousing and defending a single discipline or paradigm, it seeks to allow a broad, pluralistic range of viewpoints and models to be represented, compared, and ultimately synthesized into a richer understanding of the inherently complex systems it deals with. It represents a commitment among economists, ecologists, and other academics and practitioners to learn from each other, to explore new patterns of thinking together, and to facilitate the derivation and implementation of effective economic and environmental policies.
- Ecological economics therefore includes some aspects of neoclassical environmental economics, traditional ecology and ecological impact studies, and several other disciplinary perspectives as components, but it also encourages completely new, more integrated, ways to think about the linkages between ecological and economic systems.

'Ecological Economics' or 'Sustainability Science' is all about working with 'nature' whilst we deal with an improved quality of life for diverse sections of our society, particularly for those who still remain marginalized in the developing world context. What is heartening to perceive, however, is that the desire of humans to be close to 'nature' is emerging rather rapidly amongst all sections of the human society; this is becoming more and more evident even amongst the most industrialized parts of the world as in USA (Shutkin, 2000). Working with 'nature', rather than being delinked from it, has significant implications for conservation linked societal well-being of of marginalized sections of the society which is proving to be critical for conflict resolution, human security and peace (Ramakrishnan, 2009a).

For such an effort to take roots amongst varied sections of the society, it is important one needs to create awareness amongst all stakeholders for community-participatory efforts for conservation linked sustainable development. The experience of this author has been that this is a prerequisite for arriving an action plan - our efforts in this direction through many policy documents (Ramakrishnan, 2000b; 2006b, to cite some examples), out-reach volumes (Ramakrishnan, 2008a, b; Ramakrishnan et al. 2005; 2006) and audio-visual documentaries too ('*When the mist is lifted*' – a documentary on the Apatani cultural landscape management as evolved by this highly innovative ethnic group , to cite one of the many audio-visuals that we have done) have yielded rich dividends.

At the international level, there are very many rapidly emerging initiatives – the UNESCO's World Heritage sites and Biosphere Reserve sites are already well in place. There are now many other emerging initiatives: (i) that of the International Human Dimensions Programme on climate change dealing with knowledge systems and societal learning as the basis for coping with environmental uncertainties which implies community participation in conservation linked sustainable developmental issues; and (ii) FAO's initiative on 'Globally Important Agricultural Heritage Systems' (GIAHS) that deals with using traditional agricultural systems as learning grounds for sustainable agricultural practices, which implies conservation of agrobiodiversity for food security for the future. Conserving biodiversity endemics inclusive, as part of 'nature' in a socio-ecological sense is the key for dynamic conservation linked sustainable development, with implications for 'human well being', with benefits accruing to all sections of the society.

References

Bodeker, G. (1994). Traditional health knowledge and public policy. *Nature & Resources*, 30: 5–16.

Brauch, H.G., Spring, U.O., Grin, J., Mesjasz, C., Kameri-Mbote, P., Behera, N.C., Chourou, B. and Krummenacher, H. (Eds.). (2009). *Facing Global Environmental Change: Environmental, Human, Energy, Food, Health and Water Security Concepts, Vol. 4*. Hexagon Series on Peace and Environmental Security and Peace. Springer-Verlag, Berlin, Germany, p. 1586.

Costanza, R. (Ed.). (1991). *Ecological Economics: The Science and Management of Sustainability*. New York: Columbia Univ. Press, p. 355.

________ (2003). Ecological economics is post-autistic. Post-autistic Economics. http://www.paecon.net/PAEReview/issue20/Costanza20.htm (Review – article 2, Issue No. 20, 3).

Dube, S.C. (Ed.). (1998). *Antiquity to Modernity in Tribal India*. Vol. 1. Inter-India Publ., p. 358.

Holling, (1995). Sustainability: The cross-scale dimension. In: *Defining and Measuring Sustainability: The Biophysical Foundations*. Munasinghe, M. and Shearer, W. (Eds.). pp. 65–75. U.N. Univ. of Tokyo and The World Bank, Washington, D.C., USA.

Hughes, J.D. and Chandran, M.D.S. (1998). Sacred groves around the earth: An overview. In: *Conserving the Sacred: For Biodiversity Management*. Ramakrishnan, P.S. Saxena, K.G. and Chandrashekara, U.M. (Eds.). pp. 69–86. UNESCO and Oxford & IBH Publ., New Delhi.

Indian National Science Academy, Chinese Academy of Sciences and U.S. National Academy of Sciences, 2001. *Growing Populations, Changing Landscapes: Studies from India, China and the United States*. Wolman, M.G., Ramakrishnan, P.S., George, P.S., Kulkarni, S., Vashishtha, P.S., Shidong, Z., Qiguo, Z., Yi, Z., Long, J.F., Rosenzweig, C. and Solecki, W.D. (Eds.). National Academy Press, Washington, D.C., USA, p. 299.

Lambin, E.F., Turner II, B.L., Geist, H.J., Agbola, S., Angelsen, A., Bruce, J.W., Coomes, O., Dirzo, R., Fischer, G., Folke, C., George, P.S., Homewood, K., Imbernon, J., Leemans, R., Li, X., Moran, E.F., Mortimore, M., Ramakrishnan, P.S., Richards, J.F., Skånes, H., Steffen, W., Stone, G.D., Svedin, U., Veldkamp, T., Vogel, C., Xu, J. (2001). The causes of land-use and land-cover change: Moving beyond the myths. *Global Environmental Change*, 11: 261–269.

Ministry of Environment & Forests (1994). *Ethnobotany in India: A Status Report*. Min. of Environment & Forests, Govt. of India, New Delhi, p. 68.

National Academy of Sciences (1975). *Underexploited Tropical Plants with Promising Economic Value*. National Academy of Sciences, Washington, D.C., p. 189.

National Academy of Sciences (1979). *Tropical Legumes: Resources for the Future*. National Academy of Sciences, Washington, D.C., p. 331.

NEPED and IRRR (1999). *Building Upon Traditional Agriculture in Nagaland.* Nagaland Environ. Protection & Economic Development, Nagaland, India & International Inst. of Rural Reconstruction, Phillippines, p. 235.

Ramakrishnan, P.S. (1992a). *Shifting Agriculture and Sustainable Development: An Interdisciplinary Study from North-Eastern India.* UNESCO-MAB Series, Paris, Parthenon Publ., Carnforth, Lancs. U.K., p. 424. (republished by Oxford University Press, New Delhi, 1993).

________. (1992b). Tropical forests: Exploitation, conservation and management. *Impact of Science on Society,* 42 (No. 166): 149–162.

________. (2008a). *Ecology and Sustainable Development: Working with Knowledge Systems.* National Book Trust of India, Govt. of India. p. 217. Revised Edition of Ramakrishnan, P.S. 2001. *Ecology and Sustainable Development.* National Book Trust of India, Govt. of India, p. 198.

________. (2008b). *The Cultural Cradle of Biodiversity.* National Book Trust, India, New Delhi, p. 243.

________. (2009a). Ecological economics: From the viewpoint of the marginalized sections of the society. *Second Prof. R. Misra Birth Centenary Lecture*, 2009. National Institute of Ecology, Jaipur & New Delhi, p. 26.

________. (2009b). Linking knowledge systems for socio-ecological security. In: *Facing Global Environmental Change: Environmental, Human, Energy, Food, Health and Water Security Concepts.* Vol. 2. Peace Research and European Security Studies. p. 817–828. AFES Press, Brauch, Hans Günter; Grin, John ; Mesjasz, Czeslaw; Krummenacher, Heinz ; Behera, Navnita Chadha; Chourou, Béchir; Spring, Ursula Oswald; Kameri-Mbote, Patricia (Eds.). Germany (in press).

Ramakrishnan, P.S., Saxena, K.G. and Chandrasekhara, U. (Eds.). (1998). *Conserving the Sacred: For Biodiversity Management* (UNESCO Vol.). Oxford & IBH Publ., New Delhi, p. 480.

Ramakrishnan, P.S., Chandrashekara, U.M., Elouard,C., Guilmoto, C.Z., Maikhuri, R.K., Rao, K.S., Sankar, S., and Saxena, K.G. (Eds.). (2000a). *Mountain Biodiversity, Land Use Dynamics and Traditional Ecological Knowledge.* UNESCO & Oxford & IBH Publ., New Delhi, p. 352.

________. (2000a). *Mountain Biodiversity, Land Use Dynamics and Traditional Ecological Knowledge.* (Policy document) UNESCO & Oxford & IBH Publ., New Delhi, p. 16.

Ramakrishnan, P.S., Boojh, R., Saxena, K.G., Chandrashekara, U.M., Depommier, D., Patnaik, S., Toky, O.P., Gangawar, A.K. and Gangwar, R. 2005. *One Sun, Two Worlds: An Ecological Journey.* UNESCO and Oxford & IBH, New Delhi, p. 286. (with a Hindi translation: UNESCO, 2006 – *'Vigyan aur Paramparik Gyan: Ek Samvad'*)

Ramakrishnan, P.S., Saxena, K.G. and Rao, K.S. (2006a). *Shifting Agriculture and Sustainable Development of North-East India: Tradition in Transition.* UNESCO and Oxford & IBH, New Delhi, p. 495.

________. (2006b). *Shifting Agriculture and Sustainable Development of North-East India: Tradition in Transition.* (Policy document) UNESCO and Oxford & IBH, New Delhi, p. 27.

Raven, P. (1998). Biodiversity and the human prospect – Plenary lecture, Frontiers in biology: The challenges of biodiversity. In: *Biotechnology and Sustainable Agriculture.* Chou, C.H. and Shao, K.T. (Eds.). pp. 1–9. Academina Sinica, Taipei.

Senapati, B.K., Lavelle, P. and Ramakrishnan, P.S. 2002. Earthworm based technology application for assessment and management of traditional agroforestry systems. In: *Traditional Ecological Knowledge for Managing Biosphere Reserves in South and Central Asia.* Ramakrishnan, P.S., Rai, R.K., Katwal, R.P.S. and Mehndiratta, S. (Eds.). pp. 139–160. UNESCO and Oxford & IBH, New Delhi.

Swift, M.J. and Anderson, J.M. (1993). Biodiversity and ecosystem function in agricultural systems. In: *Biodiversity and Ecosystem Function.* Schulze, E.D. and Mooney, H. (Eds.). pp. 15–42. Springer, Berlin.

Swift, M.J., Vandermeer, J, Ramakrishnan, P.S., Anderson, J.M., Ong, C.K., Hawkins, B. 1996. Biodiversity and agroecosystem function. In: *Functional Roles of Biodiversity: A Global Perspective.* Mooney, H.A., Cushman, J.H., Medina, E., Sala, O.E., Schulze, E.D. (Eds.). pp. 261–298. SCOPE Series. John Wiley, Chichester, U.K.

UNEP (1995). *Global Biodiversity Assessment.* United Nations Environment Programme & Cambridge Univ. Press, Cambridge, U.K., p. 1140.

UNESCO and CEE (2005). *Nanda Devi Prakriti Samrakshan Geeth* A collection of environmental folk-music (on CD). UNESCO, New Delhi & Centre for Environmental Education, Lucknow.

Valiathan, M.S. (1998). Healing plants. *Current Science,* 75: 1122–1126.

4

Evolution and Diversity of Rice Genome

J. Giri, S. Tyagi and A.K. Tyagi

INTRODUCTION

Among the food grains, cereals and millets represent most important crop plants. They belong to grass family Poaceae. So far, in grass family, genome level information is available for rice (International Rice Genome Sequencing Project 2005; Yu et al. 2005), sorghum (Paterson et al. 2009), maize (Schnable et al. 2009) and *Brachypodium* (The International Brachypodium Initiative 2010). Simultaneously, limited information on physical map, transcriptome, and partial genome sequence is also available for other important members like barley and wheat. This has allowed a comparative genome analysis within grass family in relation to genome evolution and diversity. The aim of this article is to provide information on evolution and diversity of the genome and domestication genes of rice – the most important staple food grain.

EVOLUTION

In recent years, an understanding of the evolutionary events leading to the origin of rice and other cereals has been greatly helped by comparative genomics studies. Availability of genome-wide information from sequenced genomes, mapped expression sequence tags (ESTs) and conserved orthologous sequence (COS) markers has greatly facilitated the same. It has been proposed that genomes of barley, maize, rice, sorghum and wheat have diverged from an ancestral cereal genome about 50–70 million years ago (MYA) (Moore et al. 1995: Devos, 2005; Bolot et al. 2009). The reliability of the proposed synteny models depends on the identification of true orthologs distinct from paralogs since genome or segmental duplications have been crucial for evolution of plant genomes. It is evident that microcolinearity of genomes has also been influenced by genomic rearrangements. Analysis of intragenomic duplications and syntenic relationship of maize, rice, sorghum and wheat helped identify seven common duplications and five ancestral chromosomes (Salse et al. 2008). Evidence gives credence to a whole genome duplication (WGD) event from a putative ancestor with five chromosomes about 50–70 MYA. This was apparently followed by two intrachromosomal translocations and fusions leading to an intermediate ancestor with 12 chromosomes ($5 + 5 + 2 = n$). While rice appears to have retained the same chromosome number ($n = 12$), other cereal genomes evolved and diverged differently. Two chromosomal fusions paved the way for the evolution of maize and sorghum genomes ($n = 10$). In contrast to sorghum genome ($n = 10$), for maize an additional WGD event lead to 20 chromosomes which was followed by chromosomal fusion events leading to a novel genome with 10 chromosomes. For wheat or barley, five chromosomal fusions would have occurred, in a 12 chromosome ancestral genome, to yield a $n = 7$ genome. By incorporating the above in a model, it is possible to depict the evolutionary relationship between cereal genomes with a greater insight (Bolot et al. 2009). The analysis of evolutionary relationship based on comparison of *Brachypodium* genome (The International Brachypodium Initiative 2010) with wheat, rice and sorghum supports the earlier views. It seems that five chromosome *Brachypodium* genome evolved from a putative ancestral genome ($n = 5$) via an intermediate ancestor with 12 chromosomes and seven chromosomal fusion events. Interestingly, such chromosomal fusions are not shared with Triticeae and must have occurred independently.

DIVERSITY

Rice is one of the most diverse crops on earth. The genus *Oryza* has about 24 species including both wild and domesticated rice found in different geographical areas with diverse environmental conditions ranging from dry to aquatic (Vaughan et al. 2003). Four species complexes namely, *O. sativa* complex, *O. officinalis* complex, *O. ridleyi* complex and *O. granulata* complex, covering all *Oryza* species, were delimited (Vaughan et al. 2003). Cultivated rice and their close wild relatives belong to *O. sativa* complex with same haploid chromosome number (12) while members of other species complexes are diplod to allotetraploid. Members of a species complex can interbreed but this is not possible between the members of different species complexes (Sweeney and McCouch, 2007). The commonly cultivated rice belongs to two species *O. sativa* or Asian rice and *O. glaberrima* or African rice. It has been established now that *O. sativa* species (*japonica* and *indica*) was derived from wild rice *O. rufipogon* through the process of domestication which converted wild rice into a crop that serves as a staple food for more than half of the mankind. *O. rufipogon* is considered to exist in two forms, annual (*O. nivara*) and perennial (*O. rufipogon*). Which of these two forms serves as the immediate ancestor of *O. sativa* is a matter of ongoing research and *O. rufipogon* as a whole is considered as ancestor of common rice. *O. sativa* and *O. rufipogon* overlap in their geographical range and lack of strict reproductive barrier between them has given rise to diverse rice populations (Kovach et al. 2007). *O. sativa* includes two distinct major groups, namely *japonica* and *indica,* sometimes referred as two sub-species and recently five different subpopulations have been identified in *O. sativa* on the basis of genetic, biochemical and morphological evidences (Garris et al. 2005). These subpopulations encompassing diverse varieties are *indica* and *aus* (*indica* varietal group), *tropical japonica, temperate japonica*, and *aromatic* (*japonica* varietal group) (Garris et al. 2005). The subspecies of *O. sativa, japonica* and *indica,* are believed to have separated around 0.2-0.4 million years ago and still have several morphological and physiological differences despite overlap in their growing geographical areas (Vitte et al. 2004, Zhu and Ge 2005) suggesting the difference at genome level. These two species differ in leaf color, seed size, shape, texture, shattering and apiculus hair length.

Two cultivated and 22 wild species of the genes *Oryza* represented by 10 genome types could serve as useful resource to study genome evolution and domestication. The variation among these species could be identified and related to species evolution. In particular, wild species with AA genome, similar to cultivated *Oryza sativa* (rice) genome, are of great interest. Even within *O. sativa indica* and *japonica* genomes, 5% asymmetrically located and 5% subspecies-specific genes were observed (Ding et al. 2007). Two-fold size of *O. australiensis* genome in comparison to *O. sativa* has been attributed to proliferation of three LTR retrotransposons (Piegu et al. 2006). Recently, structural variation between genomes of *O. sativa* spp. *japonica* (Asian rice), *O. glaberrima* (African rice) and wild relatives of *O. sativa,* namely, *O. nivara* and *O. rufipogon* (diverged <1 MYA) has been analyzed for regional expansions, contractions and inversions. In all, 674 expansions of 25–38 Mb, 611 contractions of 4–19 Mb and 140 inversions of 14–19 Mb were observed which resulted from enrichment of TEs and LTRs or genes related to defence (Hurwitz et al. 2010). A comparative framework physical map representing ten genomes of rice in the form of BAC fingerprints and end sequences is also available (Kim et al. 2008). Such information can form the basis for understanding genes level dynamics of evolution, species formation and comparative biology.

Genome Structure and Function

The genomes of both the cultivated subspecies (*indica* and *japonica*) of *O. sativa* have been sequenced (International Rice Genome Sequencing Project 2005; Yu et al. 2005, Vij et al. 2006; Raghuvanshi et al. 2010). The 370 Mb of *japonica* rice represents a map-based sequence, while sequence of *indica* rice has been generated by whole genome shot-gun approach. Although, certain gaps, centromeric and telomeric regions remain to be sequenced, the information available represents most of the genome (>95%) and gene content. Besides, >35% genome sequence is represented as repeat elements. In addition, whole genome and segmental duplications represent almost 60% of the genome. The analysis of *japonica* rice genome also unraveled >18,000 SSRs helpful in molecular mapping and breeding. Feltus et al. (2004) have identified 408,898 SNPs/INDELs on comparison of *indica* and *japonica* rice genomes. A resequencing of about 100 Mb unique genome of 20 landraces and varieties of rice and their analysis has provided 160,000 non-redundant SNPs (McNally et al. 2009). Such resources are of great value for breeding history of cultivated rice varieties and selection of heterotic parents.

The number of predicted genes in *japonica* rice was 37,544. Continuing annotation has improved the prediction and reduced the number of genes to about 32,000. Such knowledge-based improvements in annotation of rice genes can be followed at major rice genome portals (rapdb.dna.affrc.go.jp, rice.plantbiology.msu.edu, rice.genomics.org.cn/rice, www.ncbi.nlm.nih.gov/sites/entrez?db=genome). In addition, information about transcriptome in the form of ESTs, MPSS tags, micro-array, small RNAs and miRNAs is emerging rapidly to help understand biology of rice (Raghuvanshi et al. 2010). The analysis of the function of rice genes requires a combination of several approaches including transgenics, mutagenesis, and tagging (Kathuria et al. 2007; Krishnan et al. 2009). Epigenetic studies at genome-wide level are also being carried out to investigate variability by nucleotide modification (Li et al. 2008), which positively correlates to heterochromatin formation.

Variation in Domestication Related Genes of Agronomic Significance

Wild rice (*O. rufipogon*) differs significantly from domestic rice varieties in many aspects like seed shattering, awn length, seed dormancy, pericarp colour, seed size and number, seed texture, panicle structure and plant architecture. These traits were targeted during domestication by artificial selection of rice plants for minimizing the seed shattering and dormancy and maximizing the yield and grain quality for human consumption. However, these traits do not exist as clear cut divide between wild and domesticated rice. It should be noted here that rice domestication is not a single event rather it happened as a continuous process over the period of time by selection of plants from pre-existing variations among wild cultivars. Therefore, genetic loci underlying domestication traits would display variations among the wild accessions but not in domesticated rice where they are fixed due to continuous artificial selection (Sweeney and McCouch, 2007). The genes controlling domestication traits are called domestication genes and their continuous selection has led to the present day rice varieties (Izawa, 2008).

Seed shattering

Seed shattering is the natural way of seed dispersal in plants. Wild rice show high degree of seed shattering on maturity that makes it difficult to harvest the seed. On the other hand, all domesticated rice varieties showed highly reduced degree of seed shattering indicating the selection of this trait by ancient farmers. The QTLs underlying this trait have been identified and cloned. By crossing the annual form of wild rice *O. nivara* and *indica*, Li et al. (2006) have identified three QTLs, major of them, *SH4*, accounted for 69% of total variation observed and localized to chromosome 4. The wild allele of *SH4* showed dominance and caused shattering. The candidate *SH4* gene was found to be encoding a MYB domain transcription factor and carries a SNP resulting in one amino acid change in predicted DNA binding domain in *indica*. This transcription factor was found to be involved in cell wall degradation leading to development of abscission layer, responsible for seed shattering. In *O. sativa*, development of abscission layer is incomplete resulting in reduced seed shattering. The non-shattering allele is common in all *O. sativa* accessions while dominant shattering allele is found in shattering accessions of *O. rufipogon*. Interestingly, non-shattering allele was also common among the accessions of *japonica* and *indica* indicating the origin of this mutation in *O. rufipogon* before the separation of *japonica* and *indica* (Lin et al. 2007).

Another QTL for seed shattering *SH1* was identified by QTL analysis between seed shattering *indica* (Kasalath) and non-shattering *japonica* (Nipponbare) cultivars (Konishi et al. 2006). The major QTL, *SH1* accounted for 69% of total variation and candidate gene was found to encode a BEL-type homeobox transcription factor. The responsible mutation, a SNP in regulatory region at 12 kb upstream of this gene, reduced the expression of gene only at site of potential abscission layer formation, resulting in less seed shattering in *japonica* rice. On sequence analysis in SNP region between *O. sativa* and *O. rufipogon,* it was found that SH1 allele is not fixed in all the accessions. While it was common in temperate and tropical *japonica, indica* accessions showed absence of this allele suggesting a sub-species level fixation of the mutation.

Grain size

Wild rice produces very less number of thin grains in thin panicles as compared to domesticated rice. Accessions with larger size grains would increase crop yield and would have been picked up easily by earlier farmers for successive seed sowing, leading to their domestication. Like other domestication trait, seed size is also a quantitative character and QTL analysis detected several loci for grain width using Nipponbare and Kasalath populations (Shomura et al. 2008). The major one, *qSW5,* explained 38% of total variations in F2 population. The candidate gene *qSW5* was identified and cloned, however, the nature of its molecular function could not be established. Nipponbare seeds showed higher cell number in lemma as compared to Kasalath. Increased cell number resulted in increase sink size for grain filling resulting in wider and heavier grains in Nipponbare. The NIL of Kasalath *qSW5* region in Nipponbare background showed decreased cell number indicating the role of *qSW5* gene in determining cell number in outer glume of rice flower. A deletion of 1,212 bp deletion in *qSW5* orf of Nipponbare was found to be responsible for the observed phenotype. Analysis of *qSW5* sequence in different rice landraces revealed the source of this mutation to *japonica* subpopulation. Shomura et al. (2008) also analysed the domestication pattern of a grain number QTL, *GN1a* (Ashikari et al. 2005). *Gn1a* encodes a cytokinin oxidase that degrades cytokinin in panicle. *Indica* cultivars carry a mutated *Gn1a* gene resulting in increased grain number in panicles. Further analysis revealed the prevalence of mutated *Gn1a* allele in *indica* cultivars while *japonica* cultivars carry a different form of *Gn1a* (Shomura et al. 2008). This work showed that such mutation occurred after *japonica* and *indica* divide and strengthens the view about independent domestication events in *japonica* and *indica* (Kovach et al. 2007). A QTL (*GW2*) with similar role in regulating grain width and weight was found to be encoding a E3 ubiquitin ligase (Song et al. 2007). The loss of function of *GW2* resulted in increased number of cells in rice hull providing the wider space for grain filling. Minor mutations were found to be responsible for loss of function in *japonica* cultivar, however, role of GW2 in the light of rice domestication process has not been investigated. Like grain width, grain length is also an agronomically important trait. A major QTL, *GS3*, for grain length accounts for 80–90% of total length variation and encodes a transmembrane protein (Fan et al. 2006). Tested varieties with longer grains have nonsense mutations which account for premature stop codon in *GS3* coding region while those with shorter grains lack this mutation.

Another gene for grain weight, *GIF1*, characterized from chromosome 1 has an effect on duration and rate of grain filling, contributing towards the grain weight and ultimately to grain yield (Wang et al. 2008). Mutant screening in *japonica* cultivar led to the identification of *gif1* mutation, responsible for slower grain filling and 24% less grain weight as compared to normal plants. The candidate gene encodes a cell wall invertase and has one bp deletion in orf, leading to premature stop codon in mutants. The functional *GIF1* gene regulates the sucrose unloading in ovular and stylar tissue for starch synthesis during grain filling. The analysis on the gene and regulatory sequences in *japonica* and *indica* rice as well as *O. rufipogon* indicated the artificial selection of *GIF1* mutation and its role in rice domestication.

Grain quality

Red pericarp is common in wild landraces of rice where it is supposed to provide resistance against pests and pathogens while all the modern cultivars have white colour grains. A single locus *Rc* was found to be responsible for this change from red to white pericarp (Sweeny et al. 2006). This locus was cloned by positional cloning using the population derived from a cross between *O. rufipogon* (red pericarp) and *japonica* (white pericarp) rice. *Rc* locus encodes a bHLH transcription factor and showed a 14 bp deletion in exon 6 knocking out its function, responsible for red to white pericarp in modern cultivars. A survey of wild and domesticated rice for *Rc* locus revealed that 98% of tested white pericarp accessions have 14 bp deletion while all the tested red grain accessions lack this mutation (Sweeney et al. 2007). This study further suggested that *Rc* mutation first appeared in *japonica* haplogroup and later spread to other *O. sativa* accessions. An SNP named as *Rc-s* in region near to *Rc* locus was also found to be responsible for white pericarp in *aromatic* and *aus* accessions independent of *Rc* mutation (Sweeney et al. 2007).

Grain texture is an important feature that separates *japonica* and *indica* cultivars. *Japonica* grains are low in amylose content and thus have sticky or glutinous texture while *indica* grains are high in amylose content and are nonsticky. The sticky texture is result of malfunctioning of Waxy (*Wx*) gene that encodes a granule bound starch synthase which catalyses the starch biosynthesis in grains (Wang et al. 1995). Two allele of *Wx* locus, Wx^a and Wx^b are common in *japonica* and *indica* rice, respectively. *O. rufipogon* also possesses Wx^a allele, indicating its wild type nature. The difference between two alleles is alternative splicing caused by a SNP at a donor site of the first intron within the *Wx* gene (Issiki et al. 1998). Wx^b and its derivative *wx* allele are most common in glutinous or sticky rice while *indica* type Wx^a allele has only limited dissemination (Yamanaka et al. 2004).

Plant architecture

O. rufipogon has prostrate growth during vegetative stage and produces few panicle carrying erect branches during reproductive stage. This type of plant growth occupies more space, thus inhibiting weed growth, and protects panicle from animals in their natural habitat. Domestic rice (*O. sativa*) produces erect branches throughout the life, in order to maximize plant density and photosynthesis efficiency ultimately leading to increased yield. This transition from prostrate to erect growth has been a critical event in rice domestication. Recently, a pleotropic effect gene, *PROG1* (prostrate growth 1), has been identified on chromosome 7 that controls prostrate growth in addition to grain number and yield (Tan et al. 2008). Functional *PROG1* encodes a C_2H_2 zinc-finger transcription factor protein that increases the tiller angel by increasing the number of cells at the base of tillers resulting in prostrate growth in wild rice. *O. sativa* carrying the defective *prog1* with 15 SNPs and 6 indels in coding region accounted for 23 amino acids change from *O. rufipogon PROG1*. Interestingly, SNPs and indels in 5' regulatory region of *prog1* were also found to have effect on its function. Sequence comparison in 182 *indica* cultivars with erect growth showed same SNPs and indels in *prog1* region indicating the strong selection on this allele during domestication. Out of 30 accessions of *O. rufipogon* tested, 16 have same *PROG1* allele while other showed variations, indicating the diversity at *PROG1* locus among wild accessions. This work further showed that the selection on *prog1* locus not only resulted in erect growth but also increased grain number and yield. Simultaneously, another group has also cloned *PROG1* gene from same accession and showed similar relationship between *PROG1* function and prostrate growth (Jin et al. 2008). Although orf length and mutation sites of *prog1* differ between these two studies, sequence conservation was seen among the tested wild and domestic rice cultivars. Panicle architecture has major influence on plant yield. *Japonica* cultivars have more erect and dense panicle branching as compared to *indica* (curved loosely packed panicles). The QTL responsible for this trait was mapped and named as *DEP1* (dense and erect panicle1). The underlying gene encodes a PEBP (phosphatidylethanolamine-binding protein) like domain protein and acts as a negative regulator of panicle branching and grain number (Huang et al. 2009). Mutant plants with nucleotide deletions in exon of *dep1* allele account for pre-mature stop codon, have increased degree of panicle branching and grain number. This allele was common in tested *japonica* accessions while absent in *indica* indicating the *japonica* specific fixation of this allele. Two independent groups characterized *OsSPL14*, a gene involved in regulation of tiller number and panicle branching in rice resulting in higher yield (Jiao et al. 2010; Miura et al. 2010). OsSPL14 (rice Squamosa Promoter binding Protein-Like), a transcription factor, is regulated post-transcriptionally by microRNA-156. A point mutation in complementary region of *OsSPL14* alters its repression by miR156, as a result plants have water lodging resistant tillers, bearing highly branched panicle. A detailed investigation on variability of *OsSPL14* locus would help in understanding its role in domestication process.

During the course of domestication, ancient farmers selected and bred wild accessions with increase yield and better grain quality. These major phenotypic variations were derived from minor changes at genome level and lead to multiple domestication events (Izawa et al. 2009). Thousands of years of breeding and selection has fixed those mutations in the genome of present day rice while wild rice still represents the source of genetic variability.

References

Ashikari, M., Sakakibara, H. and Lin, S. et al. (2005). Cytokinin oxidase regulates rice grain production. *Science*, 309. 741–745.

Bolot, S., Abrouk, M. and Masood-Quraishi, U. et al. (2009). The 'inner circle' of the cereal genome. *Curr. Opin. Plant Biol.*, 12: 119–125.

Devos, K.M. (2005). Updating the 'crop circle'. *Curr. Opin. Plant Biol.,* 8: 155–162.

Ding, J., Araki, H. and Wang, Q. et al. (2007). Highly asymmetric rice genomes. *BMC Genomics,* 8: 154.

Fan, C., Xing, Y. and Mao, H. et al. (2006). GS3, a major QTL for grain length and weight and minor QTL for grain width and thickness in rice, encodes a putative transmembrane protein. *Theor. Appl. Genet.,* 112: 1164–1171.

Feltus, F.A., Wan, J. and Schulze, S.R. et al. (2004). An SNP resource for rice genetics and breeding based on subspecies indica and japonica genome alignments. *Genome Res.,* 14: 1812–1819.

Garris, A.J., Tai, T.H. and Coburn, J. et al. (2005). Genetic structure and diversity in *Oryza sativa* L. *Genetics,* 169: 1631–1638.

Huang, X., Qian, Q. and Liu, Z. et al. (2009). Natural variation at the DEP1 locus enhances grain yield in rice. *Nat. Genet.,* 41: 494–497.

Hurwitz, B.L., Kudrna, D. and Yu, Y. et al. (2010). Rice structural variation: A comparative analysis of structural variation between rice and three of its closest relatives in the genus *Oryza*. *Plant J.,* 63: 990–1003.

International Rice Genome sequencing Project (2005). The map-based sequence of rice genome. *Nature,* 436: 793–800.

Isshiki, M., Morino, K. and Nakajima, M. et al. (1998). A naturally occurring functional allele of the rice waxy locus has a GT to TT mutation at the 5' splice site of the first intron. *Plant J.,* 15: 133–138.

Izawa, T. (2008). The process of rice domestication. *Rice,* 1: 127–134.

Izawa, T., Konishi, S. and Shomura, A. et al. (2009). DNA changes tell us about rice domestication. *Curr. Opin. Plant Biol.,* 12: 185–192.

Jiao, Y., Wang, Y. and Xue, D. et al. (2010). Regulation of OsSPL14 by OsmiR156 defines ideal plant architecture in rice. *Nat. Genet.,* 42: 541–544.

Jin, J., Huang, W. and Gao, J.P. et al. (2008). Genetic control of rice plant architecture under domestication. *Nat. Genet.,* 40: 1365–1369.

Kathuria, H., Giri, J. and Tyagi, H. et al. (2007). Advances in transgenic rice biotechnology. *Crit. Rev. Plant Sci.,* 26: 65–103.

Kim, H., Hurwitz, B. and Yu, Y. et al. (2008). Construction, alignment and analysis of twelve framework physical maps that represent the ten genome types of the genus *Oryza*. *Genome Biol.,* 9: R45.

Konishi, S., Izawa, T. and Lin, S.Y. et al. (2006). An SNP caused loss of seed shattering during rice domestication. *Science,* 312: 1392–1396.

Kovach, M.J., Sweeney, M.T. and McCouch, S.R. (2007). New insights into the history of rice domestication. *Trends Genet.,* 23: 578–587.

Krishnan, A., Guiderdoni, E. and An, G. et al. (2009). Mutant resources in rice for functional genomics of the grasses. *Plant Physiol.,* 149: 165–170.

Li, C., Zhou, A. and Sang, T. (2006). Rice domestication by reducing shattering. *Science,* 311: 1936–1939.

Li, X., Wang, X. and He, K. et al. (2008). High-resolution mapping of epigenetic modifications of rice genome uncovers interplay between DNA methylation, histone methylation and gene expression. *Plant Cell,* 20: 259–276.

Lin, Z., Griffith, M.E. and Li, X. et al. (2007). Origin of seed shattering in rice (*Oryza sativa* L.). *Planta,* 226: 11–20.

McNally, K.L., Childs, K.L. and Bohnert, R. et al. (2009). Genomewide SNP variation reveals relationships among landraces and modern varieties of rice. *Proc. Nat. Acad. Sci., USA.,* 106: 12273–12278.

Miura, K., Ikeda, M. and Matsubara, A. et al. (2010). OsSPL14 promotes panicle branching and higher grain productivity in rice. *Nat. Genet.,* 42: 545–549.

Moore, G., Devos, K.M. and Wang, Z. et al. (1995). Cereal genome evolution: Grasses, line up and form a circle. *Curr. Biol.,* 5: 737–739.

Paterson, A.H., Bowers, J.E. and Bruggmann, R. et al. (2009). The *Sorghum bicolor* genome and the diversification of grasses. *Nature,* 457: 551–556.

Piegu, B., Guyot, R. and Picault, N. et al. (2006). Doubling genome size without polyploidization: Dynamics of retrotransposon-driven genomic expansions in *Oryza australiensis*, a wild relative of rice. *Genome Res.,* 16: 1262–1269.

Raghuvanshi , S., Kapoor, M. and Tyagi, S. et al. (2010). Rice genomics moves ahead. *Mol. Breed.,* 26: 257–273.

Salse, J., Bolot, S. and Throude, M. et al. (2008). Identification and characterization of conserved duplications between rice and wheat provide new insight into grass genome evolution. *Plant Cell,* 20: 11–24.

Schnable, P.S., Ware, D. and Fulton, R.S. et al. (2009). The B73 maize genome: Complexity, diversity and dynamics. *Science,* 326: 1112–1115.

Shomura, A., Izawa, T. and Ebana, K. et al. (2008). Deletion in a gene associated with grain size increased yields during rice domestication. *Nat. Genet.,* 40: 1023–1028.

Song, X.J., Huang, W. and Shi, M. et al. (2007). A QTL for rice grain width and weight encodes a previously unknown RING-type E3 ubiquitin ligase. *Nat. Genet.,* 39: 623–630.

Sweeney, M. and McCouch, S. (2007). The complex history of the domestication of rice. *Ann. Bot.,* 100: 951–957.

Sweeney, M.T., Thomson, M.J. and Cho, Y.G. et al. (2007). Global dissemination of a single mutation conferring white pericarp in rice. *PLoS Genet.,* 3, e133.

Sweeney, M.T., Thomson, M.J. and Pfeil, B.E. et al. (2006). Caught red-handed: Rc encodes a basic helix-loop-helix protein conditioning red pericarp in rice. *Plant Cell,* 18: 283–294.

Tan, L., Li, X. and Liu, F. et al. (2008). Control of a key transition from prostrate to erect growth in rice domestication. *Nat Genet.,* 40: 1360–1364.

The International Brachypodium Initiative (2010). Genome sequencing and analysis of the model grass *Brachypodium distachyon. Nature,* 463: 763–768.

Vaughan, D.A., Morishima, H. and Kadowaki, K. (2003). Diversity in the *Oryza* genus. *Curr. Opin. Plant Biol.,* 6: 139–146.

Vij, S., Gupta, V. and Kumar, D. et al. (2006). Decoding the rice genome. *Bio. Essays,* 28: 421–432.

Vitte, C., Ishii, T. and Lamy, F. et al. (2004). Genomic paleontology provides evidence for two distinct origins of Asian rice (*Oryza sativa* L.). *Mol. Genet. Genomics,* 272: 504–511.

Wang, E., Wang, J. and Zhu, X. et al. (2008). Control of rice grain-filling and yield by a gene with a potential signature of domestication. *Nat. Genet.,* 40: 1370–1374.

Wang, Z., Zheng, Y. and Shen, F.Q., et al. (1995). The amylose content in rice endosperm is related to the post-transcriptional regulation of the waxy gene. *Plant J.,* 7: 613–622.

Yamanaka, S., Nakamura, I. and Watanabe, K.N. et al. (2004). Identification of SNPs in the waxy gene among glutinous rice cultivars and their evolutionary significance during the domestication process of rice. *Theor. Appl. Genet.,* 108: 1200–1204.

Yu, J., Wang, J. and Lin, W. et al. (2005). The genome of *Oryza sativa*: A history of duplications. *PLoS Biol.,* 3, e38.

Zhu, Q., Ge, S. (2005). Phylogenetic relationships among A-genome species of the genus *Oryza* revealed by intron sequences of four nuclear genes. *New Phytol.,* 167: 249–265.

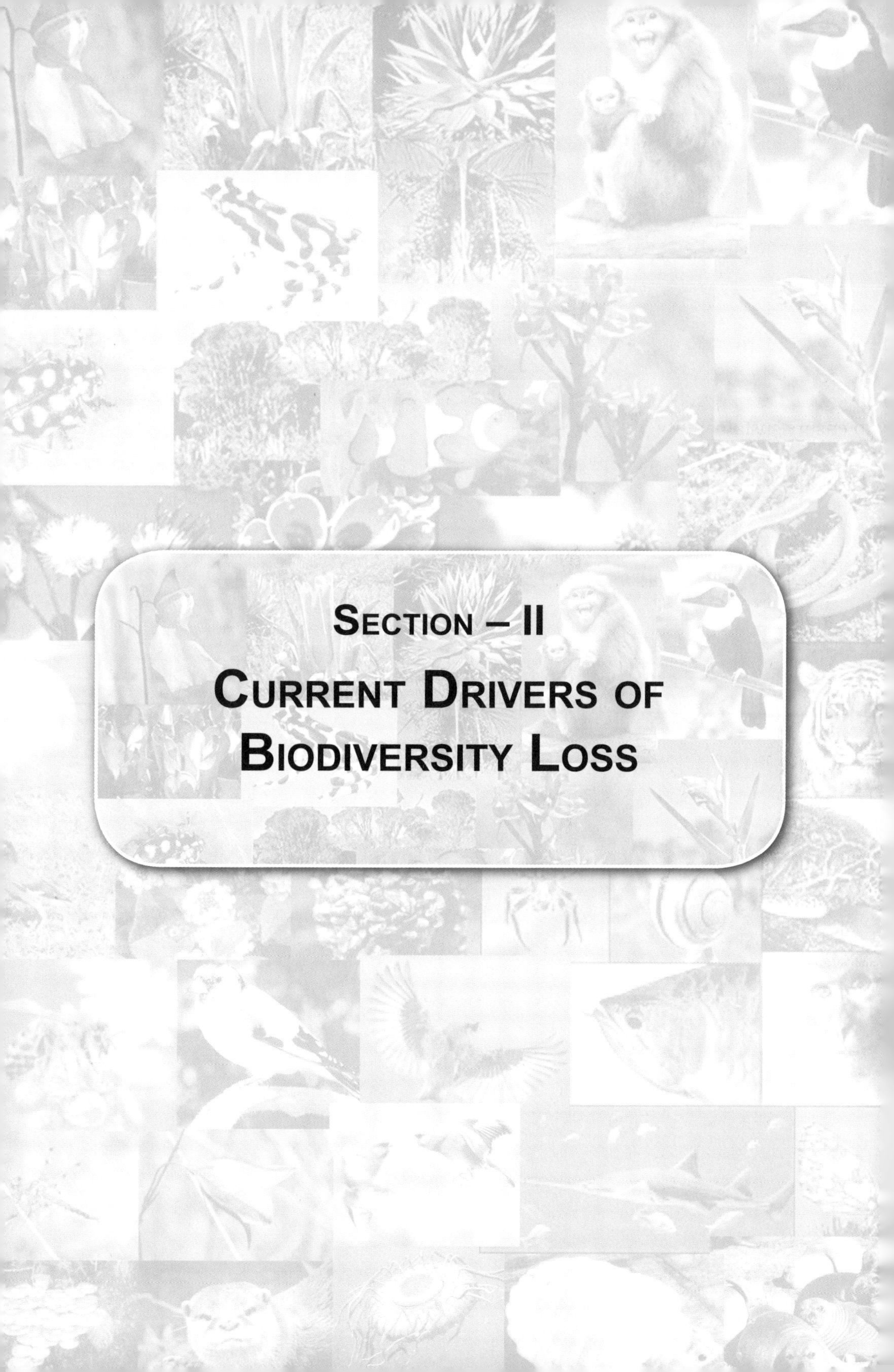

SECTION – II

CURRENT DRIVERS OF BIODIVERSITY LOSS

5

Ecology of Plant Invasion with Particular Reference to India

Gyan P. Sharma, Hema Singh and J.S. Singh

INTRODUCTION

Organisms immigrating to new localities and their descendants have been referred to as alien, adventive, exotic, introduced or non-indigenous (Mack et al. 2000) Species whose native status and origin are not clear are called cryptogenic species (Carlton, 1996). A taxon can be considered successfully naturalized in a new area after it has overcome geographical, environmental and reproduction barriers. An invasive species requires, in addition, to overcome dispersal barrier also within its new region (Richardson et al. 2000). According to Rejmanek (2000), invasive taxa represent a subset of naturalized taxa. Invasion is usually discovered once the plant has already naturalized. The biotic invaders tend to establish a new range in which they proliferate, spread and persist to the detriment of the environment. Although defined variously by different authors, we consider the definition of alien invasive species given by GISP (2003), as most pertinent to the present discussion: "an alien species which becomes established in natural or semi-natural ecosystems or habitat, is an agent of change, threatens native biological diversity and harm the environment, economies, or human health". Thus establishment and spread of these species threatens ecosystems, habitats, or species with economic/environmental harm (GISP, 2000). Invasion of exotic species is among the most important global scale problems experienced by natural ecosystems. The growing human population and improved transcontinental transport have increased the scales of movement of non-indigenous organisms, and the current enhanced rate of invasion constitutes one of the most important effects that humans have had on the earth ecosystem. In the past, many of the irretrievable losses of native biodiversity have gone unrecorded, but today there is an increasing realization of the ecological costs of biological invasion.

Over 40% of the species on the list of threatened and endangered species are due to invasive species (Wilcove et al. 19980. In many continental areas 20% or more of the plant species are now non-indigenous. On many islands the proportion of non-indigenous plant species is 50% or more (Rejmaneek and Randall, 1994). As many as 10% of the 260,000 vascular plant species are estimated to be potential invaders (Rapoport, 1991). About 18% of the Indian flora constitutes adventives aliens, of which 55% are American, 10% Asian, 20% Asian and Malaysian, and 15% European and Central Asian species (Nayar, 1977). All non-native species are, however, not harmful. For example, over 70% of the world's food comes from just nine crops (wheat, maize, rice, potato, barley, cassava, soyabean, sugarcane and oats), each of which is cultivated far beyond its place of origin (Prescott and Prescott, 1990). In New Zealand, 95% of export earning derives from alien species (New Zealand Department of Statistics, 1996). Nevertheless, the invasive species are the second largest threat to biodiversity globally after habitat destruction and constitute the number one cause of species extinctions in most island states (Schei, 1996; Gurevitch and Padilla, 2004; Fritts and Rodda, 1998). In India, *Cytisus scoparius*, *Chromolaena odorata*, *Eupatorium adenophorum*, *Hyptis suaveolens*, *Lantana camara*, *Mikania micrantha*, *Mimosa diplotricha*, *Parthenium hysterophorus* and *Prosopis juliflora* among terrestrial exotics, and *Eichhornia crassipes* and *Pistia stratiotes* among aquatics, have posed serious threat to the native flora. The problem seems to be alarming because many of the biological invasions are effectively irreversible (Sharma et al. 2005a).

In this communication, we provide an overview of the invasion process, hypotheses propounded to explain the success of invaders, a list of important invasive species (including non-plants) in India, brief information on the biology and ecology of 10 most troublesome invasive plants, and knowledge gaps and research needs in the area of invasion ecology in the country.

INVASION PROCESS

The essential first step in invasion by an alien plant is its introduction to an area beyond its previous geographical range. Introduction of non-native species may occur through (i) accidental introduction, (ii) import for a limited purpose and subsequent escape, or deliberate introduction on a large scale (Levine, 1998). Once introduced, the invader colonizes the new habitat, produces new self-perpetuating populations and is naturalized by getting incorporated into the resident flora. This is followed by spread to new locations. Recently, Wilson et al. (2009), categorized extra-range dispersal into six types of dispersal pathway *viz.* leading edge dispersal, corridor, jump dispersal, extreme long distance dispersal, mass dispersal and cultivation. The invader is likely to exist for some time as a single or small-localized population. The concept of dormant invader has been applied to species that are present in an area for an extended period of time before becoming a significant invader. At some point of time, however, the invader will enter a period of rapid expansion both in terms of total population size, and the number and size of individual infestations. Finally, an invader will reach a stage at which it will be a major problem (Fig. 1). The invasion is not necessarily a smooth process; major episodes of population expansion may be punctuated by uneventful periods. A statistical rule, known as the 10th rule, holds that 1 in 10 of imported species become introduced, 1 in 10 of those introduced become established, and 1 in 10 of those established become pests (Sharma et al. 2005a and Williamson, 1993).

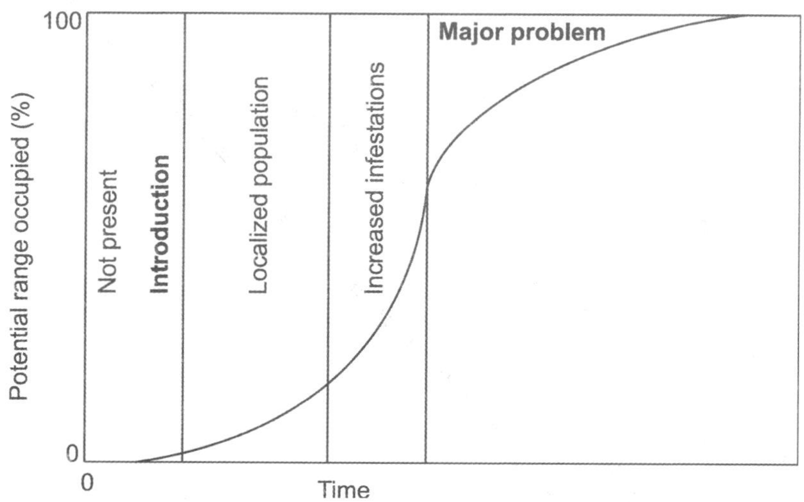

Fig. 1. Diagrammatic representation of generalized phases of invasion (adapted from Sharma et al. 2005a).

Species and Community Invasibility

Susceptibility to invasion varies from habitat to habitat and from community to community. Capacity to invade also varies among alien species. In a classical '*naturalization-invasion continuum*' model, (Richnadson and Pysek,

2006), visualize various barriers that a plant must overcome to become alien, casual, naturalized or invasive (see Fig. 2 for explanation) in a new environment, and the model highlights the dynamics of plant invasions requiring insights on traits of the plant (elements of species invasiveness) and features of the environment (components of community invasibility), and emphasizes that none of the components can be fully assessed without reference to the other (Fig. 2). So to understand plant invasion in totality one must consider the complimentarity of the species invasibility and community invasibility.

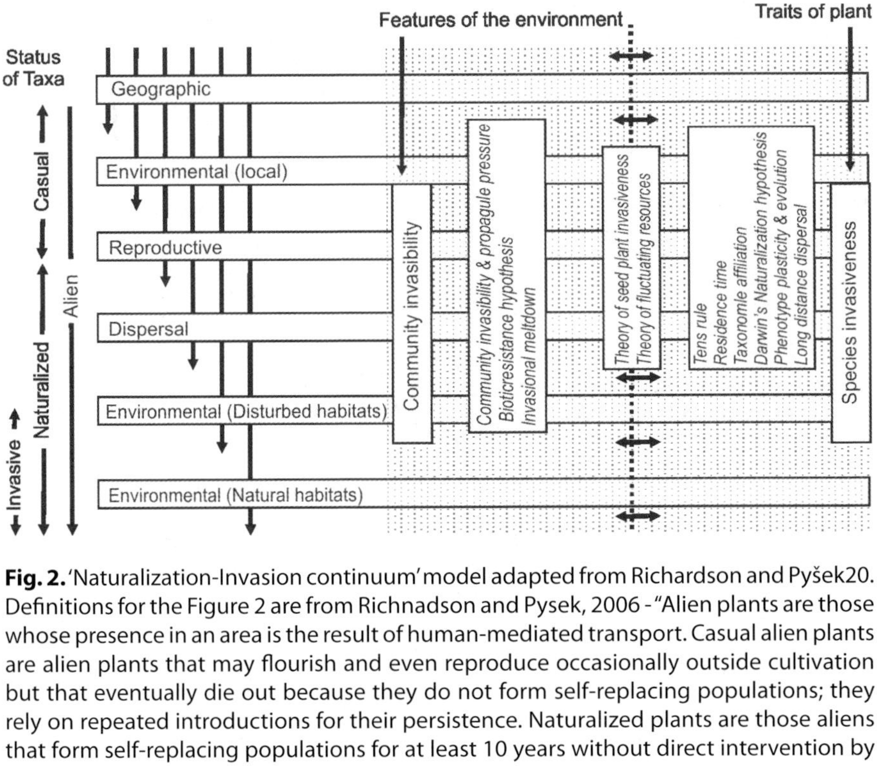

Fig. 2. 'Naturalization-Invasion continuum' model adapted from Richardson and Pyšek20. Definitions for the Figure 2 are from Richnadson and Pysek, 2006 - "Alien plants are those whose presence in an area is the result of human-mediated transport. Casual alien plants are alien plants that may flourish and even reproduce occasionally outside cultivation but that eventually die out because they do not form self-replacing populations; they rely on repeated introductions for their persistence. Naturalized plants are those aliens that form self-replacing populations for at least 10 years without direct intervention by people (or despite human intervention) by recruitment from seeds or ramets capable of independent growth. Invasive plants are a subset of naturalized plants that produce reproductive offspring, often in large numbers, at considerable distances from parent plants, and thus have the potential to spread over a large area".

Hypotheses Regarding Success of Invaders

Based on the analysis of the traits of invasive species, a number of hypotheses have been proposed as a mechanism conferring invasiveness to a plant.

Fluctuating Resource Hypothesis (FRH) (Davis et al. 2000) states that a "plant community becomes more susceptible to invasion whenever there is an increase in the amount of unused resources". This could happen subsequent to disturbance in a habitat. Meanwhile, the Presence of Empty Niche (PEN) hypothesis (Mack et al. 2000) asserts that a species poor habitat is more susceptible to invasion than a species-rich one, as new niche opportunities are created following the disturbance. The Enemy Release Hypothesis (ERH) (Keane & Crawley, 2002) suggests that invaders perform better in their introduced range than in their native range because they lose their enemies (often, but not always, parasites) during the process of colonization. The Evolution of Increased Competitive Ability (EICA) (Muller, 2004) hypotheses is based on the fact that exotics, when liberated from their native specialized enemies, can allocate resources, otherwise used for costly traits that helped them resist those enemies, to the development of traits that provide greater competitive advantage. Whereas, Allelopathic Advantage Against Resident Species (AAARS) or Novel Weapon (NW) hypotheses (Callaway & Ridenour, 2004) suggests that invasive plant species possess novel biochemical weapons that function as unusually powerful allelopathic agents or as mediators of new plant–soil microbial interactions. Once the species gets naturalized, it can spread through interaction with the faunal diversity of the region and through the possession of its vegetative traits (Sharma et al. 2005a). Sharma et al. (2005b) suggested that no single theory is responsible for the success of invaders as different theories operate at different stages of the invasion process. For example, after the arrival of the propagules, FRH and PEN play a role in species establishment. Once the species gets established, ERH helps in colonization. Subsequently, EICA, through developed AARS (NW), gives a supportive advantage to naturalize, and faunal interaction and vegetative traits encourage the spread of invasive species.Therefore, trait analysis of invasive species may indicate that the invasiveness of invaders can only be explained when a set of hypotheses is applied in complimentarity (Sharma et al. (2005b).

INVASIVE SPECIES IN INDIA

We used the invasive alien species data base of the Invasive Species Specialist Group of the IUCN (World Conservation Union) which has compiled the most geographically comprehensive database on invasive species worldwide, the Global Invasive Species Database (GISD, 2005). It includes 227 countries and profiles on 357 invasive alien species across all taxa that are significant threats to native biodiversity (GISD, 2005). Global Invasive Species Database for India includes 197 invasive species (important invaders compiled in Table 1). There are also certain regional lists compiled by different authors. Khanna and Singh (2006), enumerated 60 invasive alien species for the Chhattisgarh region, Rao and Murugan (2006) compiled a list of 37 dominant alien weeds which have affected the biodiversity in India. Shukla et al. (2009), documented 106 alien plant species belonging to 36 families in the Achanakmar-Amarkantak Biosphere Reserve, which represent almost 10% of the reserve flora. A comprehensive inventory of the invasive alien flora of Uttar Pradesh, revealed 152 species from 109 genera and 44 Families (Reddy et al. 2008). In this list, Dicots represented 137 species and monocots 15 species. About 73% of these alien species were introduced from tropical America including South America, followed by tropical Africa (10.5%). All flowering families were not equally represented in the invasive flora. Maximum number of species (30) were from the family Asteraceae, followed by Fabaceae (12 species). A comprehensive list has been put together by Reddy et al. (2008), which includes 173 alien plant species, under 44 families, representing 1% of the Indian flora. Gupta and Khetarpal (2006), argued that about 40% of the species in the Indian flora are alien, of which 25% are invasive. A complete, authentic list of invasives for India is still not available. In our opinion there is an urgent need to develop such a list which we may use to assess the degree to which the invasive alien species impact biodiversity, ecology and economy of a region.

Table 1. Some noxious invasive species (plants and non-plants) in India and their impact (compiled from GISD 2005 database searched on 7 March 2010; www.issg.org/database).

S. No.	*Species*	*Nativity*	*Impact*
1.	*Acacia farnesiana* (tree, shrub)	Tropical America	Chokes streams
2.	*Acacia mearnsii* (tree, shrub)	Australia	Reducing native biodiversity
3.	*Acacia melanoxylon* (tree)	Eastern Australia	Timber plantations
4.	*Achatina fulica* (mollusc)	Africa	Native fauna
5.	*Alternanthera philoxeroides* (aquatic plant, herb)	South America	Waterways
6.	*Ambrosia artemisiifolia* (herb)	Temperate North America	Pollen allergy

7.	*Anoplolepis gracilipes* (insect)	West Africa	Interfere in the reproduction of a variety of arthropods, reptiles, birds and mammals
8.	*Aristichthys nobilis* (fish)	Asia	Interfere with the zooplankton populations thus alter the food web
9.	*Ascidiella aspersa* (tunicate)	Norway	Displace native fauna
10.	*Asparagus densiflorus* (herb)	South Africa	Forest regeneration
11.	*Bemisia tabaci* (insect)	Asia	Crop
12.	*Bugula neritina* (bryozoan)		Under water structures
13.	*Cabomba caroliniana* (aquatic plant)	America	Clog water bodies
14.	*Carassius auratus* (fish)		Aquatic community
15.	*Chromolaena odorata* (herb)	South America	Compete with native plants
16.	*Cirsium arvense* (herb)	Canada	Grassland communities
17.	*Columba livia* (bird)	Europe	Monument corrode by droppings
18.	*Cryphonectria parasitica* (fungus)	Japan	Forest regeneration
19.	*Cryptostegia grandiflora* (vine, climber)	Madagascar	Chokes water access and biodiversity loss
20.	*Ctenopharyngodon idella* (fish)	China	Aquatic community
21.	*Cyprinus carpio* (fish)	Europe	Reduce water clarity
22.	*Eichhornia crassipes* (aquatic plant)	South America	Clog water bodies and aquatic community
23.	*Elaeagnus umbellata* (tree, shrub)	Asia	Biodiversity loss
24.	*Equus asinus* (mammal)	Africa	Flora and fauna
25.	*Eugenia uniflora* (tree, shrub)	South America	Biodiversity loss
26.	*Euglandina rosea* (mollusc)	Southeastern US	Biodiversity loss
27.	*Eupatorium cannabinum* (herb)	United States	Alter the nutrient structure

28	*Gambusia affinis* (fish)	United States	Pest of waterways
29.	*Gambusia holbrooki* (fish)	Southern US	Native fish and amphibian
30.	*Gymnocoronis spilanthoides* (aquatic plant)	Mexico to Argentina	Clog water bodies and aquatic community
31.	*Hypophthalmichthys molitrix* (fish)	Asia	Biodiversity loss
32.	*Kappaphycus spp.* (algae)	Malaysia	Clog water bodies and aquatic community
33.	*Lantana camara* (shrub)	Mexico	Ecosystem structure and function
34.	*Leucaena leucocephala* (tree)	Southern Mexico	Native plants
35.	*Limnocharis flava* (aquatic plant)	US	Aquatic community
36.	*Ludwigia peruviana* (aquatic plant)	US	Clog water bodies and shoreline community
37.	*Macfadyena unguis-cati* (vine, climber)	Central and south America	Biodiversity loss
38.	*Mikania micrantha* (vine, climber)	Central and south America	Biodiversity loss
39.	*Mimosa diplotricha* (vine, shrub)	Tropical America	Biodiversity loss
40.	*Mimosa pudica* (herb)	South America	Biodiversity loss`
41.	*Monomorium pharaonis* (insect)	Africa	Increase infection rates through pathogens
42.	*Mytilopsis sallei* (mollusc)	Gulf of Mexico	Biodiversity loss
43.	*Oncorhynchus mykiss* (fish)	Western North America	Native fish diversity
44.	*Oreochromis mossambicus* (fish)	Mozambique	Native fish diversity
45.	*Oreochromis niloticus* (fish)	Africa	Aquatic community
46.	*Oreochromis spp.* (fish)	Africa	Native fish diversity
47.	*Orthezia insignis* (insect)	South America	Feeding on phloem of host plant
48.	*Parthenium hysterophorus* (herb)	Mexico	Agricultural yield and biodiversity loss
49.	*Pennisetum clandestinum* (grass)	Eastern Africa	Reduce native biodiversity

50.	*Physalis peruviana* (shrub)	Brazil	Agriculture
51.	*Pittosporum undulatum* (tree, shrub)	South-eastern Australia	Reduce native biodiversity
52.	*Prosopis* spp. (tree, shrub)	America	Reduce native biodiversity
53.	*Pycnonotus jocosus* (bird)	Asia	Spread invasive plants
54.	*Quadrastichus erythrinae* (insect)	Taiwan	Affect growth of coral trees
55.	*Rattus exulans* (mammal)	Asia	Native insects and lizards
56.	*Ricinus communis* (tree, shrub)	Asia	Displaces native vegetation
57.	*Salmo trutta* (fish)	South East Asia	Native fishes
58.	*Salvelinus fontinalis* (fish)	America	Native fishes and quality of water
59.	*Salvinia molesta* (aquatic plant, herb)	South America	Chokes water access and biodiversity loss
60.	*Solanum mauritianum* (tree, shrub	South America	Shade out native species
61.	*Solanum sisymbriifolium* (herb)	South America	Displaces native vegetation
62.	*Solanum viarum* (shrub)	Brazil	Displaces native vegetation
63.	*Solenopsis geminata* (insect)	America	Agriculture and coffee plantations
64.	*Spartina alterniflora* (grass)	Tropical America	Displaces native vegetation
65.	*Technomyrmex albipes* (insect)	Asia	House walls
66.	*Tinca tinca* (fish)	Europe	Native fishes
67.	*Triadica sebifera* (tree)	China	Alter nutrient cycle
68.	*Vibrio cholerae* (micro-organism)	–	Cause epidemic with poor sanitation
69.	*Zosterops japonicus* (bird)	Pacific region of Asia	Native bird species

Biology and ecology of the most troublesome invasive plants in the country

Lantana camara

Lantana camara (Verbenaceae) is a multi-stemmed bushy shrub, usually ≤ 1m in diameter (Palmer & Pullen, 1995). It is commonly known as wild sage and is native to tropical America. In its naturalized range, it often forms dense mono-specific thickets, 1–4m high and ≈ 1–4m in diameter (Swarbrick et al. 1998), although some varieties climb trees and reach heights of 8–15m (Matthews, 2004). The plant usually flowers as early as the first growing season after its establishment and, in most places, flowers all the year round if adequate moisture and light are available (Gujral & Vasudevan, 1983 and Duggin & Gentle, 1998). Inflorescences are produced in pairs in leaf axils. Wide variations in flower colour, size, and shape have been reported (Van Oosterhout et al. 2004). The fruit is a round, fleshy, drupe, ≈ 0.5 cm wide, initially green, but turning purple then blue-black. Each drupe contains 1–2 seeds that are ≈ 0.1–0.2 cm long. *Lantana* seeds germinate at any time of the year with sufficient soil moisture, light, and temperature (Duggin & Gentle, 1998; Van Oosterhout et al. 2004 & Parsons & Cuthbertson 2001). It is a prolific seed producer; heavy infestations can yield up to ≈ 3000–4000 seeds per m^2. Once established, through rapid vegetative growth lantana forms large, impenetrable clumps (Sharma and Esler, 2010). In India, densities of 1500–1600 individuals per hectare have been reported with fruit production reaching 0.6-0.7 million per hectare in pure stands of *Lantana* (Sharma et al. 2005b). Pollination in *Lantana* results in 85% fruit-set, with an Australian study showing that each inflorescence bears about eight fruits (Barrows, 1976), whereas, in India, ≈ 25–28 fruits are produced (Sharma et al. 2005b), showing intra-specific variation (Thorp, 2000). Because of its broad environmental tolerance, high seed production, fast development to maturity, as well as ability to re-sprout vigorously after damage, *Lantana* is definitely a serious problem species.

It is not clear how it arrived in India, although there are anecdotal references to the introduction of *Lantana* as an ornamental in early 19th century. *Lantana* might have been introduced in India on multiple occasions in the late 1800's - as an ornamental in Coorg (Hiremath & Sundaram, 2005) and as a bio-hedge in Calcutta in the early 19th century (Hakimuddin, 1929; Pereira, 1919). Hakimuddin (1929), gives an account of its introduction and spread in the Nainital district of India around 1905, where it was confined to hedges and had a sparse distribution until 1911, but by 1929, *Lantana* had spread from point of introduction forming dense thickets covering farms, pastures, fallows and forest areas. It has invaded the majority of Indian pasture lands (13.2 million ha) besides other areas and the cost of its control is estimated at US$ 70 per hectare (Sing, 1996). Therefore, the damage from *Lantana* to pastures alone is estimated to be US$ 924 million per year (Pimentel et al. 2001), this does not include vast tracts of disturbed forest colonized by *Lantana*.

Studies revealed that *Lantana* invasion in the dry deciduous forest can impact tree species regeneration, decrease the herb layer diversity due to shading effect, and also increase the rate of nutrient cycling that may benefit its growth (Sharma et al. 2005a, 2005b; Sharma & Raghubanshi, 2005, 2006, 2007, 2009, 2010, and Raizada et al. 2008). Removal of *Lantana* bushes showed a positive effect on herb–shrub richness in the livestock-free blocks, although no effect was reported on tree seedlings for the livestock-free and livestock blocks of the Bandipur Tiger Reserve in southern India (Prasad, 2010). There exists a close relation between the presence of *Lantana* and livestock grazing which in turn affect the understory plant species composition (Prasad, 2010). Sahu and Singh (2008), compared the structural attributes of vegetation in 20 plots with differential *Lantana* intensity in the Achanakmar–Amarkantak Biosphere Reserve, central India and reveal that there has been no adverse effect of *Lantana* invasion on vegetation at community level but the threat can build up with time because at individual species level there was measurable effect. Babu et al. (2009), conducted a classic study in the Corbett Tiger Reserve for restoring *Lantana* infested landscapes, and developed an eradication technology using knowledge on its ecology.

Parthenium hysterophorus

It is an erect herb of Asteraceae family native of Mexico. *Parthenium* is considered among the serious invaders affecting the indigenous ecosystems all over the world especially in the urban areas (Kohli et al. 2006). Many people believe that *Parthenium* arrived in India in appreciable amounts with the contaminated grains of PL-480 wheat scheme during 1950–1960. Now the species is wide spread in open unshaded wastelands all across the India forming mono-specific stands.

P. hysterophorus has alternate, deeply dissected leaves and can grow up to 2m with much branched inflorescence bearing white flower heads and numerous obovoid, smooth and black achenes. Plants start producing seed often within three weeks of germination. Generally a mature plant alone can produce over 15000 seeds in a growing season. Seeds are light and produced in large quantities (Smith, 2002). Seeds are dispersed by wind, water, as a contaminant of harvest materials and by vehicles, machinery or animal. The allelopathic potential of *P. hysterophorus* is well established and results from release of phytotoxic substances such as ferulic, caffeic, vanillic, chlorogenic, p-coumaric and p-hydroxybenzoic acids, parthenin, ambrosin and coronopilin, which inhibit the germination and growth of several crop plants and multi-purpose trees (Swaminathan et al. 1990). The factors responsible for the invasive success of the species range from allelopathy, persistent seed bank, longevity of seed bank, faster and year round germination, early flowering and fruiting, extensive colonization capacity and efficient nutrient extraction potential. *Parthenium hysterophorus* shows plastic response to soil quality leading to contrasting strategies that contribute to its success as an invader, allowing range expansion and establishment, through production of lighter and heavier seeds, respectively (Annapuran and Singh, 2003a, 2003b).

Hyptis suaveolens

Hyptis suaveolens (Lamiaceae) is an erect annual woody herb, commonly 1m in height (maximum height = 1.5m) and reproduces by seed and root stocks (Willis, 1973). It is commonly known as horehound weed or mint weed, is native to the tropical America and is now a pantropic weed (Hickey and King, 1988; Hutchinson and Dalziel, 1963). The leaves are ovate, 4 to 9cm long, pointed at the tip, somewhat heart-shaped at the base, and toothed on the margins. The flowers occur in the axils of the leaves in long stalks. The calyx, in flower, is hairy and about 4mm long but soon enlarges in fruit to about 1cm long, with erect, stiff teeth. The corolla is blue and about 8mm long, with a limb 5mm in diameter, nutlets are about 1.2–1.5mm long (Stone, 1970). The seeds are flat and characteristically dimorphic. This dimorphism is reflected in the seed size (Wulff, 1973). Seed polymorphism may help in establishment at different microsites or safe sites (Harper et al. 1961)[68], *H. suaveolens* is a prolific seed producer and dense infestations can yield up to ≈ 3000 seeds per m^2, forming persistent propagule banks within a short period. The seeds are slightly notched and protected by spined burr which help in its dispersal through animal fur (Persons & Cuthbertson, 2001 and Stubbs, 1999). *H. suaveolens* can germinate under a wide variety of temperature conditions and tolerates dry periods well (Felippe, 1983). It is found on a variety of habitats like rail tracks, roadsides, foothills of open forests, forest clearings and can heavily invade wastelands, particularly on arid and rocky substrates (Verma and Mishra, 1992; Mudgel, 1977). *H. suaveolens* was not recorded in the Vindhyan highlands in 1903 during a large survey (Verma and Mishra, 1992). However, Agrawal (2002) reported heavy infestations of *H. suaveolens*, 2km north of the super thermal power plant at Singrauli, where as recently as 10 years ago the species was not prevalent in the area. *H. suaveolens* has spread widely in open areas of this forest and can be classified as a major common and wide spread invader (Sharma et al. 2007). Sharma et al. (2009), have suggested that for conserving biodiversity in the dry deciduous forests invaded by *H. suaveolens*, management agencies should focus on areas with a high level of rockiness as these are the niches especially dominated by *H. suaveolens* that harbour typical dry deciduous flora and fauna.

Ageratum spp.

Ageratum conyzoides (Asteraceae) is an erect annual herb, commonly 45-70cm in height (maximum height 1.4m), that reproduces by seed. It is commonly known as Billygoat-weed and is native of tropical America (Holm et al. 1977). *A. conyzoides* leaves are 2–10cm long and 1–5.6cm wide, ovate to rhombic-ovate, acute at apex, and rounded at base. Flower heads are in flat-topped clusters, bluish lavender to white in colour. Seeds are tiny, rough achene (dry, single-seed fruit) with achenes 1.5–2mm long, nearly glabrous. *A. conyzoides* may produce ≈ 40,000 seeds per plant and in some areas one-half of the seeds will germinate shortly after they are shed. Seeds are mainly spread by wind and water and can germinate under a wide range of conditions (Holm et al. 1977). *A. conyzoides* is a weed of cultivated fields, but also invades other ecosystems such as pastures, grasslands, wastelands and even forest areas. In the agricultural fields, it is very troublesome, interferes with the growth and establishment of crops and reduces their yield (Holm et al. 1977). The successful invasion of *A.conyzoides* is attributed to its wide range of environmental adaptability, higher reproductive potential and allelopathy. Seeds reportedly stick in the throats of livestock and can choke them to death (Motooka et al. 2003).

Ageratum houstonianum (Asteraceae) is also an erect annual herb, commonly 30–90cm in height that reproduces by seed. It is commonly known as flossflower and is native to Central America and adjacent parts of

Mexico. *A. houstonianum* leaves are 2–7cm long and 1.5–6cm wide, ovate to deltate, and margins are toothed (Anonymous, 2007). Flower heads are in flat-topped clusters and ray florets are threadlike, bluish lavender to rarely white in colour. Seeds are tiny, rough achenes, (dry, single-seed fruit) 0.8–2mm long, black and nearly glabrous. *A. houstonianum* may also produce ≈ 40,000 seeds per plant. It has become an invasive weed in the eastern USA, Australia, Europe, Africa, China, India, Japan and New Zealand (Anonymous, 2007). *A. houstonianum* is toxic to grazing animals, causing liver lesions. A recent study by Singh et al. (2010), revealed that *A. houstonianum* is a more potent invader than *A. conyzoides* in the eastern Indo-Gangetic plains, as the trade-off among the eco-physiological and morphological traits gives advantage to *A. houstonianum*.

Calotropis procera

Calotropis procera (Asclepiadaceae) is an erect spreading shrub, commonly 1–2m in height (maximum 5.5m), that reproduces by seed. It is commonly known as Giant milkweed /Apple of Sodom and is native of Israel (Rahman and Wilcock, 1991). *C. procera* leaves are opposite oblong-ovate to nearly orbicular, short-pointed to blunt at the apex with heart-shaped base, 7 to 18cm long and 5 to 13cm broad, slightly leathery, and have a fine coat of soft hairs. Flower clusters are umbelliform cymes that grow at or near the ends of twigs. Flowers are 1 to 1.5cm long, fleshy and variable in colour from white to pink, often spotted or tinged with purple. The fruits are 8 to 12cm long, inflated, obliquely ovoid follicles that split and invert when mature to release flat, brown seeds with a tuft of white hairs at one end (Howard, 1989; Liogier, 1995; Little et al. 1974). Flowering and fruiting takes place throughout the year (Little et al. 1974). Hundreds to thousands of seeds may be produced per plant each year. Seeds are dispersed by wind and may fly for several hundred metres in gentle breezes. The successful invasion of *C. procera* can be attributed to its wide range of environmental adaptability and higher reproductive potential. It thrives best in disturbed urban lots and favours open habitat. Sharma et al. (2010), reported that *C. procera* distribution is directly related to road use type and distance from the urban vicinity in the eastern Indo-Gangetic plain of India.

Mikania micrantha

M. micrantha, native of Central and South America, is a perennial creeping climber characterized by vigorous and rampant growth. It grows best where fertility, organic matter, soil moisture and humidity are all high. *M. micrantha* was introduced in India after the Second World War to camouflage airfields and is one of the most widespread and problematic weeds in the Pacific region[26].

M. micrantha is a branched, slender-stemmed perennial vine. The leaves are arranged in opposite pairs along the stems and are heart-shaped or triangular with an acute tip and a broad base. Leaves may be 4–13cm long. The flowers are 3–5mm long and are arranged in dense terminal or axillary corymbs. Individual florets are white to greenish-white. The seed is black, linear-oblong, five-angled and about 2mm long. Each seed has a terminal pappus of white bristles that facilitates dispersal by wind or on the hair of animals. It reproduces sexually by seed, and vegetatively by rooting at nodes. A single plant may cover over 25 square metres within a few months, and release as many as 40,000 viable seeds every year. In some locations flowering and seed production are during short days only (GISD, 2005).

Once established, *M. micrantha* spreads at an alarming rate, readily climbing and twining on any vertical support, including crops, bushes, trees, walls and fences. Its shoots have been reported to grow up to 27mm a day. Vegetative reproduction is also efficient and vigorous. It is intolerant of heavy shade and readily colonizes gaps (GISD, 2005). *M. micrantha* damages or kills other plants by attenuating light and smothering them. It also competes for water and nutrients, and perhaps releases substances that inhibit the growth of other plants. *M. micrantha* is considered one of the three worst weeds of tea in India. It also causes serious problems in oil palm, banana, and forestry crops, and in pastures.

Chromolaena odorata

Chromolaena odorata, native to North America, from Florida and Texas to Mexico and the Caribbean, is a shrub of the family Asteraceae. It has been introduced to tropical Asia, west Africa, and parts of Australia. Commonly known as Siam weed or Floss flower, it was earlier taxonomically classified under the genus *Eupatorium*, but is now considered more closely related to other genera in the tribe Eupatorieae (Schmidt and Schilling 2000).

C. odorata is an herbaceous perennial that forms dense tangled bushes 1.5–2.0m in height. As a climber on other plants, it may reach a height of 6m. Its stems branch freely, with lateral branches developing in pairs from the axillary buds. The root system is fibrous and does not penetrate beyond 20–30cm in most soils. The flower heads are borne in terminal corymbs of 20 to 60 heads on all stems and branches. The flowers are white or pale bluish-lilac, and form masses covering the whole surface of the bush (Cruttwell and McFadyen, 1989). Florets are all alike (disc-florets) with pale purple to dull off-white, the styles extending about 4mm beyond the apex of the involucre, spreading radiately; florets are about 20–30 or a few more, 10–12mm long; ovarian portion 4mm long; corolla slender trumpet form; pappus of dull white hairs 5mm long; achenes glabrous (Stone, 1970). The seeds are small (3-5mm long, ~1mm wide, and weigh about 2.5mg) (Cruttwell & McFadyen, 1989). Seed production is prolific (up to 260,000m^{-2}) with about 20–46% of seeds being viable (Vanderwoude et al. 2005 and Witkowski, 2002). Plants can germinate and set seed within a 12-month period.

C. odorata grows on a wide range of soils, in varied vegetation types, e.g. forests (annual rainfall 1500mm), grassland and arid bushveld (annual rainfall less than 500mm) (Goodall and Erasmus, 1996). In arid areas, it is restricted to riverbanks. It becomes invasive in the frost-free areas of medium to arid woodland which are not water-stressed in the growing season (Honu and Dang, 2000). Seedlings grew best at the relative humidity in the range of 60–70%; at values higher than 80% the growth performance was poor (Vanderwoude et al. 2005). Experiments show that seedlings grew well at 30°C and even better on mulched soils at 25°C (Vanderwoude et al. 2005). In heavy shade, the plant will not seed. It has a negative relationship with tree canopy cover and appears to be most abundant on the edge of forested areas (Vanderwoude et al. 2005). Witkowski (2002), reports that in north-eastern India, the plant is regarded as a nutrient-demanding early successional species. It takes advantage of the flush of nutrients after a disturbance, such as fire or land clearing for agriculture, and exhibits relatively high foliar N, P and K contents (Saxena and Ramakrishnan, 1983). *C. odorata* forms dense stands preventing establishment of other species, both due to competition and allelopathic effects. The dried shoots may promote wild bushfires. It is a major weed in plantations and croplands, including plantations of rubber, oil palm, forestry and coffee plants. *Chromolaena* can replace other weeds such as *Lantana camara* and *Imperata cylindrica* and become the single dominant species in an area (Lai, 2006). *C. odorata* directly impacts humans as it may cause skin complaints and asthma in allergy-prone people.

Mimosa diplotricha

Belonging to the family Fabaceae (subfamily Mimosaceae) and introduced from tropical America, the plant occurs in three varieties (*diplotricha*, *odibilis* and *inermis*). In India it occurs throughout Kerala and in certain parts of the northeast. Species is a nitrogen-fixer and a drought resistant, fast growing annual shrub or climber forming dense thickets over other plants, smothering and killing the native flora and preventing their regeneration. It causes heavy damage to forest, and crops like sugarcane, rubber and tea. The weed is sun-loving and susceptible to dense shade. Leaves are feathery, each leaf with about 20 pairs of small leaflets. Flowers from August-February, on short-stalked clustered inflorensces, subsequently producing flat pods covered with prickles, and sets seeds in September-February. Seeds are also spiny, >8000 per m^2, and are dispersed through running water, animal furs, or as contaminant of crop seeds. This weed also checks the spread of *Chromolaena* and *Mikania* and replaces them.

Prosopis spp.

Prosopis, commonly known as mesquite or algarrobo, includes about 44 defined species and many hybrids. This leads to problems with identification. Native to America, the genus belongs to the family Fabaceae and subfamily Mimosoideae. *Prosopis* species are fast growing, nitrogen fixing and salt and drought tolerant shrubs or trees (www.weeds.org.au/docs/msqstrat.). They are phreatophytes with deep tap roots. This enables trees to remain green during droughts as they can access the water table. Lateral roots draw on surface water during the rains and, in common with other desert specialized plants, leaf adaptations reduce water loss (Pasiecznik, 1999). Feral animals are considered to be the most effective dispersers, and high seed germination occurs after passing through the animal gut. This interaction with the animals exacerbates the spread of mesquites.

The four main species that have presented problems as weeds world-wide are *P. glandulosa* and *P. velutina* in more subtropical regions and *P. juliflora* and *P. pallida* in the truly tropical zone. Among all the species *P. juliflora* is the most dominant invader in India. It is one among the 44 species in the genus, mainly found in the arid and semi-arid regions of the world (Asfaw and Thulin, 1989; Mohamed, 1997). Due to its wide ecological amplitude, *Prosopis* occurs on a large variety of soils and over a wide range of altitudes. The climate of its natural range is characterized by temperature averaging 20°C. It endures temperature as high as 50°C, and resists occasional frost of -12°C (Mohamed, 1997). Many species of *Prosopis* are known to nodulate and fix nitrogen in an endo-symbiotic relationship with root nodule bacteria (Kanzaria and Varshney, 1999).

P. juliflora is a tree or shrub, which reaches 15m height and can attain a diameter of 1m at breast height. It is armed with stipular spines 0.5–5cm long. Its leaves consist of 1–4 pairs of pinnae and 6–29 glabrous leaflet pairs measuring 1.5–5.5mm. The flowers are yellowish and develop on 5–15cm long spiciform racemes. The flowers have 10 stamens and anthers with an often-inconspicuous apical gland. The pod is pale yellow, linear and/or straightly curved measuring 0.8–1.7cm. It is compressed, internally septate between the seeds, indehiscent and with a sugary-pulpy mesocarp (Asfaw and Thulin, 1989).

Eradication of *P. juliflora* has proven to be extremely difficult or impossible. Better management, different land use strategies and the exploitation as a resource may reduce its invasiveness in some regions as well as improving local economies (Pasiecznik, 1999). Destroying seed, by the collection and use of pods for stall feeding or processing, reduces re-invasion.

Eichhornia crassipes

Originally from South America, *Eichhornia crassipes* is one of the worst aquatic weeds in the world (Holm et al. 1977). Beautiful, large purple and violet flowers lend ornamental value to the plant. It is now found in more than 50 countries on five continents. Water hyacinth is a very fast growing plant, with populations known to double in as little as 12 days. Infestations of this weed block waterways, limiting boat traffic, swimming and fishing. Water hyacinth also prevents sunlight and oxygen from reaching the water column and submerged plants. The shading and crowding effects dramatically reduce biological diversity in aquatic bodies.

E. crassipes is a free-floating aquatic macrophyte growing generally to 0.5m in height but to nearly 1m in height in some Southeast Asian locations (Gopal, 1987). *E. crassipes* may form dense floating mats. Its leaves are thick, waxy, rounded, and glossy and rise well above the water surface on stalks. They are broadly ovate to circular, 10–20cm in diameter, with gently incurved, often undulate sides. Leaf stalks are bulbous and spongy. The stalk is erect, up to 50cm long, and carries at the top a single spike of 8–15 showy flowers. The flowers have six petals, purplish blue or lavender to pinkish, the uppermost petal with a yellow, blue-bordered central splotch. Its roots are purplish black and feathery (Gopal, 1987). Water hyacinth reproduces both vegetatively and sexually (Penfound and Earle, 1948; Gopal and Sharma, 1981). Vegetative reproduction is more important. Stolon buds develop that bear offshoots from axillary buds and stolons are readily distributed by water currents, winds and boat traffic. The plant flowers year-round in mild climates, producing abundant amounts of long-lived seeds (Penfound and Earle, 1948; Sculthorpe, 1985). Maximum fruiting occurs in 90% humidity and at 22.5°C to 35°C (Gopal, 1987). Several species of bee pollinate the flowers and several researchers report a high level of self-compatibility. High light intensity and altering high and low temperatures (5°C to 40°C) favour germination.

E. crassipes forms thick mats that cover rice paddies, clog irrigation channels, impede navigation, halt fishing, sweep away buildings during floods and foster breeding of disease-transmitting mosquitoes (Room and Fernando, 1992). Doubling in biomass every 6 to 18 days, the exact time being dependent on location and time of year (Lindsey and Hirt, 1999), this weed rapidly invades water ways and has caused problems for people around the globe (Julien and Orapa, 2001). Invasive plant theory predicts that a release from environmental constraints due to altered hydrology can often lead to a successful invasion (Galatowitsch et al. 1999). Disruptions of wetland ecosystems involving irrigation canals, hydroelectric projects and construction of artificial lakes have made areas particularly susceptible to invasion by water hyacinth (Barrett, 1989; Toft, 2000). The dense mats reduce dissolved oxygen levels and light, significantly altering ecosystems and plant and animal communities. Low oxygen levels harm fish populations and fish spawning areas may be reduced, as well as critical waterfowl habitat degraded (Schmitz, et al. 1993).

KNOWLEDGE GAPS AND RESEARCH NEEDS

Permanent Plot Study

Researches focusing on the impact of invasives on the ecosystem structure and functioning can be best performed at a temporal scale using a permanent plot study. Use of permanent plots to study the long term impact of invasives on various life-forms is lacking in the country. Studies should be designed using some of the invasive species to study their impacts in isolation and in combination with other invasive species at a permanent plot scale. Looking to the present scenario this calls for an urgent attention to the funding agencies to deliver funds pertaining to such ideas.

Niche evolution, local adaptation, genetic variability, and geographical expansion

Niche requirements will determine which species can invade which regions and where they will spread within those regions (Peterson, 2003; Peterson and Vieglais, 2001). If their fundamental niches are conserved, species will only be able to invade regions that have a climate similar to that of their native range. Wiens and Graham (2005), argued that the correspondence between the native climate of few exotic species and their introduced range seems striking. Invasive species offer an intriguing system to study the ecological and evolutionary causes of biogeographic patterns. Invasive species in particular represent biogeographic experiments that can be used to test if species niche is conserved or it evolves to the new environmental conditions. If niche conservatism determines the range limits of species, then we should expect to see consistent parallels between their climatic distribution in their native and introduced ranges. Conversely, if introduced sub tropical species routinely spread into tropical regions or vice versa, then competition (or other biotic factors) may be more important in setting geographic range limits in their native ranges. Members of some families are often over-represented in the invasive flora of a region, but we do not know why. Research is needed to find out the explanation for the over-representation of certain genera/families compared to others in the invasive flora of a region.

The "Enemy Release" hypothesis (Davis et al. 2000), suggests that invaders perform better in their introduced range than in their native range because they lose their natural enemies during the process of colonization. A corollary of this hypothesis is the "Evolution of increased competitive ability" (Muller-Scharer, 2004), hypothesis which assumes that exotics, when liberated from their native specialized enemies, can allocate resources otherwise used for costly traits to resist those enemies, for the development of traits that provide greater competitive advantage. Either of the two variables, the abiotic environment and competition have the potential to shape the evolution of traits that confer invasiveness. However, the extent to which each driver becomes more important to explain invasive success is an issue that remains to be explored.

Many organisms can respond with considerable flexibility to a changing environment generating a range of characteristics depending on the environment in which the developing organism finds itself. A new species, when introduced into an alien environment, must become established within the constraints of site characteristics and competition with resident vegetation. A comparative approach has been useful in generating hypotheses as to what makes invasive species so successful (Daehler, 2003). Differences among the invader populations might be attributed to rapid adaptive evolution giving rise to plastic response of species which ultimately favours invasiveness. One approach is to pick up phenotypic variation among populations and then to explore the basis of such variation using an experimental protocol in an attempt to elucidate if the variation is indeed a product of local adaptation or plasticity.

The success of invasive species depends upon whether colonization events are facilitated by adaptive evolution or are limited to sites where a species is pre-adapted to thrive (Dlugosch and Parker, 2008). In particular, successful colonization is most likely under conditions requiring little adaptive change (pre-adaptation); however, as spread proceeds, recipient habitats deviate more strongly from the original habitats; in this case larger numbers of founders and more substantial adaptive evolution are required for successful expansion (Holt et al. 2005). Some of the world top invaders do not succeed in one go in the novel environment, but they migrated gradually following a number of non-adaptive and adaptive phases (Cronk and Fuller, 1995).

The ability of an organism to express different phenotypes in response to environmental cues, i.e., phenotypic plasticity, is generally interpreted as an adaptation/acclimation for dealing with variability in the environment. New habitat challenges may cause the expression of potentially beneficial traits through plasticity. Wide spread colonizing species are often characterized by high phenotypic plasticity allowing them to inhabit diverse new sites without undergoing local genetic adaptation through natural selection. Work on several colonizing species has confirmed that their population across broad geographic and environmental ranges consisted of genetically similar populations of highly plastic genotypes. Thus, being plastic may allow a population of invaders to quickly adapt to the new environment and to maintain higher fitness relative to native taxa. To measure the extent of phenotypic plasticity (Bradshaw, 1965), genetically identical organisms need to be exposed in a range of different, but relevant, environments. We need to address the difficult key question of whether plasticity enhances fitness (Scheiner, 1993 and Via et al. 1995) and thus invasiveness. One approach would be to compare levels of plasticity within and among populations, specifically for some of its vegetative and reproductive traits. By focusing on particular traits that allow invasion into different environments, one could also directly test the link between invasiveness and plasticity at the genotype level by using manipulative approaches (Schmitt et al. 1999). However, the spatial scale over which invasive plant populations are adapted is not well understood and most certainly varies with population and the degree of environmental heterogeneity in its invaded range. Studies of plasticity in its ecological context provide a rich opportunity for interdisciplinary collaborations, to reveal not only patterns and mechanisms of environmental response, but also the direct and indirect effects of those responses on the relationship of organisms to their abiotic environments and, consequently on higher-level ecological processes. Such study design can be potentially utilized to test the theory of "niche conservatism" and can provide opportunity to investigate the role of genetic changes during invasions, so that we can begin to make more robust generalizations about the long-term risks of species introductions. Such study designs can give a bigger picture of rapid plant evolution with invasive species as a novel system to observe real time evolution.

References

Agrawal, M. (2002). Atmospheric deposition and budgeting of the nutrients and trace elements around industrial sites at Singrauli region. Final technical report. Ministry of Environment and Forest, Government of India, New Delhi. Report No. 19/93/95-RE.

Agriculture & Resource Management Council of Australia & New Zealand, Australian & New Zealand Environment & Conservation Council and Forestry Ministers (2000). *Weeds of National Significance Mesquite* (*Prosopis* species) *Strategic Plan*. National Weeds Strategy Executive Committee, Launceston. Available from: http://www.weeds.org.au/docs/msqstrat.pdf.

Annapurna, C. and Singh, J.S., (2003a). Variation of *Parthenium hysterophorus* in response to soil quality: Implications for invasiveness. *Weed Res.*, 43: 190–198.

_________. (2003b). Phenotypic plasticity and plant invasiveness: Case study of congress grass. *Curr. Sci.*, 85: 197–201.

Anonymous (2007). *Global Compendium of Weeds Ageratum Houstonianum* (Asteraceae). Retrieved from URL: http://www.hear.org/gcw/species/ageratum_houstonianum dated: 25 August 2008.

Asfaw, H. and Thulin, M. (1989). In: *Flora of Ethiopia: Pittosporaceaeto Araliaceae, Mimosoideae.* Vol. 3. Hedberg, I. and Edwards, S. (Eds.) Addis Ababa University, Addis Ababa and Uppsala University, Sweden.

Babu, S., Love A. and Babu C.R. (2009). Ecological Restoration of *Lantana*-Invaded Landscapes in Corbett Tiger Reserve, India. *Ecol. Rest.*, 27(4): 467–477.

Barrett, S.C.H. (1989). Waterweed invasions. *Sci. Am.*, 264: 90–97.

Barrows, E.M. (1976). Nectar robbing and pollination of *Lantana camara* (Verbenaceae). *Biotropica,* 8: 132–135.

Bradshaw, A.D. (1965). Evolutionary Significance of Phenotypic Plasticity in Plants. *Advances in Genetics,* 13: 115–155.

Callaway, R.M. and Ridenour, W.M. (2004). Novel weapons: Invasive success and the evolution of increased competitive ability. *Front. Ecol. Environ.*, 2: 436–443.

Carlton, J.T. (1996). Biological invasions and cryptogenic species. *Ecology,* 77: 1653–1655.

Cronk, Q.C.B. and Fuller, J.L. (1995). *Plant Invaders,* Chapman and Hall.

Cruttwell, R.E. and McFadyen, R.C. (1989). Siam weed: A new threat to Australia's north. *Plant Prot., Q.,* 4: 37.

Daehler, C.C. (2003). Performance comparisons of co-occurring native and alien invasive plants: Implications for conservation and restoration. *Ann. Rev. Ecol. Evol. Syst.,* 34: 183–211.

Davis, M.A., Grime J.P. and Thompson, K. (2000). Fluctuating resource in plant communities: A general theory of invasibility. *J. Ecol.,* 88: 528–534.

Dlugosch, K.M. and Parker, L.M. (2008). Invading populations of an ornamental shrub show rapid life history evolution despite genetic bottlenecks. *Ecol. Letters,* 11: 701–709.

Duggin, J.A. and Gentle, C.B. (1998). Experimental evidence on the importance of disturbance intensity for invasion of Lantana camara L. in dry rainforest–open forest ecotones in north-eastern NSW, Australia. *For. Ecol. Manag.*, 109: 279–292.

Duthie, J.F. (1903). *Flora of the Upper Gangetic Plain,* Vol. 1 and 2. Government of India, Calcutta.

Felippe, G.M., Polo, M., Cardosa, V.J.M. and Figeiredo-Ribeiro, R.C.L. (1983). Germinacao da unidade de dispersao de ervainvasora *Hyptis suaveolens. Ann. Sem. Reg. Ecol.*, 3: 245–261.

Fritts, T.H. and Rodda, G.H. (1998). The role of introduced species in the degradation of island ecosystems: A case study of Guam. *Annu. Rev. Ecol. Syst.*, 29: 113–140.

Galatowitsch, S.M., Anderson N.O., and Ascher P.D. (1999). Invasiveness in wetland plants in temperate North America. *Wetlands*, 19: 733–755.

Gentle, C.B. and Duggin, J.A. (1997). *Lantana camara* L. invasions in dry rainforest–open forest ecotones: The role of disturbances associated with fire and cattle grazing. *Austr. J. Ecol.*, 22: 298–306.

GISD, (2005). Global invasive species data. Database retrieved from URL: www.issg.org/database on 7 March 2010.

GISP, (2000). *Global strategy on invasive alien species.* Global Invasive Species Programme Workshop, September 2000, Cape Town, South Africa.

GISP, (2003). *The IAS Problem.* The global invasive species programme, http://www.gisp.org/about/IAS.asp.

Goodall, J.M. and Erasmus, D.J. (1996). Review of the status and integrated control of the invasive alien weed, *Chromolaena odorata*, in South Africa. *Agri. Ecosyst. Environ.*, 56: 151–164.

Gopal, B. and Sharma, H.P. (1981). *Water-Hyacinth*, Hindasia Publication, Delhi, India.

Gopal, B. (1987). *Water Hyacinth.* Elsevier Science Publishers, Amsterdam, p. 471.

Gujral, G.S. and Vasudevan, P. (1983). *Lantana camara* L., a problem weed. *J. Sci. Indust. Res.*, 42: 281–286.

Gupta, K. and Khetarpal, S. (2006). Regulatory measures dealing with invasive alien species: Global and national scenario. In: *Invasive Alien Species and Biodiversity in India.* Rai, L.C. and J.P. Gaur, (Eds.). pp. 169–185. Banaras Hindu University Press, Varanasi, India

Gurevitch, J. and Padilla, D.K. (2004). Are invasive species a major cause of extinctions? *Trends Ecol. Evol.*, 19: 470–474.

Hakimuddin, M. (1929). *Lantana* in northern India as a pest and its probable utility in solving the cowdung problem. *Ind. For.*, 56: 405–410.

Harper, J.L., Clatworthy, J.N., McNaughton, I.H. and Sagar, G.R. (1961). The evolution and ecology of closely related species living in the same area. *Evolution,* 15: 209–227.

Hickey, M. and King, C. (1988). *100 Families of Flowering Plants.* Cambridge University Press, Cambridge.

Hiremath, A.J. and B. Sundaram, (2005). The fire-lantana cycle hypothesis in Indian forests. *Conserv. Soc.,* 3: 26–42.

Holm, L., Plucknett, Pancho, D.J. and Herberger, J. (1977). *The World's Worst Weeds: Distribution and Biology, Honolulu*: University of Hawaii Press, p. 609.

Holt, R.D., Barfield, M. and Gomulkiewicz, R. (2005). Theories of niche conservatism and evolution. Could exotic species be potential tests? In: *Species Invasions. Insights into Ecology, Evolution and Biogeography.* Sax D.F., Stachwicz, J.J. and Gaines, S.D. (Eds.). Sinauers Associates Inc., Publishers, Sunderland; Massachusetts, pp. 259–290.

Honu, Y.A.K., Dang, Q.L. (2000). Response of tree seedlings to the removal of *Chromolaenaodorata* Linn. in a degraded forest in Ghana. *For. Ecol. Manag.,* 137: 75–82.

Howard, R.A. (1989). *Flora of the Lesser Antilles, Leeward and Windward Islands.* Dicotyledoneae. Part 3, Vol. 6, p. 658. Jamaica Plain, MA: Arnold Arboretum, Harvard University.

Hutchinson, J. and Dalziel, J.M. (1963). *Flora of West Tropical Africa.* Crown Agents, London.

Julien, M.H. and Orapa, W. (2001). Insects used for biological control of the aquatic weed water hyacinth in Papua New Guinea. *Papua New Guinea Journal of Agriculture, Forestry and Fisheries,* 44: 49–60.

Kanzaria, M.K. and Varshney, A.K. (1999). *Prosopis juliflora*—Its use. State Forest Department, India, pp. 113–115.

Keane, R.M. and Crawley, M.J. (2002). Exotic plant invasions and the enemy release hypotheses. *Trends Ecol. Evol.,* 17: 164–170.

Khanna, K.K. and Singh, K.P. (2006). Invasive Alien Species of Chhattisgarh. In: *Invasive Alien Species and Biodiversity in India.* Rai, L.C. and Gaur, J.P. (Eds.). pp. 35–42. Banaras Hindu University Press, Varanasi, India.

Kohli, R.K., Batish D.R., Singh, H.P. and Dogra, K.S. (2006). Status, invasiveness and environmental threats of three tropical American invasive weeds (*Parthenium hysterophorus* L., *Ageratum conyzoides* L., *Lantana camara* L.) in India. *Biol. Inv.,* 8: 1501–1510.

Lai, P.Y., Muniappan, R., Wang, T.H. and Wu, C.J. (2006). Distribution of *Chromolaena odorata* and its biological control in Taiwan. *Proc. Hawaiian Entomol. Soc.,* 38: 119–122.

Levine, S.A. (1989). Analysis of risk for invasions and control program. In: *Biological Invasion: A Global Perspective,* SCOPE 37 Drake, J.A. (Eds.). et al. John Wiley Chichester, UK., pp. 425–435.

Lindsey, K. and Hirt, M.H. (1999). *Use Water Hyacinth, A Practical Handbook for the Water Hyacinth from Across the World.* Action for Natural Medicines, Winnenden, Germany.

Liogier, H.A. (1995). *Descriptive flora of Puerto Rico and adjacent islands,* San Juan, PR: Editorial de la Universidad de Puerto Rico. 4: 617.

Little, E.L., Jr., Woodbury, R.O. and Wadsworth, F.H. (1974). *Trees of Puerto Rico and the Virgin Islands,* Vol. 2, p. 1024. Agriculture Handbook 449. Washington, D.C.: U.S. Department of Agriculture.

Mack, R.N., Simberloff, D., Lonsdale, W.M., Evans, H., Clout, M. and Bazzaz, F.A. (2000). Biotic invasions: Causes, epidemiology, global consequences and control. *Ecol. Appl.,* 10: 689–710.

Matthews, S. (2004). Tropical Asia invaded: The growing danger of invasive alien species Global Invasive Species Programme. Available from: http://www.gisp.org/casestudies/showcasestudy.asp?id=233&MyMenuItem=casestudies&worldmap=&country.

Mohamed, A.E.F., (1997). In: *Tropical Forestry Report: Management of Prosopis juliflora for Use in Agroforestry Systems in the Sudan.* Luukkanen, O. (Ed.). Ph.D. Thesis, University of Helsinki, Helsinki, Finland.

Motooka, Philip, Luisa Castro, Duane Nelson, Guy Nagai, and Lincoln Ching. (2003). *Weeds of Hawaii's Pastures and Natural Areas: An Identification and Management Guide.* College of Tropical Agriculture and Human Resources (University of Hawaii--Manoa), Honolulu, p. 184. color illus. ISBN 1-929325-14-2.

Mudgal, V., Khanna, K.K. and Hazra, P.K. (1997). *Flora of Madhya Pradesh II.* Botanical Survey of India, Kolkata, India.

Muller-Scharer H., Schaffner U. and Steinger, T. (2004). Evolution in invasive plants: Implication for biological control. *Trends Ecol. Evol.,* 19: 417–422.

Nayar, M.P. (1977). Changing patterns of the Indian flora. *Bull. Bot. Surv. India,* 19: 145–154.

New Zealand Department of Statistics (1996). *New Zealand Official Yearbook,* Wellington, New Zealand, 99th edn.

Palmer, W.A. and Pullen, K.R. (1995). The phytophagous arthropods associated with *Lantana camara, L. hirsuta, L. urticifolia and L. urticoides* (Verbenaceae) in North America. *Biol. Control,* 5: 54–72.

Parsons, W.T. and Cuthbertson, E.G. (2001). Common *Lantana.* In: *Noxious Weeds of Australia.* CSIRO, Melbourne, 627–632.

Pasiecznik, N. (1999). Prosopis – Pest or providence, weed or wonder tree? In: *European Tropical Forest Research Network News 28*. W. Brinkman, (Ed.). The Tropenbos Foundation (ETFRN). Available from: http://www.etfrn.org/etfrn/newsletter/frames/nl28.html.

Penfound, W.T. and Earle, T.T. (1948). The biology of the hyacinth. *Ecol. Monogr.*, 18: 448–472.

Pereira, W.E. (1919). *Lantana* in the Math working circle in Savantavadi state forest. *Ind. For.,* 46: 188–193.

Peterson, A.T. and Vieglais, D.A. (2001). Predicting species invasions using ecological niche modeling. *Bio. Science,* 51: 363–71.

Peterson, A.T. (2003). Predicting the geography of species' invasions via ecological niche modeling. *Q. Rev. Biol.,* 78: 419–33.

Pimentel, D., McNair, S., Janecka, J., Wightman J., Simmonds, C. O'Connell, C. Wong E., Russel L., Zern J., Aquino T. and Tsomondo, T. (2001). Economic and environmental threats of alien plant, animal, and microbe invasions. *Agri. Ecosyst. Environ.,* 84: 1–20.

Prasad, A.E. (2010). Effects of an Exotic Plant Invasion on Native Understory Plants in a Tropical Dry Forest. *Conserv. Biol.,* 24: 747–757.

Prescott-Allen, R. and Prescott-Allen, E. (1990). How many plants feed the world? *Conserv. Biol.,* 4: 365–374.

Rahman, M.A. and Wilcock, C.C. (1991). A taxonomic revision of *Calotropis* (Asclepiadaceae). *Nordic J. Bot.,* 11: 301–308.

Raizada, P., Sharma, G.P. and Raghubanshi, A.S. (2008). Ingress of *Lantana* in dry tropical forest fragments: Edge and shade effects. *Curr. Sci.,* 94(2): 180–182.

Rao, R.R. and Murugan, R. (2006). Impact of exotic adventive weeds on native biodiversity in India: Implication for conservation. In: *Invasive Alien Species and Biodiversity in India.* Rai, L.C. and Gaur, J.P. (Eds.). pp. 93–109. Banaras Hindu University Press, Varanasi, India.

Rapoport, E.H. (1991). Tropical versus temperate weeds: A glance into the present and future. In: *Ecology of Biological Invasion in the Tropics.* Ramakrishnan, P.S. (Ed.). International Scientific Publications, New Delhi, pp. 441–451.

Reddy, C.S., Bagyanarayana, G., Reddy, K.N. and Raju, V.S. (2008). Invasive alien flora of india. Available on line at: http://www.gisinetwork.org/IndiaInvasivePlants/documents/assessment_Invasive_India_jan08.pdf.

Rejmanek, M. (2000). Invasive plants: Approaches and predictions. *Aust. J. Ecol.*, 25: 497–506.

Rejmanek, M. and Randall, J.M. (1994). Invasive alien plants in California: 1993 summary and comparison with other areas in North America. *Madrono*, 41: 161–177.

Richardson, D.M. and Pyšek P. (2006). Plant invasions: Merging the concepts of species invasiveness and community invasibility. *Progress in Physical Geography,* 30: 409–431.

Richardson, D.M., Pysek, P., Rejmanek, M., Barbour, M.G., Panetta, F.D. and West, C.J. (2000). Naturalization and invasion of alien plants: Concept and definitions. *Divers. Dist.*, 6: 93–107.

Room, P.M. and Fernando, I.V.S. (1992). Weed invasions countered by biological control: *Salviniamolesta* and *Eichhorniacrassipes* in Sri Lanka. *Aquatic Bot.*, 42: 99–107.

Sahu, P.K. and Singh, J.S. (2008). Structural attributes of *Lantana*-invaded forest plots in Achanakmar–Amarkantak Biosphere Reserve, Central India. *Curr. Sci.*, 94: 494–500.

Saxena, K.G. and Ramakrishnan, P.S. (1983). Growth, resource allocation patterns and nutritional status of some dominant annual weeds of shifting agriculture (jhum) in north-eastern India, *ActaOecol.*, 4: 323–333.

Schei, P.J. (1996). Conclusions and recommendations from the UN/Norway Conference on Alien species. *Sci. Int.*, 63: 32.

Scheiner, S.M. (1993). Genetics and evolution of phenotypic plasticity. *Ann. Rev. Ecol. Evol. Syst.*, 24: 35–68.

Schmidt, G.J. and Schilling, E.E. (2000). "Phylogeny and Biogeography of Eupatorium (Asteraceae: Eupatorieae) Based on Nuclear ITS Sequence". *Am. J. Bot.*, (*Bot. Soc. Am.*) 87: 716–726.

Schmitt, J., Dudley, S.A. and Pigliucci, M. (1999). Manipulative approaches to testing adaptive plasticity: Phytochrome-mediated shade-avoidance responses in plants. *Am. Nat.*, 154: 43–54.

Schmitz, D.C., Schardt, J.D., Leslie, S.J., Dray, F.A. Jr., Osborne, J.A. and Nelson, B.V. (1993). The ecological impact and management history of three invasive alien aquatic plant species in Florida. In: *Biological Pollution: The Control and Impact of Invasive Exotic Species*. McKnight, B.N. (Ed.). Indiana Academy of Science. Indianapolis, IN.

Sculthorpe, C.D. (1985). *The biology of aquatic vascular plants.* Koeltz Scientific Books, Konigstein, West Germany.

Sharma, G.P. and Esler, K.J. (2010). Performance of invasive shrub *Lantana camara* L. in two of its invaded ranges: India and South Africa. Biological Invasion. *J. Biogeo.*, (in press).

Sharma, G.P. and Raghubanshi, A.S. (2005). A comparative account of theories related to plant invasions. *Range. Manag. Agrofor.*, 26: 37–42.

________. (2006). Tree population structure, regeneration and expected future composition at different levels of *Lantana camara* L. invasion in the Vindhyan tropical dry deciduous forest of India. *Lyonia,* 11(1): 25–37.

________. (2007). Effect of *Lantana camara* L. cover on plant species depletion in the Vindhyan tropical dry deciduous forest of India. *Appl. Ecol. Environ. Res.*, 5: 109–121.

________. (2009). How *Lantana* invades a dry deciduous forest: A case study from Vindhyan highlands, India. *Trop. Ecol.,* 51(2S): 305–316.

________. (2010). *Lantana* invasion alters the soil pools and processes: a case study with special reference to nitrogen dynamics in the tropical dry deciduous forest of India. *Appl. Soil Ecol.,* 42: 134–140.

Sharma, G.P., Kumar, M. and Raghubanshi, A.S. (2010). Urbanization and road-use determines Calotropisprocera distribution in the eastern Indo-Gangetic plain, India. *Ambio.,* 39: 194–197.

Sharma, G.P., Raghubanshi, A.S. and Singh, J.S. (2005b). *Lantana* invasion: An overview. *Weed Biol. Manag.,* 5: 157–167.

Sharma, G.P., Raizada, P. and Raghubanshi, A.S. (2007). *Hyptis suaveolens* – a threat to the flora of the Vindhyan Highlands, India. *Newsletter of the Asia-Pacific Forest Invasive Species Network* (APFISN), 7: 4–5.

Sharma, G.P., Raizada, P. and Raghubanshi, A.S. (2009). *Hyptis suaveolens*: An emerging invader of Vindhyan plateau, India. *Weed Biol. Manag.*, 9: 185–191.

Sharma, G.P., Singh J.S. and Raghubanshi, A.S. (2005a). Plant invasions: Emerging trends and future implications. *Curr. Sci.,* 88: 726–734.

Shukla, A.N., Singh, K.P. and Singh, J.S. (2009). Invasive alien species of Achanakmar-Amarkantak Biosphere Reserve, Central India. *Proc. Nat. Acad. Sci.* India, Sect. B, 79: 384–392.

Singh, K.P., Shukla, A.N. and Singh, J.S. (2010). State-level inventory of invasive alien plants, their source regions and use potential. *Cur. Sci.*, 99(1): 107–114.

Singh, S.P. (1996). Biological control. In: *50 Years of Crop Science Research in India*. Paroda, R.S., Chadha, K.L. (Eds.). Indian Council of Agricultural Research, New Delhi, 88–116.

Singh, V., Singh, H., Sharma, G.P. and Raghubanshi, A.S. (2010). Eco-physiological performance of two invasive weed congeners (*Ageratum conyzoides* and *Ageratum houstonianum*) in the Indo-Gangetic plains of India. *Environ. Monit. Ass.* (on line first). doi: 10.1007/s10661-010-1700–4.

Smith, N.M. (2002). *Weeds of the Wet/Dry Tropics of Australia – A Field Guide*. Environment Centre NT, Inc., p. 112.

Stone, B.C. (1970). The flora of Guam. *Micronesica,* 6: 1–659.

Stubbs, C.J. (1999). *Feral burro Removal: New Solutions to an Old Problem*. Natural Resource Year in Review: publication D-1346.

Swaminathan, C., Rai, R.S. and Suresh, K.K. (1990). Allelopathic effects of *Parthenium hysterophorus* on germination and growth of a few multi purpose trees and arable crops. *Int. Tree Crops J.*, 6: 143–50.

Swarbrick, J.T., Willson, B.W. and Hannan-Jones, M.A. (1998). *Lantana camara* L. In: *The Biology of Australian Weeds.* Panetta F.D., Groves R.H. and Shepherd R.C.H. (Ed.). R.G. & F.J. Richardson, Melbourne, 119–136.

Thorp, J.R. (2000). Weeds of national significance. *Lantana* (*Lantana camara*) Strategic plan. National Weed Strategy Executive Committee, Launceston, Tas., Australia. Available from URL: http://www.weeds.org.au/docs/lanstrat.pdf. Accessed 12 July 2005.

Toft, J.D. (2000). Community effects of the non-indigenous aquatic plant water hyacinth (*Eichhornia crassipes*) in the Sacramento/San Joaquin Delta, California. Masters Thesis, University of Washington, Seattle, Washington.

Van Oosterhout, E., Clark, A., Day, M.D. and Menzies, E. (2004). *Lantana* Control Manual. Current Management and Control. Options for *Lantana* (*Lantana camara*) in Australian State of Queensland. Department of Natural Resources, Mines and Enegry, Brisbane, Qld, Australia. Available from URL: http://www.nrm.qld.gov.au/pests/wons/Lantana. Accessed 23 November 2004.

Vanderwoude, C., Scanlan, J.C., Davis , B. and Funkhouser, S. (2005). *Plan for National Delimiting Survey for Siam Weed.* Natural Resources and Mines Land Protection Services: Queensland Government.

Verma, B.K. and Mishra, B.K. (1992). *Flora of Allahabad District UP India.* B. Singh & M.P. Singh Publishers, Allahabad, India.

Via, S., Gomulkiewicz, R., de Jong, G., Scheiner, S., Schlichting, C.D. and van Tienderen, P.H. (1995). Adaptive phenotypic plasticity: Concensus and controversy. *Trends Ecol. Evol.,* 10: 212–217.

Wiens, J.J. and Graham, C.H. (2005). Niche conservatism: Integrating evolution, ecology, and conservation biology. *Annu. Rev. Ecol. Evol. Syst.,* 36: 519–539.

Wilcove, D.S., Rothstein, D., Dubow, J., Phillips, A. and Losos, E. (1998). Quantifying threats to imperiled species in the United States. *Bioscience,* 48: 607–615.

Williamson, M. (1993). Invaders, weeds and risk from genetically modified organisms. *Experientia,* 49: 219–224.

Willis, J.C. (1973). *Dictionary of Flowering Plants and Ferns* (rev. by Shaw, A.K.). Cambridge University Press, London.

Wilson, J.R.U., Dormontt, E.E., Prentis, P.J., Lowe, A.J. and Richardson, D.M. (2009). Something in the way you move: Dispersal pathways affect invasion success. *Trends Ecol. Evol.,* 24: 136–144.

Witkowski, E.T.F. (2002). Invasion intensity and regeneration potential of the non-native invasive plant *Chromolaenaodorata* at St Lucia, South Africa. In: *Proceedings of the Fifth International Workshop on Biological Control and Management of Chromolaenaodorata, Durban, South Africa, October 2000.* Costas Zachariades, R. Muniappan and Lorraine W. Strathie, (Eds.). 34–39. Teeanem Pietermaritzburg.

Wulff, R.D. (1973). Intrapopulational variation in the germination of seeds in Hyptis suaveolens. *Ecology,* 54: 646–649.

6

Our Changing Geo-Biosphere

V.M. Meher-Homji

INTRODUCTION

The year 2010 has been dedicated as the year of Biodiversity by the United Nations and it is heartening to learn that the Biodiversity Board of West Bengal is bringing out a publication on Biodiversity with reference to India on this occasion when efforts are on to conserve whatever remains of our floral and faunal heritage.

The discipline that scans the biodiversity is the science of Biogeography. Encompassing Life Sciences and Earth Sciences, it links the past, present and future in a three dimensional framework: Present distribution pattern of an organism or a community, their past i.e., evolution in time, and application to meet future challenges. Maps showing location of individual species or vegetation (physiognomies like forest, thicket, grassland) best indicate distribution in space i.e., chorology; chronology or distribution in time on long-term basis (evolution) is accessed through fossil records; palynology is a modern contribution to palaeobotany. On a short-term basis, the succession of communities is assessed. Explanations for the existence of a species or community (vegetation) are sought in terms of climate, edaphic and biotic (mainly anthropogenic) factors. It is pertinent to mention that the vegetation is the integrator of environmental factors as any change in environmental factors is reflected in its structure and composition. Besides, vegetation plays a fundamental role as the habitat for animals.

HISTORY OF THE EARTH

At this juncture a brief review of the history of the Geo-Biosphere would expose the vicissitudes and their repercussions. Formed 5 to 7 billion years ago, life originated on earth in its simplest form 3600 million years (my) ago.

The history of past vegetation has been summarized in Table 1 along with the environmental features since the Pre-Cambrian age, revealing marked climatic changes. For example, part of Europe experiencing cold temperate climate to day was under tropical regime during the Palaeogene 50 my ago as shown by the fossils. The fossilized fruit of the palm *Nypa* growing in the mangroves of the Sundarbans and Sri Lanka has been discovered in the London clay geological formation. During the Quaternary, 3 Ice Ages alternated with 3 warm interglacial phases, the last Ice Age terminated 10,000 years B.P.

Table 1. Vegetation History

Period	*Age from to day in millions of years*	*Plant fossil record*	*Indication of environment*
Middle-Upper Cambrian	600	Cutinized spores probably belonging to land plants	A landscape of green vegetation mainly composed of ferns and Pteridosperms
Paleozoic	300–500	Several fern groups prevailed; Carboniferous members were trees	
Lower Gondwana (Upper Carboniferous to Permian)	270–350	*Glossopteris* flora	Glacial to cold Temperate climate (Peninsular India at 65° latitude south)
Mesozoic-Middle Gondwana (Triassic)	230	*Dicroidium* flora. Some Conifers made their appearance in the Upper Triassic	Warm, semi-arid climate
Upper Gondwana (Jurassic to Lower Cretaceous)	130–180	Golden age of the Indian gymnosperms	Warm, humid climate
Mid to Late Cretaceous	65–130	*Gymnosperms* begin to decline. *Mesembrioxylon,* only gymnosperm fossil to be recorded in Upper Cretaceous (in Tiruchirapalli district)	Volcanic activity

Cenozoic-Eocene Miocene	26–54 5–25	Fossils of tropical evergreen forest species recorded from Coromandel in the South to Shivalik in the north, from Bengal basin in the east to Rajasthan-Kutch in the west.	Uniformly warm, humid conditions throughout Peninsular India
Pliocene	5	Present day pattern of vegetation established	The Himalayas and the Western Ghats attain maximum elevation; establishment of the monsoon regime as known today.
Quaternary	1	Temperate species including conifers colonize the Himalayas	Three glacial phases alternating with three interglacials
Holocene	10,000 years B.P.	Man's interference with vegetation begins in the upper Holocene	

The climate of the Indian Peninsula was uniformly warm and humid throughout its length and breadth in the Eocene-Miocene. The continuing northward drift of the Indian plate after its separation from Gondwana brought about a major change in climate. In the late Cretaceous or early tertiary, India occupied a southerly position astride the equator (Frakes and Kemp, 1972). As it crossed the equator, it was deprived of the equatorial climate characterized by daily rainfall. In the wake of its collision with the Asian landmass, the Himalaya was uplifted and the monsoon pattern was established. The Western Ghats too were elevated in the Pliocene, introducing a rainshadow on the leeward side reducing rainfall of the Deccan and the Coromandel coast, changing the vegetation from evergreen forests to thorny scrubs. Elsewhere in the Peninsular forests, the evergreen species were replaced by deciduous ones everywhere except in the Western Ghats, pockets of the Eastern Ghats and in the North-East where rainfall continued to remain favourable (Meher-Homji, 1977, 1990) Climate-vegetation changes during the Quaternary suggested by palynological studies have been reviewed by Meher-Homji and Gupta, 1999.

The main vegetation types in India are determined by macro-climatic factors like temperature, especially the mean of the coldest month, rainfall quantum and distribution i.e., the length of the dry period (number of dry months) and the regime i.e., the season of occurrence of rains, and soils.

Homocene, the age characterized by the interference of Homo sapiens with his environment

Introduction of agriculture in the Neogene (for India estimated 4500 years B.P.) marked the beginning of interference with the natural vegetation. Fire was a very effective tool in the possession of man even before the use of metal implements. Large tracts of forests were cleared to make room for cultivation. Woodlands have been mainly confined to hills and hillocks. Fires promoted spread of grasslands and felling of trees accompanied by grazing created thicket type of physiognomy with dominance of thorny, spiny shrubs and stragglers.

Vegetation, the integrator of environmental factors

Mosaics of vegetation comprise floristic diversities grouped in varied physiognomies or landscapes like forest, thicket, grassland. Vegetation maps published by the French Institute, Puducherry with the co-operation of the State Forest Departments and the ICAR (Gaussen et al. 1961–1978; Pascal et al. 1982) depict forest types (series) with essential floristic composition, physiognomy, links with environmental factors, successional trend and cropping patterns. The distribution is revealed at a glance as also the fragmentation of the habitat through the disruption of vegetation.

The forest area in India is estimated at about 677,000 sq. kms or 20% of the geographical area. Even then only about 10% is under dense cover, the rest being open woodlands and degraded vegetation (savannas, thickets). Using the vegetation maps of the Peninsula south of the latitudinal belt Bikaner-Kolkata (Gaussen et al. 1961–1978) and other documents like Forest Atlas of Das Gupta (1976), maps of Bellan (1985) and NRSA, an attempt was made to assess the percentage of physiognomic forest area in each of the vegetation (Forest) type with reference to the potential area that the type could cover. The percentage of the area under degraded vegetation (savanna, thickets) for each vegetation type (with reference to the potential area) has also been noted. As expected no forest (in physiognomic form) is left in the thorny vegetation types of the arid and semi-arid zones. The deciduous vegetation types have mostly less than 23% forest cover. The semi-evergreen and evergreen types fare somewhat better.

In spite of this alarming scenario, the forests that remain are under a threat in view of the demand for land for various projects. Another point of concern is that the worth of a forest is judged from the timber species it contains. The thickets and scrub-jungles of the Coromandel coast under the "dry evergreen type" are dovoid of timber species and are therefore considered worthless and felled or replaced by the plantations of the exotic Eucalyptus. However, these jungles contain valuable medicinal plants. One example is that of the climber *Ormocarpum sennoides*, reputed for its bone healing property. The use is known only to the villagers, not to the medical world. With the disappearance of the thickets a rich source of nature's wealth is lost. With extinction of one species, it is feared that 6 to 7 more species may disappear because of interlinks among plants and animals. After extermination of the endemic bird Dodo, the tree *Calvaria* stopped regeneration as its seeds could only germinate after being crushed in the gizzard of the Dodos.

Global Warming vis-à-vis the Forest Ecosystem

The climate change mainly concerns the rising temperature and secondarily the rainfall. However, associated factors like the length of the dry (rainless) period and the time (season) of occurrence of rains are left out of the discussion in spite of their crucial role in governing the vegetation types.

Further, rainfall amount and its spread over the year present curious contradictions. Kochi in Kerala receives 2 to 3m of rains in 10 months, further north Agubme in Karnataka, supposed to be Cherrapunji of South India, receives 8m but in 6–7 months only. Mumbai has annual average of 2000mm with dry season of 8 months. Bangalore and Mysore receive 800mm only but over 6 months. Forests in the vicinity of these 3 stations, however, are clothed with deciduous forests as from forest point of view there seems to be some sort of compensation between the rainfall amount and distribution.

Similarly, the season of occurrence of rains seems to be a critical issue. The tropical dry evergreen "forest" type is strictly confined to the Coromandel-Circar coastal area of Tamil Nadu, Andhra Pradesh and Orissa where the rainfall regime is dissymmetric (rains mainly from the depressions and cyclones in the months of October to January from the North-East monsoon). Where the regime is typically tropical (rains from June to September from the southwest monsoon), the resultant vegetation is a deciduous forest. In central India, the eastern part is dominated by the

Sal (*Shorea robusta*) forests whereas the western counterpart by the Teak (*Tectona grandis*). The reason is that the seeds of Sal remain viable for only about 10 days or so after falling on the ground and it should rain during this critical period to favour germination. The Bay of Bengal branch of the South-West monsoon being earlier than the Arabian Sea branch, the growth of Sal is promoted in the eastern half of the Peninsula, leaving the dominance to the teak in the western half. This example shows the complicated relationship between the distribution of species and the climate. Some additional points also need consideration while in the topic of global warming and forest eco-system:

1. The highest expression of vegetation is a Rain-Forest which occurs under high rainfall and high temperature. One tends to associate deserts with high temperatures but in fact it is the lack of water that determines the arid zone ecosystem. Besides deserts like the thar experience low temperatures in winters', not to speak of cold deserts, like Leh, Kargil.
2. The expected rise in temperature is estimated at 2.5° to 3.9°C. However, a place in Andhra Pradesh like Warangal and another in U.P., say Varanasi will have a difference of 3°C in the mean minimum of the coldest month, which is the limiting factor for the vegetation rather than the average or maximum temperature. In physiognomy and floristic composition, vicinity of both these places bear a close similarity *viz.* a deciduous forest. India with its large latitudinal expanse from the equatorial level up to 36° North offers ample opportunity to the flora to migrate northwards in case of increase in temperature, south being warmer than the north. But as the plains have been cleared of forests for cultivation and settlements (woodlands being relegated to hills), what may come in the way of migration of species through the dispersal of diaspores would be the anthropogenic landscapes. The Himalayan heights may likewise accommodate the species from the lower altitudes to the higher following warming, from the tropical belt to the sub-tropical which would be gradually turning tropical; sub-tropical to temperate to sub-alpine and alpine. Of course this is a theoretical scenario that could be altered by multifarious human activities.

MONTANE AND COASTAL ECOSYSTEMS

Whereas the rate of retreat of glaciers in the Himalayas has been a debated issue calling for further research, the coastal erosion merits a discussion on factors other than global warming. The islands in the Bay of Bengal are no doubt getting submerged as reported by the School of Oceanographic Studies of Jadhavpur University. Both the West Coast and the East Coast of the country have been facing littoral erosion when the monsoon is at its peak and a storm is blowing. With the fall in the atmospheric pressure by 1mb, the sea-level rises by 1cm. Therefore, during a severe storm or a cyclone, the sea-level may rise considerably. As waves are generated by winds, the fury of cyclonic winds produces giant waves. The resultant damage is to a greater extent if the storms take place on the full-moon or no-moon days when the tides are the longest.

Any obstruction, natural or man-made, coming in the path of waves, leads to long shore drift with accumulation of sand at one end and severe erosion at the other. After Chennai harbour came up, the Marina beach to the South received heavy deposits of sand but the suburb of Ennore to the North started eroding. Among other harmful practices may be mentioned mining of beach sand for collection of radio-active minerals like monazite in Kerala, or for gems or yet for construction purposes that aggravate sea-erosion.

CONCLUDING REMARKS

Among the green-house gases, carbon dioxide has received maximum attention, particularly with reference to the use of fossil fuels. Tropical deforestation has not received the attention that it deserves and so also the emission of methane, much more potential than CO_2 and of prime concern for countries like New Zealand, Australia with large livestock population.

The relationship between forests and water-cycle merits further investigation. Rainfall is generally associated with the monsoons and cyclones in India. However, there are other phenomena that generate rains. Orographic rainfall is due to mountain effect and convectional rains are linked to a dense forest cover. The Amazon basin receives more than 50% of rainfall due to convection thanks to the dense evergreen forests recycling rains (Salati et al. 1979). Meher-Homji (1989, 1991) enumerates the following factors that are involved in the link between vegetation-hydrological cycle: Evapo-transpiration, albedo (Charney et al. 1977), mechanical friction effect of aerodynamically rough, undulating forest canopy and organic debris and pollen grains floating over a dense forest serving as effective condensation nuclei (Glantz, 1987). However, the most drastic and immediate effect of deforestation is felt at the level of micro-climate or eco-climate. In the Palni hills of South India, the temperature recorded by thermographs in the open grassland was as low as minus 9°C in winter nights. At the same time, the temperature registered in a nearby *shola* (montane forest) was above 0°C. The low temperature eliminates whatever seedlings of the *shola* that may have germinated during favourable periods. Also during the prevalence of the dry season, the relative humidity remains very low in the grassland favouring the spread of fires and once again destroying the seedlings of the *shola* species. The conclusion is that once cleared, the *shola* species find it extremely difficult to stage a come back in the open grassland.

The current tendency to blame global warming for any single extreme or freak meteorological event like drought, flood, cloudburst or cyclone may not be fully justified because a scrutiny of statistical records would verify their occurrence in the past decades and last century. Sea-levels too have been fluctuating. The casualty due to a cyclone may be higher in the last decade because of the burgeoning population and consequent migration to the marginal areas prone to inundation and fury of winds and waves.

Whereas climate change is an issue not new to the 21st century, the rate of change accelerated by human impact degenerating the delicate environment of the Geosphere is a cause of concern. The biodiversity is under a greater threat not only from the warming up phenomenon but also from a multitude of anthropogenic disturbances.

References

Bellan, M.F., (1985). *Eco-floristic maps of Peninsular India*. Carte Intern. du Tapis Vegetal, Toulouse.

Charney, J.C., Quirk, W.J., Chow, S.H. and Kongfield, J. (1977). A comparative study of the effect of albedo change in semi-arid region. *J. Atmos. Sci.*, 34: 1366–1385.

Das Gupta, S.P. (Ed.). (1976). *Forest Atlas of India*. Nat. Atlas Organisation, Calcutta.

Frakes, L.A. and Kemp, E.M. (1972). Influence of Continental positions on early Tertiary climates. *Nature,* 24: 79–100.

Gaussen, H., Legris, P., Viart, M. and Meher-Homji, V.M. (1961–1978). *Maps of Vegetation and Environmental Conditions.* (12 sheets of Peninsular India with their explanatory booklets). French Institute, Pondicherry & ICAR, New Delhi.

Glantz, M. (1987). La Secheresse en Afrique. *Pour la Science,* Aout, 1987: 18–25.

Kawosa, M.A. (1983). *Vegetation map of Himalaya.* 7 parts. Lieth, H. and Esser. (Eds.). Biologie, Universitat Osnabruck, Germany.

Meher-Homji, V.M. (1977). Tropical dry deciduous forests of Peninsular India. *Feddes Repertorium,* 88: 113–134.

________. (1989). Deforestation and probable climatic change: A review. In: *Global Warming and Climate Change. Perspective from Developing Countries.* Gupta, S. and Pachauri, R.K. (Eds.). Tata Energy Research Institute, New Delhi.

________. (1990). Evolutionary studies in plant geography. *Biology Education,* 7(11): 262–277.

________. (1991). Probable impact of deforestation on hydrological processes. *Climate Change,* 19: 163–173.

Meher-Homji, V.M. and Gupta, H.P. (1999). A critical appraisal of vegetation and climate changes during the Quaternary in the Indian region. *PINSA* B 65(5): 205–244.

Pascal, J.P., Shyam Sundar, S. and Meher-Homji, V.M. (1982). *Forest Maps of South India.* French Institute, Pondicherry.

Salati, E., Dall Olio, A., Natsui, E. and Gat, J.R. (1979). Recycling of water in the Amazon basin: An isotopic study. *Water Resources Res.,* 15: 1250–1258.

Schweinfurth, U. (1957). Die horizontale und vertikale verbreitung der vegetation im Himalaya. Bonner Geogr. Abhandlungen Heft 20. Ferd. Dumlers Verlag, Bonn.

7

Aquatic Invasive Species in India

Ramakrishna

INTRODUCTION

Alien or exotic species, have locally become dominant and invade natural communities, are referred as **Invasive species.** The invasive species could be a plant, animal or microbes. Invasive species normally cause changes in ecological services of the species profile by disturbing the operation of the hydrological cycle, including flood control and water supply, waste assimilation, recycling of nutrients, conservation and regeneration of soils, pollination of crops and seed dispersal. The invasive species in its introduced habitat system tend to displace native plants and animals, disrupt food webs and alter fundamental natural environmental processes. Such services have both current use value and option value. On `1`islands and isolated ecosystems, invasive alien species are the main threat to native biodiversity and people's livelihood. IUCN also defines alien invasive species, as a species that become established in natural or semi natural ecosystem or habitat, is an agent of change and threatens biological diversity. Invasive species are so much important in the present scenario that article 8(h) of the Biodiversity Convention asks for measures "*to prevent the introduction, control or even the eradication of those alien species which threaten ecosystems, habitats or species*".

Biological invasions though the introduction and spread of exotic organisms in regions outside of their native range has emerged as a major environmental problem to our indigenous fauna and many of these exotics being hardy species compete with native species for food and space or modify the ecosystems making it unsuitable for the native species or impair the ability of natural managed ecosystems to provide services of significant economic value. Further, they bring new parasites and diseases, may prey on eggs and larvae of the species, ultimately eliminating them. Invasive species are either accidentally introduced or they are introduced by man to fulfill his needs. After introduction, they expand their population and create monospecific thickets. Recent studies have revealed that exotic organisms constitute the second biggest threat to global biodiversity after habitat destruction and degradation but far above pollution and overharvesting. Of 256 vertebrate extinctions with an identifiable cause, 109 are known to be due to biological invaders, while 70 are known to be caused by human exploitation (Cox, 1993). Intentional introductions of non native species raise a distinct set of economic and policy issues. Among them are the economics of intra seasonal pest management (Carlson and Wetzstein, 1993), the economics of resistance (Laxminarayan 2002) and the economics of infectious disease control in humans (Philipson 2000). Many invasive species introductions occur as a result of trade. International Trade Agreements recognise that it is an important for individual country to "have right to take sanitary and phytosanitory measures necessary for the protection of human, animal or plant life or health" (Article 2 of World Trade Organisation Agreement on the Application of Sanitary and Phytosanitory Measures; SPS Agreement). The main objective of the non tariff or technical barriers to trade is to correct externalities or market inefficiencies caused by invasive species in the production and distribution and consumption of goods.

Exotic species of fishes were either purposefully introduced world over either to enhance fisheries, for sport fishing, as aquarium species or for bio-control measures. Welcome (1998) reported that 168 species of fishes belonging to 37 families have been introduced outside their natural range of distribution the world over, with most of them establishing themselves in different water bodies and several of them turning out to be pests. Catastrophic impacts of exotic introductions have been common phenomena. A typical case is that of Lake Victoria, the World's largest lake also known for Cichlid fauna, until the introduction of exotic Nile Perch – *Lates niloticus*. In 1970's there were 300 endemic Cichlid species, representing 99 percent of the lake's fish species which were mostly wiped out with the introduction of Nile Perch disseminating the Cichlid species which now form only one percent of the lake's fish diversity.

INDIAN SCENARIO

In the Indian context, freshwater ecosystem represents only one percent of the earth's surface, the loss of aquatic biodiversity is severe in freshwater ecosystems since they harbor 40% of the fish species – estimated to be 29,977 valid species (Nelson, 2006). The Indian subcontinent is a confluence of three bio-geographic realms viz., the Palearctic, Afro-Tropical and Indo-Malayan and exhibits a great variety of ecological habitats, harbouring rich ichthyofaunal diversity, constituting nearly 8.9% of the known fish species of the world. India ranks eighth in the world and third in Asia in terms of freshwater fish diversity (Kottelat & Whitten, 1996). During the last several decades, over 300 exotic species have been introduced for various purposes and also unintentionally into the Indian waters (Biju Kumar, 2000, Lakra et al. 2006). Several of these exotic fishes are now well established in the natural water bodies. Though studies on the impact of the species have not been carried out, reports of loss of indigenous diversity are often reported where these species abound. The species introduced into India for various purposes are listed below.

Selected Exotic Fishes Introduced into India

A. Game Fishes (Reservoirs, Streams and Lakes)

1. Brown Trout (*Salmo trutta fario*) from England and Japan during 1863 – 1908 at Nilgiris (Tamil Nadu), Kerala and Harwan (Kashmir)
2. Loach Level Trout (*Salmon levensis*) from England during 1863 at Nilgiris
3. Steelhead Trout (*Salmo gairdneri irideus*) during 1867 – 1940 from England, Sri Lanka at Kashmir, Kerala and Nilgiris
4. Steelhead Trout (*Salmo giardneri shasta*) from England during 1941 at Kerala
5. Atlantic Salmon (*Salmo salar*) during 1968 from North America to Kashmir
6. Lake Brook Trout (*Salvelinus fontinalis*) during 1959 from Canada at Kerala
7. Splake Trout (*Salvelinus namaykush*) during 1968 from Japan at Nilgiris
8. Rainbow Trout (*Oncorhynchus mykiss*) from Sri Lanka, Germany, New Zealand at Nilgiris
9. Sockeye Salmon (*Oncorhynchus nerka*) during 1968-70 from Canada at Nilgiris

B. Food Fishes (experimental culture)

1. Golden Carp (*Carassius carassius*) during 1974 from England at Nilgiris

2. Tench (*Tincta tincta*) during 1870 from England
3. Gourami (*Osphronemus goramy*) during 1856 from Jawa to Calcutta, West Bengal
4. Silver Barb (*Barbodes gonionotus*) during 1972 from Indonesia to West Bengal
5. Mosambique Tilapia (*Oreochromis mossambicus*) during 1952, 1962, 1985 from Indonesia, Bongkok, Sri Lanka, Bangladesh, Nepal to West Bengal, Tamil Nadu and Rajasthan
6. Red Tilapia (Hybrid strain of *O. niloticus*) during 1980 from Phillipines, Israel, to Chennai, Calcutta, CCMB, Hyderabad
7. Common Scale Carp (*Cyprinus carpio communis*) during 1939, 1957 from Sri Lanka and Bongkok to Cuttack (Orissa)
8. Mirror Carp (*Cyprinus carpio specularis*) during 1939, 1957 from Sri Lanka and Bongkok to Nilgiris
9. Leather Carp (*Cyprinus caprio nudus*) during 1937, 1957 from Sri Lanka, Bongkok to Cuttack, Orissa
10. Grass Carp (*Ctenopharyngodon idella*) during 1959 from Japan and Hongkong to Cuttack, Orissa.
11. Silver Carp (*Hypopthalmichthys molitrix*) during 1959 from Japan and Honkong to Cuttack, Orissa

C. Larvicidal Fishes

1. Guppy (*Poecilia reticulata*) during 1908 from South America
2. Top Minnow (*Gambusia affinis*) during 1928 from Italy

D. Weedicide fish

1. Tilapia (*Oreochromis zilli*) during 1986 from Thailand to Indira Gandhi Canal, Rajasthan

E. Ornamental Fishes (Aquarium Keeping)

1. Live bearers (27 species) from different countries
2. Egg Layers (261 species) from different countries
3. Armoured Catfish (Plecostomus multiradiatus, Pterigophlichthyes sp., Pseudacanthicus sp.)

F. Unauthorised Introductions

1. Bighead Carp (*Aristichthyes nobilis*) during 1980 possibly from Hongkong, Bangladesh and Nepal to West Bengal

2. African Catfish (*Clarius gariepinus*) from neighbouring countries to West Bengal and Tamil Nadu
3. Piyasi (*Pangasius sutchi*) from Neigbouring countries to West Bengal and Tamil Nadu
4. Nile Tilapia (*Oreochromis niloticus*) from Thailand, Israel, to Tamil Nadu, West Bengal and Rajasthan
5. Tilapia (*Oreochromis urolepis*) to coastal waters of Andaman and Nicobar
6. Red Piranha (*Serrasalmus natteriri*)

Approximately 350 species of exotic species of fishes, under various categories are reported from India. They include 7 species of game/sport fishes, 12 species of food fishes, 4 species of larvicidal fishes and as many as 327 species of ornamental fishes. Of these only a minor share of 12 species are known to be the potential exotic species which have, to a reasonable to great extent, attained the status of Invasive Alien Species in the country. A scrutiny of the ichthyofaunal compilation of Talwar and Jingran (1991) and Jayaram (1999), including the deletion of the introduced species as well as the addition of newly described species in the most recent times, as many as 784 species of freshwater species are known to occur in the country. Thus the presence of potential invasive fish species in the country constitute only 1.5 percent of the total freshwater fish diversity of the country, including 12 invasive alien species.

The first record of introduction of the English Trout, *Salmo trutta fario* Linnaeus to the country dates back to 1863. The introduction of exotics like *Oriochromis mossambica*, *Cyprinus caprio* have no doubt contributed much to the country's fish production, but they have also had a negative impact much to the country's fish production in the form of competition, predation, environmental modification etc. The unprecedented spread and colonization of the hardy *O. mosssambica* (introduced in to the country in 1952 from its native place of East coast of Africa), has been posing severe threat to our indigenous fishes competing with the latter for space and food and quickly replacing them. The popular aquaculture and angling species *C. carpio* causes harm to other species by stirring up bottom sediments and creating turbid conditions in impoundments. In the Kashmir Valley this species has replaced the indigenous Schizothoracinae fishes and in the Loktak lake in Manipur it has eliminated *Osteobrama belangeri*, and the large indigenous *Hypselobarbus periyarensis* from Periyar lake, Kerala (Menon, 1989 – 92). Another species – *Hypopthalmichthys molitrix* has replaced

Catla catla in Govind Sagar Reservoir, Himachal Pradesh. There are reports of occurrence of different life history stages of several ornamental species such as *Xiphophorus helleri* and armoured catfish in the wild, especially in Pamba river and Kol wetlands of Trichur district of Kerala.

Though the introduction of sport fishes has been considered non problematic in Indian waters (Shetty et al. 1989), sport fishes such as the Rainbow Trout are found to be a major predator on the eggs and young ones of native species. Likewise the introduction of larvicidal fishes *Gambusia affinis* and *Poecilia reticulata* has almost replaced the indigenous *Aplocheilus parvus* and *Oryzias carnaticus* and *O. dancena. The Gourami Osphronemus gorami* (Lacepede) introduced from its natural habitats in Indonesia, Thailand, Malaya and Vietnam also has become naturalized in many water bodies. These introduced mosquito fishes are known to be prolific breeders and are capable of encroaching in micro niches of its Indian counterpart and are also predatory in nature. Considering its negative impact the mosquito fish has been branded a "fish destroyer" by Myers (1965) and reports of IUCN (1986) implicate the species as being responsible for extinctions world over. *Carrassius auratus Linnaeus* and *Carrassius carrassius,* the Gold fishes, introduced from native land of China, Korea, Japan, Taiwan, Europe, Siberia etc., are now ubiquitous aquarium pets in the country. It is believed that the intentional and unintentional releases of this species caused formation of wild populations in many locations. The major concern about the impacts of goldfish on the aquatic community are its potential for increasing turbidity, predation of native species and helping to facilitate algal blooms. The Gold fishes and the Tench or Shoemaker, *Tinca tinca* (Linnaeus), introduced into the country from freshwaters of Europe and Western Siberia, can be seen in the naturalized conditions in Ooty Lake and also in the Nilgiri waters (Tamil Nadu). Other illegally imported species such as the African catfish *Clarius gariepinus* and aquarium fish the Red Piranha *Serrasalmus nattereri* are carnivorous and voracious feeders are the worst invasive species wreck havoc to indigenous fish fauna. *C. gariepinus* an omnivorous in its natural range feeding from plant material to plankton, arthropods, mollusks, fishes, amphibians, reptiles. The species is ill reputed to be a voracious predatory carnivore that consumes a wide variety of prey objects. The species though a freshwater in its native habitat, is known to occur in brackish waters of upper estuaries, but poorly oxygenated waters. It has attained the status of potential invasive species in India especially in southern states. A recent disturbing observation is the introduction and widespread culture of the African catfish in several tanks and ponds whereever they have been introduced. Species introduced for ornamental purposes are now well established in the wild as the Cichlid the flower horn or *Cichlosoma* species and the scaled Loricarid catfish species. Transportation within the country as in the release of Gangetic carps into the peninsular rivers has also affected the indigenous carps of the peninsula. These introductions generally cause zoogeographic pollution, loss of genetic identity of local populations, a high level of hybridization and extinction or reduction of local communities of endemic species.

Ctenopharyngodon idella (Grass carp), an introduced species known worldwide as a biological control agent of aquatic vegetation and as a food fish, is a voracious feeder and incredibly efficient for removing aquatic weeds; however, results in widespread ecological effects as they carry parasites. *Cyprinus carpio* has been introduced as a food and ornamental fish throughout the world is now considered as a pest because of its abundance and its tendency to reduce water clarity and destroy, uproot the aquatic vegetation used as a habitat by a variety of fish species. *Gambusia affinis*, native of United States, has become a pest throughout the world as they predated on and endangers rare indigenous fish and invertebrate species. *Hypopthalmichthys molitrix*, native to Asia introduced throughout for aquaculture and control of phytoplankton in natural waters, alters food web by depleting the zooplankton population. It is also reported to carry and transmit the *Salmonella typhimurium*. *Salmo trutta*, *Tinca tinca* and *Salvelinus fontinalis* popular angling sports fishes are blamed for reducing the hill stream fishes in the cold and temperate habitat through predation, displacement and food competition. *Oreochromis niloticus*, highly invasive species, has effective mouth breeding reproductive strategy which allows it to increase in numbers at a rate that not only crowds native species, but pollutes and unbalance the water column.

CONDITIONS CAUSING INVASION

A combination of mechanisms causes an invasive situation in a new habitat; in general, most introduced animals/plants do not become invasive. A competition among species often appears to be a major characterstic mechanism, while all species compete to survive, invasive species appears to have specific traits or combination of traits that allow them to compete native species. They have the ability to grow and reproduce more rapidly than native species and attain, often, early sexual maturity with high reproductive output and the ability to disperse widely, including tolerance of a broad range of environmental conditions. Often the characterstics of invasive potential involves complex interaction not easily catagorised (Kolar and Lodge, 2001). Typically an introduced species may survive at low population densities before it becomes invasive in a new location. In a new habitat the invasive species out compete native species for resources such as nutrients, light, space, water or food. Some species when introduced into a new environment are at an advantage lacking the competition and predation they had undergone earlier in their native environments, making them to proliferate quickly.

ECOSYSTEM BASED CHANGES

Every species has a role of occupying and feeding facilitating them thriving in its native ecosystem. These roles are known as niches, while some species fill large and varied roles, others are highly specialized. Some invading species are able to fill niches that are not utilized by native species and they can create niches that did not exist. The best example is the establishment of Tilapia in our country. When changes occur to ecosystems, conditions change in the dynamics of the species interaction and niche development. This could often cause a species to replace other species, because they become capable of utilizing the better available resources that did not exist before. This mechanism describes a situation where the ecosystem in question has suffered disturbance of some sort, which changes the fundamental nature of the ecosystem (Byers, 2002). All species go through periods of increasing and decreasing population numbers in many cases accompanied by expansion and contraction of range. Human "alterations" on landscapes are especially significant, thus anthropogenic alterations of an environment may enable the expansion of a species into a geographical area where it had not been seen before and thus species could be described invasive.

IMPACT OF INVASION

Biological species invasion alter ecological systems in a multitude of ways. Naturally evolved, region specific pure bred wild species can be threatened with extinction by an introduced plant or animal through a process of genetic pollution i.e., uncontrolled hybridization, introgression and genetic swamping which leads to homogenization or replacement of local genotypes as result of introduced plant or animal (Mooney and Cleland, 2003). Economic costs from invasive species can be direct through natural bioproduction loss and biodiversity loss of native species itself, and management costs. Economic cost of invasions, calculated as bio-production loss and management costs, are low when the environmental damages are not usually considered. If monetary values are assigned to the extinction of the species, loss in biodiversity and loss of ecosystem services, costs from impacts of invasive species would be drastically increase in invaluable measures.

INVASIVENESS OF EXOTIC FISH SPECIES AND QUARANTINE MEASURES

Exotic aquatic organisms are introduced for various purposes, primarily for aquaculture, increasing productivity of an ecosystem, aquarium trade, sport fishing and biological control. Unintentional introductions may happen due to discharge of ballast water, common water boundaries or connection between water bodies such as manmade canals within the country. Every introduction has both negative and positive effects. The negative impact is when any alien or exotic species become invasive. The "invasiveness" or when exotic species become invasive alien species, their detrimental effects are both biological and socio-economic (Lakra et al. 2006). They include :

- (a) Spread of exotic diseases
- (b) Loss of biodiversity through competition, predation and habitat degradation
- (c) Destruction of aquatic fauna/flora
- (d) Genetic deterioration of wild population through hybridization and introgression

Condition responsible for alien species

- (a) Lack of effective implementation of the quarantine procedures
- (b) Lack of seed of indigenous species to meet specific local needs
- (c) Short-term or long term production superiority of exotic varieties
- (d) Lack of awareness of the local people/agencies about the harmful effects

CONCLUSION

Over the last decade there has been a great demand by the aquaculture entrepreneurs/farmers for new fish strains and varieties in view of competition in the world market as well as in the domestic market. However, environmental, socio-economic and biodiversity issues are important considerations which have been seriously lacking in India. In India, many species have been introduced legally or illegally and several have become well established in our natural waters.

There is a growing awareness of the environmental problems caused by the introduction of invasive alien species, described by IUCN as one of the major threats to biological diversity. Despite difficulties in defining and identification, international regimes have increasingly attempted to tackle the issue, with over 40 treaties referring to the regulation of invasive species. One mechanism to assist in the responsible use of introduced species is the development of *code of practice*, which has now been developed by the International Council for Exploitation of Sea and the European Inland Fishery Advisory Commission and other organization. In India, similar guidelines and plan for exotic fish introductions – "National Strategic Plan for Aquatic Exotics and Quarantine" – have been developed by the National Bureau of Fish Genetic Resources in consultation with experts and stake holders in the field. Quarantine controls, though originally designed to protect human health and agricultural commerce, could play a critical role as a country's first line of defense in containing the spread of invasive species. Further, scientific studies on the impact of alien species on indigenous biodiversity and socio-economic aspects and their potential invasiveness are of paramount importance.

ACKNOWLEDGEMENT

The author wish to express his deepfelt gratitude to the Ministry of Environment and Forests, Government of India for facilities and support. Also wish to express thanks to my colleague Dr. Rema Devi and Dr. Gopi for the input.

REFERENCES

Biju Kumar, A. (2000). Exotic fishes and freshwater fish diversity. *Zoos' Print Journal*, 15(11): 363–367.

Byers, J.E. (2002). Impact of non-indigenous species on natives enhanced by anthropogenic alteration of selection regimes. *Oikos*, 97(3): 449–458. IUCN, 1986. Red Data Book: *Freshwater Fishes.*

Carlson, G.A., and Wetzstein, M.E. (1993). Pesticides and pest management. In: *Agricultural and Environmental Resource Economics.* Carlson, G.A., Zilberman, D. and Miranowski, J.A. (Eds.). New York: Oxford University Press.

Cox, G.W. (1993). The ecology of extinction. In: *Conservation Ecology*. Cox, G.W. (Ed.). W.C. Brown. Publishers, Dubuque, Iowa, pp. 13–22.

Jayaram, K.C. (2006). *Catfishes of India.* Narendra Publishing House, Delhi (India): xv+ 383, 11.

Kolar, C.S., and Lodge, D.M. (2001). Predicting invaders: Response from Kolar and Lodge. *Trends in Ecology and Evolution,* Vol. 16, Issue 10, p. 546.

________. (2001). Progress in invasion biology: Predicting invaders. *Trends in Ecology and Evolution,* 16(4): 199–204.

Kottelat, M. and Whitten, T. (1996). Freshwater biodiversity in Asia with special reference to fish. *World Bank Technical Paper* No. 343, Washington, p. 59.

Lakra, W.S., Rehana Abidi, Singh, A.K., Sood, N., Rathore, G. and Swaminathan, T.R. (2006). *Fish Introductions and Quarantine: Indian Perspective.* National Bureau of Fish Genetic Resources, Lucknow, p. 198.

Laxminarayan, R. (2002). How broad should the scope of antibiotics patents be? *American Journal of Agricultural Economics,* 84(5): 1287–92.

Mooney, H.A. and Cleland, E.E. (2001). Hybridization and introgression; Extinctions from the evolutionary impact of invasive species. *Proc. Natl. Academy Sci.,* 98(10): 5446–5451, The National Academy of Sciences, USA.

Moyle, P.B. and Leidy, R.A. (1992). Loss of biodiversity in aquatic systems: Evidence from fish fauna. In: *Conservation Biology: The Theory and Practice of Nature Conservation, Preservation and Management.* Fiedler, P.L. and Jain, S.K. (Eds.). pp. 127–169. Chapman and Hall, New York.

Myers, G.S. (1965). Gambusia, the fish destroyer. *Tropical Fish Hobbyist,* 13: 31–32; 53–54.

Nelson, J.S. (2006). *Fishes of the World.* John Wiley and Sons, New Jersey, p. 601.

Philipson, Tomas, (2000). Economic epidemiology and infectious diseases, Handbook of Health Economics. In: *Handbook of Health Economics,*

Culyer, A.J. and Newhouse, J.P. (Eds.). Edition 1, Vol. 1, Chapter 33, pp. 1761–1799.

Ramakrishna and Alfred, J.R.B. (2008). *Faunal Resources in India,* Edited. Director, Zoological Survey of India.

Shetty, H.P.C., Nandeesha, M.C. and Jhingran, A.G., *Exotic Aquatic Organisms in Asia.* De Silva, S.S. (Ed.). Asian Fisheries Society Special Publication, 1989, Vol. 3, pp. 45–55.

Shetty, H.P., Nandeesha, M.C. and Jhingran, A.G. (1989). Impact of exotic aquatic species in inland waters. *In: Exotic Aquatic Organisms in Asia. Proceedings of the Workshop on Introduction of Exotic Organisms in Asia.* De Silva, S.S. (Ed.). pp. 45–55. Asian Fisheries Society, Spec. Publ. No. 3. p. 154.

Talwar, P. K. and Jhingran, A.G. (1991). *Inland fishes of India and adjacent countries.* Oxford and IBH Publishing Co. Pvt. Ltd., New Delhi, Vol. 2. xix + 1158.

Welcomme, R.L. (1988). International introduction of inland aquatic species. *FAO Fish. Techn. Pap.*, 294: 1–318.

8

Mass Extinction in Anthropocene

Debal Ray

INTRODUCTION

About 99.9% of all species that ever existed on earth are now extinct. The extinction of species is, thus, inevitable in the evolutionary history of species. Yet, such a common phenomenon is rarely observed to happen in nature. This is because extinction of species is seldom an abrupt event. This has been eloquently described by Wilson (1992) in his now classic book "*The Diversity of Life*" and is quoted below.

"*Extinction is the most obscure and local of all biological processes. We don't see the last butterfly of the species snatched from the air by a bird or the last orchid of a certain kind killed by the collapse of its supporting tree in some distant mountain forest. We hear that a certain plant or animal is on the edge, perhaps already gone. We return to the last known locality to search, and when no individuals are encountered there year after year, we pronounce the species extinct.*"

WHAT IS MASS EXTINCTION

The majority of extinctions that took place in the geological past were in the form of natural (background) extinction. Only about small percentage of species went extinct during the periods of accelerated species loss. Yet these episodes of accelerated species extinction, or mass extinction as they are commonly called, received great attention from scientific world and triggered imagination of general public. There are certain features that distinguish mass extinction from background extinction events. Many species, generally more than 40 to 50% of those existing, became extinct at a time. In end-Permian extinction (251 Mya), about 96% of the species went extinct. The extinct forms span a broad range of ecologies, and they typically include marine and non-marine forms, plants and animals, microscopic and macroscopic forms. The extinctions all happened within a short time, geologically speaking. Extinction events were supposed to have been triggered by a single cause or a cluster of interlinked causes. The end-Cretaceous extinction (65 Mya), for example, was caused by bolide impact at Yucatan peninsula in Mexico and/or lava outpour at Deccan trap, India.

TRIGGERS OF MASS EXTINCTION

Mass extinctions of the past are believed to have been triggered by one or more of the following agencies.

1. Bolides impacts. The end-Cretaceous Mass extinction is largely believed to have been caused by asteroid impact on earth. The "impact of a large earth crossing asteroid would inject 60 times the object's mass into the atmosphere; a fraction of this dust would stay in the stratosphere for several years and be distributed worldwide. The resulting darkness would suppress photosynthesis"(Alvarez et al. 1980). Other indirect effects of bolide impact would be stratospheric ozone depletion, nitric acid formation, anoxia, fire and destruction of food chain.
2. Volcanic eruption. Volcanic ash enters on entering into earth's atmosphere block sunlight and heat radiation. This directly affects primary production and consequently the entire food web. Volcanic eruption is also linked to anoxia.
3. Fluctuations in rate of sea floor spreading
4. Collision formation of mountain belts
5. Opening vs. closing of equatorial ocean passages
6. Sea level fluctuations related to glaciation and de-glaciation .There has been several episodes temperature extres experienced by earth. For example ice ages were experienced by earth 750 Mya, 640 Mya and agin 590 Mya. Glaciation binds water and lowers sea levels. Life adapted to warm temperature and continental shelf during high sea level would find it hard to survive and go extinct. Reverse pattern would be observed during the period of de-glaciations. As glaciers melt, chemicals, metals, ions and gasses would also be released creating hostile condition for life.
7. Anoxia in oceans and atmosphere. Large scale volcanic deposition of iron in oceans, during Cenomanian/Turonian times resulted in

algal boom which on decomposition led to anoxic condition in water and death of biota (Elewa & Joseph, 2009)

8. Global cooling. It is one of the primary cause of extinction on earth. Cooling could be the result of biological factors (Joseph, 2009), geological factors (Raup, 1992) or even an exploding comet (Firestone, 2007). Global cooling during end-Ordovician, for example, is believed to have been caused by volcanic weathering driven lowering of CO_2 level
9. Reversals of polarity in the Earth's magnetic field. Although the precise cause of it is still unknown, the reversal do take place every tens of thousands to millions of years. Last reversal took place 780,000 years ago. The earth's magnetic field is weakening at the moment and we may be living the early stages of next reversal. (Gribbin & Gribbin, 2001). Reversal of polarity affects earth's biota profoundly leading to extinction substantial number of species.
10. Gamma-ray bursts from some nearby supernova explosions. There have been 8 to 10 such episodes in earth's history. Ordovician mass extinction is believed to have been caused by such gamma-ray burst (Melott et al. 2004)

PAST EXTINCTION EVENTS

It is customary to look into the causes and patterns of past extinctions and recovery therefrom to understand the ongoing mass extinction. Barring the present, the biota of earth has undergone five major extinction events. A summary of these extinction events is given below.

The **end-Ordovician** (≈439 mya) extinction event was second largest in terms of extinction of life on earth. This event extinguishes nearly all bryozoans and brachiopods in a two pulse extinction episode. The cause of this mass extinction event is believed to be global cooling caused by passage of Gondwana supercontinent over north pole.

The **end-Devonian** (≈364 Mya) extinction event marked disappearance of 70% of all existing taxa. The cause of the event is another cooling event possibly brought about by bolide impact (Joachimski and Buggisch, 2002)

The **end-Permian** (≈251 Mya) extinction, the biggest mass extinction of all, was triggered by volcanisms that created Siberian Traps. An estimated 3 million cubic kilometer of basalt covered Siberia at a time coinciding with end Permian extinction event. Such a huge out pour of basalt is believed to have changed the chemistry of ocean to brightness of the sky with anything in between. The end Permian extinction resulted in death of nearly 96% of the species. It was assumed till recently that the end Permian event caused extinction in random and the surviving species represent the so called lucky ones. Apparently indiscriminate killing led a group of authors to believe that it is sheer chance rather than fitness of the species that determine its survival. In fact Raup (1992) famously raised the issue whether "Bad luck or bad genes?" more important in determining extinction. The oft-quoted example of terrestrial synapsid (mammal like reptile) *Lystrosaurus* illustrate the point. This simple animal survived the great end-Permian extinction and formed monospecific assemblage in Gondowanaland, although scientists could not find any great adaptive feature in it.

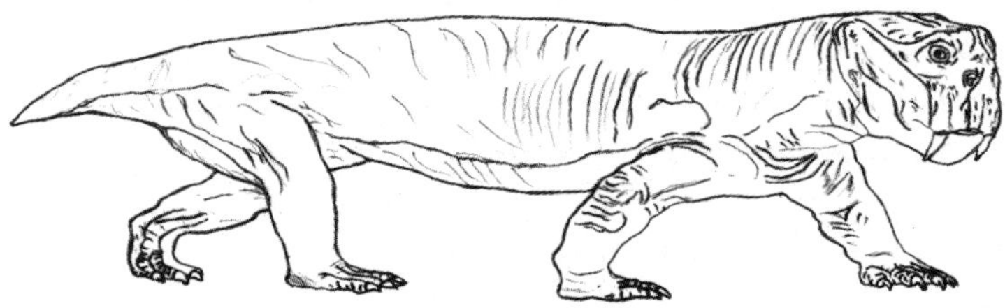

Fig. 1. *Lystrosaurus* (Redrawn from Benton, 2003)

It is known with certainty that global temperature during this event rose by 6°C. The volcanic CO_2 initially raised the temperature which in turn destabilized the methane hydrates stored in ocean and these greenhouse gases raised the ambient temperature. But, the temperature alone can not be the only cause of death. Large scale extinction of plants as well as survival of some animal species with high metabolic rates are cited as weakness of this theory (Lane, 2007).

Destruction of ozone layer by the volcanic hydrochloric acid and consequent greater penetration of ultraviolet rays have been suggested as alternate killing mechanism. Paleo palynological studies found mutant spores and pollens at sediments corresponding to this time. But, massive extinction in seas, where ultraviolet ray penetration is low, has been cited as weakness of ozone collapse hypothesis as sole killing mechanism.

End Permian and early Triassic was the time when the oxygen concentration (13%) of the atmosphere was lowest in geological history. Apart from, methane hydrates released from sea floor, huge quantity of thermogenic methane was released in Siberia Trap as lava passed through the coal layer transforming large hydrocarbons into methane. The latter burned in air depleting already low oxygen level. Interestingly, this period was preceded by Carboniferous period when the oxygen concentration of air was highest (35%) in geological history. Thus, the organisms evolved in Carboniferous had a difficult time to adjust in Permian. The low oxygen concentration restricted altitudinal ranges of many species, thereby fragmenting the population. The prevailing high temperature further reduced the oxygen concentration of sea water causing distress to marine animals. It is widely accepted that such low oxygen concentration made life on earth intolerable. But it can not be viewed as ultimate killing machine. Sedentary animals like stationary filter feeders have low metabolic rates and are expected to survive better in such conditions. Yet they were amongst worst sufferers.

Lane (2007) has summarized the condition of end Permian world, "...with dangerously low atmospheric oxygen levels, ecosystems were compressed and fragmented. The deep oceans were largely uninhabitable. Land plants were dying back in the arid greenhouse climate, making food hard to come by. And then came the hammer blows of fate, the greatest volcanic outpourings in the history of our planet, releasing vast quantities of methane and carbon dioxide, raising global temperatures by 6°C..." Life on earth came very close to getting extinguished.

Looy et al. (2001) analyzing end Permian micro flora of Greenland reported two interesting phenomena that are relevant to current biodiversity crisis: (1) there is a time lag between ecosystem disturbance and selective extinction of gymnosperms and (2) following initial perturbation, the diversity of the ecosystem increased instead of decreasing. The former phenomena can be explained in terms of extinction debt. The latter term has two components: a direct time lag between environmental perturbation and the consequent extinction of species and the idea that among species going extinct the strong competitor/weak colonizers go extinct first. As a result of delayed extinction of competitive taxa, the biological diversity is not affected immediately following perturbation. During this period, opportunistic invasion in the ecosystem by alien invasive species will result in overall increase of diversity. This explains the second phenomena.

The **end-Triassic** (≈199-214 Mya) came at a comparatively rapid succession following the great end Permian extinction. It is now believed it is actually a combination of two extinction events separated by 12 million year interval. Possibly half of the marine genera went extinct in this event. The huge supercontinent Pangaea was broken up during this period possibly as a result of bolide impact. There were large scale volcanic eruptions all over the earth. Increase in CO_2 concentration, anoxia and continental flood plunged the planet into a difficult place to live in.

The best known mass extinction event is the most recent one that took place some 65 Mya at the end of **Cretaceous**. It is now widely believed to have been caused by bolide impact at Chicxulub site at Yucatan peninsula in Mexico and lava outpour at Deccan trap, India. The evidence for bolide impact, which were earlier dismissed by Geologists as figment of imagination, was discovered as recently as 1984 by a team of physicist (Alvarez et al. 1984) working on the evidence of occurrence iridium spike at K-T sediment layer. A possible scenario of earth after impact has been given by many authors. The following extract is taken from Kring (2000).

Regional Effects

In the immediate vicinity of the crater, the shock wave, air blast and heat produced by the impact explosion killed many plants and animals. The air blast, for example, flattened any forests within a 1000-2000 km diameter.... Tsunamis were 100-300 m high as they crashed onto the gulf coast and ripped up seafloor sediments down to depths of 500m.

Global Effects

The most significant environmental perturbations were the direct and indirect result of ejected debris that rained through the atmosphere. The material was carried in a vapour-rich plume that rose through the atmosphere into space. Once above the atmosphere, it expanded on ballistic trajectories, enveloping the whole earth as it fell back into the atmosphere.... model calculations suggest that dust may have made it too dark to see for one to six months and too dark for photosynthesis for two months to one year, seriously disrupting marine and continental food chains and decreasing continental surface temperature.

The Deccan Trap represent massive outflow of lava covering an area equal to half of Europe. Geologists believe that such vast amount of flood basalt volcanoes could send clouds of dust rich in iridium to atmosphere. The later would encircle the globe cutting off sunlight and heat.

RELEVANCE FOR PRESENT

Great fluctuation of sea level was associated with some of the mass extinction events, e.g. the end Ordovician and end Devonian extinction event. In late

Maastrichtian the sea level dropped by 150 to 200 metres exposing the continental shelf and destroying the ecosystem therein completely. Just below the K-T boundary, the sea level suddenly rose. The earth today faces a similar sea level rise of uncertain magnitude due to rapid melting of polar ice caps and thermal expansion of water. In addition to physical drowning of habitat, lowering of the salinity in ocean could be the reason for extinction of marine invertebrates (Lantzy et al. 1977). Melting glaciers release chemicals, metals ions, gases creating a toxic condition in the ocean creating a hostile condition for life (Joseph, 2009).

It has been observed that propensity of a particular taxon to background extinction determine its likelihood of getting affected by mass extinction. The tribolites with higher background rate of extinction than articulate brachiopods, suffered more loss at end Ordovician extinction event. Similarly, brachiopods with higher background rate of extinction than bivalves suffered more at end Permian extinction Thus, rates of extinction measured during the background times at a higher taxon level (e.g. class) give a good prediction of magnitude of extinction at major event (Sepkoski, 2001). It is likely that groups like amphibians, mammals, gymnosperms, which have higher rate of background extinction at present will suffer more at the next big extinction event.

Fewcett et al. (2009) in his pioneering work has shown the role of polyploidy in survival of plants at the time of K-T extinction event. Using a method of dating, he has shown that polyploidy or genomic duplication took place in many unrelated plant species about 65 mya, suggesting a common causal factor for it. It is well known that polyploidy confer many advantages upon the plant like new gene functions, epigenetic changes, altered gene expression, change in mating system, plant-pollinator interaction and speciation (sympatric) (Soltis and Burleigh, 2009) and some of them might have conferred survival advantage at the time of biotic crisis. Soltis and Burleigh (2009) raised the question "if polyploidy allowed certain plant lineages to survive and adapt during the tremendous global changes of the K-T period, does polyploidy confer a similar advantage in the current period of global warming?"

All mass extinction events were followed by diversification of biodiversity. The rise of mammals following extinction of dinosaurs and diversification of angiosperms in vacated niches of gymnosperms are two such end cretaceous diversification event. The extinction of reef-builders at end Ordovician extinction event helped to release huge amount of calcium from the reefs into the oceans. The development of species with bony structure in Devonian was made possible due to abundant supply of this calcium. Many more examples of post extinction burst of diversification exists. In fact, after every extinction event, the level of biodiversity rebounds to a level similar or higher than the pre-extinction level. However, there exists a significant time lag between the extinction event and recovery following it, particularly at a human time scale. Recovery takes nearly 5 myr for third order extinction event, about 10 to 15 myr for second order extinction event and some 30 myr for first order extinction (end Permian) event. Applying the results of past post-extinction diversification to sixth mass extinction event, we can only arrive at discomforting results. The factors that drive the present biotic crisis are unlikely to be mitigated soon enough to initiate diversification /recovery of biodiversity. On the contrary, in a human time scale and involving a scenario that does not envisage sudden and early mass extinction of human, the stressors acting on biodiversity will get only worse. The prospect of rediversification, under this condition, is therefore limited (Myers & Knoll, 2001). It is also to be noted that mass extinction events invariably removed the most successful species existing on earth at that point of time. All these point to a grim future for mankind in the event of sixth extinction.

Analyzing the past extinction events Jablonski (2001) summarized the following lessons that are applicable to similar events irrespective of specific mechanism, intensity and participants.

1. Mass extinctions happen. The fossil record provides ample evidence that even the more widespread and species-rich clades, ecosystems and biogeographic provinces are not infinitely resilient.
2. Survivorship during mass extinctions need not be closely related to many aspects of biological success as measured during 'background' times.
3. Extinction itself promotes biotic interchange. Asymmetries in ancient biotic interchange generally appear to reflect geographic differences in extinction intensity.
4. The evolutionary response to mass extinction is slow on human time scales, difficult to predict owing to contingencies of post extinction conditions, including the identity and evolutionary dynamics of the survivors and geographically heterogeneous.

EXTINCTIONS MEDIATED BY HUMAN

Extinction Spasms

The sixth mass extinction started and is continuing in two extinction spasms, both mediated by human beings.

Stage One started about 50,000 years ago when humans started dispersal out of Africa. As they came in contact with various species on their migration path, human beings left a trail of killings of native biota. This together with some adverse climatic factors contributed to massive extinction of mega fauna during Pleistocene.

Stage Two was initiated about 13,000 years ago with the beginning of agriculture. This initiated the process of taking control of the immediate environment which kept on increasing in magnitude with time. The situation reached a whole new level when heightened industrial and commercial activity, following Industrial Revolution, required huge extraction of natural resources. The discovery of life saving drugs and more assured food supply helped in exponential rise of human population. These factors, in turn, triggered unprecedented level of extinction through habitat fragmentation, pollution, alteration of biogeochemical cycle, climate change, biotic homogenization etc.

PLEISTOCENE MEGAFAUNA EXTINCTION

Pleistocene megafauna (biomass >44 kg) extinction event took place at different time period in different continents, but it was mostly concentrated during the time period 50 to 7 kya. The intensity of extinction also varied across continents: North America lost 72% of species, South America lost 83% species and Australia lost 88% of species. Eurasia suffered a relatively modest level (35%) of megafauna loss and Africa was least affected of all with only 21% loss. (Baronsky, 2008)

The Pleistocene Megafauna Extinction is the only mega extinction event that took place at a time in the geological past when human beings were present and climate of the earth was warming rapidly. As such, events that caused the extinction as well as its consequences are of special significance to us. Paleontologists analyzing along multiple lines of evidence now believe that following factors, working independently or synergistically, could have resulted in extinction in Pleistocene.

1. Blitzkrieg (rapid overkill)
2. Climate Change
3. Sitzkrieg (fire, habitat fragmentation, introduction of exotic species, disease)

Several lines of evidence suggest overkill hypothesis. One of the strongest arguments is relatively low level of extinction observed in Africa and Central Eurasia. Human beings coevolved with the megafauna in this part of the world and the later got time to evolve its defences against the former. The human beings, on the other hand, were invasive species in Australia and Americas (Baronsky et al. 2004). This explains the high level of extinctions in these two continents when native species came in contact with humans. The timing of extinction pulse also suggest strong role of human mediated overkill. The extinction of megafauna started within thousands of years after human arrived in the continents of Australia, Eurasia and Central North America. Till recently, it was commonly believed that climate change was the main reason for species extinction in New Zealand and the Polynesians were thought to live in harmony with the nature. Two lines of evidence cast doubt to the climate change assumption. As Diamond (1992) has pointed out the moas died with their gizzard full with food and they experienced unusually stable climate in tens of thousands of years when the wave of extinction struck them. Secondly, the remains at living sites of Polynesian colonies show that when the humans arrived the moas were still very much alive. Remains of the bird found in these sites show that the Polynesians used moas as foods, its skin for clothing, bones for jewellery and fishhook and eggshells as water container. Clearly, the climate was not the culprit.

However, changes in climate did play role Pleistocene Megafauna Extinction. Extinction in Eurasia took place in two pulses: first pulse corresponded to cooling during Last Glacial Maximum and the second to Younger Dryas cooling followed by Holocene warming. In South America, human presence was recorded about 14.6 kya but no significant extinction could be recorded till Holocene warming stated about 11kya.

Disease brought by human pets like dogs, rats and birds are also indicated in Pleistocene Megafauna Extinction (Huggett, 2003). It is believed that one or more disease brought by these animals accompanying migrating humans affected the megafauna weakening their immune response. The weakened animals succumbed to the disease itself or to climatic stressors or to a combination of both. The population reduced to such a level that it could not recover again. The disease hypothesis has however, been criticized for the lack of suitable such disease with the possible exception of leptospirosis and rabies. The present understanding about Quaternary extinction is that human contributed to it through hunting and habitat alteration but it was exacerbated by climate change.

Baronsky (2008) examined the Quaternary Megafauna Extinction from the viewpoint of biomass tradeoff between human and other mammalian species. The megafauna biomass distributed among many species in pre crash state became concentrated in a few species in post crash state with human cornering the biggest gain. There was massive appropriation of ecological niches accompanying this biomass tradeoff. The diverse megafauna of Quaternary occupied narrow niches and following the extinction event the vacant niches were occupied mostly by a single species with broad ecological niche- human. As a result the potential of evolution of new species in the niches vacated by extinct species was not realized as long as humans were so abundant.

SELECTIVITY OF EXTINCTION

Background extinction does not strike at random. However, as the intensity of the trigger of extinction increases, the selectivity decreases and abrupt high intensity triggers affect almost all taxa, albeit differentially.

The scientific research on extinction is primarily an effort of identification of taxonomic selectivity affected by the event. Since families/genera/species belonging to a taxa share some common physiological, anatomical and ecological attributes, identification of vulnerable taxa gives an idea about attributes that render a taxon likely to go extinct. Based on such analysis, certain traits have been shown to render taxa prone to extinction or confer survival value. However, the results are far from conclusive and should be taken with some amount of caution. A list of traits hypothesized to impact extinction rate at species level during 'background extinction' time is given below.

Table 1. Impact of Traits on Rates of Extinction (Modified from Jablonski, 2008)

Serial	*Species level trait*	*Impact on extinction*
1.	Geographic range size	Low rate of extinction for widespread species
2.	Patchiness of population within range	Low rate of extinction for low patchiness

3.	Genetic population structure	Low rate of extinction with weak structure
4.	Sex ratio	High rate of extinction with biased sex ratio
5.	Population size	Low rate of extinction with large size.
6.	Population density/local abundance	Low extinction rate with high abundance.
7.	Variability in population size	Low rate of extinction with high variability.
8.	Intraspecific variation	Low rates of extinction with copious variation.
9.	Mode of intraspecific competition	Low rates of extinction with scramble competition (as opposed to contest competition)
10.	Evolvability or versatility	Low rates of extinction with high evolvability
11.	Social organization	Low rates of extinction with large social groups
12.	Ecological specialization/ environmental tolerance	Low rate of extinction in generalized and tolerant taxa.
13.	Competitive ability	Low rate of extinction with high competitive ability.
14.	Host specificity	Low rates of extinction in generalized taxa.
15.	Intensity of sexual selection, sexual dimorphism or sexual conflict	Low rates of extinction with weak selection. Low rates of extinction where males dominate mate choice.
16.	Asexual versus sexual reproduction	Low rates of extinction in sexual species.
17.	Semelparity versus iteroparity	Low rates of extinction in iteroparous taxa.
18.	Mating system	Low rates of extinction in polyandrous and polygynous taxa.
19.	Allee effect	Low rates of extinction with Allee effect.
20.	Dispersal rates	Low level of extinction with high dispersal rates.
21.	Generation time	Low rates of extinction with long generation time.
22.	Home range of individual	Low rates of extinction with small home range.
23.	Floral symmetry	Low rates of extinction with radial symmetry.
24.	Trophic level	Low rates of extinction at lower trophic level.
25.	Colonial invertebrate polymorphism	Low rates of extinction with polymorphic colonies.

EXTRINSIC DRIVERS OF EXTINCTION

The extrinsic factors like habitat fragmentation, invasive alien species, pollution, biotic homogenization, overkill are mainly responsible for present biodiversity crisis. Wilson (1992) ranked the drivers in terms of their propensity to cause species extinction.

Table 2. Drivers of Species Extinction

Serial	*Drivers*	*% of species extinction caused*
1.	Destruction of physical habitat	73
2.	Displacement by introduced species	68
3.	Alteration of habitat by chemical pollutant	38
4.	Hybridization with other species or sub species	38
5.	Overharvesting	15

Wilson did not mention about climate change; possibly importance of this factor was not properly recognized during that time.

Millennium Biodiversity Assessment (2005) identified a larger set of drivers that are acting on biodiversity. The following figure (Figure 2) taken from the report illustrate the relative importance of drivers as well as the trend.

Climate Change has emerged as newest threat to extinction. It will act primarily by exacerbating the other stressors acting on biotic world. Extinction of amphibians in Costa Rica is due to disease aggravated by climate change. The warming of climate resulted in altitudinal rise of moisture laden clouds, which made the climatic conditions favourable for the growth of the chytrid fungi *Batrachochytrium dendrobatidis*. The chytrid fungus causes dermal infection of the frogs causing death. The worldwide decline of corals is also due to recent climate change. The corals are highly sensitive to sea surface temperature and a rise of mere 1°C above long term summer temperature is enough to cause coral bleaching. Corals and its symbiotic zooxanthellae algae live at a temperature 1-2°C below their maximum tolerance limit. When the sea temperature increases, the algal symbiont photosynthesizes at a greater rate. This produces more oxygen which in turn through a series of metabolic steps produce substances toxic for the corals. The corals eject the symbiotic zooxanthellae algae in order to survive, losing colour and source of nourishment in the process. Large scale coral bleaching was observed in Indonesia during El Nino event of 1980s.

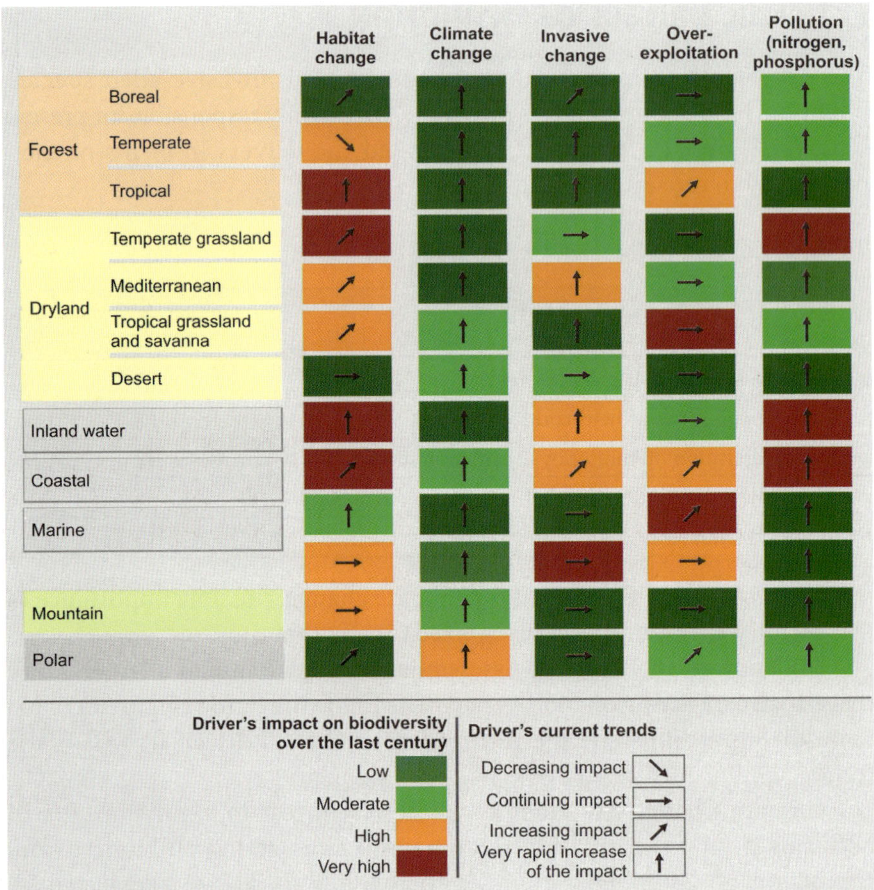

Fig. 2. Impact on Biodiversity by different Drivers
(Source: Millennium Ecosystem Assessment, 2005)

It has been estimated that 20% of corals of the world are bleached. Higher levels of CO_2 in sea water make it acidic which in turn affects calcification. Thus, higher levels of CO_2 in water weakens coral skeleton, impedes growth and erodes reefs. The IUCN report (2009) indicates that at 560 ppm CO_2 concentration calcification rates and coral growth will be reduced by 40%.

Species invasion causes extinction by predation, alteration of habitat, competition for food and other resources. Invasive alien species can be found in almost any taxa of plant and animal kingdom: *Phytopthora cinnamomi, Batrachochytrium dendrobatidis* among macro fungi, *Eichhornia crassipes, Caulerpa taxifolia* among aquatic plants *Lantana camara, Leucaena leucocephala, Rubus ellipticus* among land plants, golden apple snail (*Pomacea canaliculata*), zebra mussel (*Dreissena polymorpha*) among aquatic invertebrates, Argentine ant (*Linepithema humile*), Asian tiger mosquito (*Aedes albopictus*) among terrestrial vertebrates, Nile perch (*Lates niloticus*), Bull frog (*Rana catesbeiana*), Indian Myna bird (*Acridotheres tristis*), domestic cat (*Felis catus*) and rabbit (*Oryctolagus cuniculus*) among vertebrates. A few contemporary examples of extinction due to species invasion is given below (from Lowe et al. 2000) Miconia, a highly ornamental tree was introduced in Tahiti botanical garden in 1937. It spread into wild through fruit eating birds and now half of the island is invaded by it. The tree forms a dense canopy beneath few species grow. Scientists estimate that several of endemic plants of islands face extinction due to habitat loss caused by this species. Crazy ants are one of the fiercest invader of ecosystem in Hawaii, Seychelles and Zanzibar. The crazy ants attack and decimate red land crabs interfering with litter breakdown. They also prey on and interfere with the reproduction of a variety of arthropods, reptiles, birds and mammals on the forest floor. Scientists estimate that the endangered bird, Abbott's booby, which nests only in Christmas Island, will go extinct in near future due to predation and habitat alteration. Small Indian mongoose was introduced in Fiji, Mauritius, West Indies and Hawaii to control rats. This predator decimated native island fauna which had no defence against this agile predator. It has caused extinction of a number of endemic birds, reptiles and amphibians.

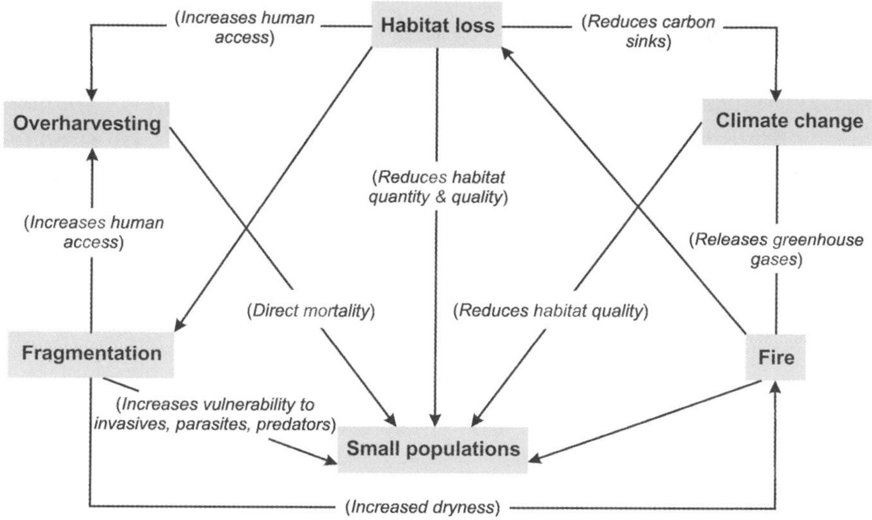

Fig. 3. Interrelationship among extinction drivers (Redrawn from Brooke et al. 2008)

The extinction drivers seldom act independently. Rather, the drivers interact with each other creating extinction pressure in excess of their individual effects. This synergistic effects of the drivers is illustrated in the figure 3.

Background extinctions are likely when the perturbation was not experienced in the evolutionary history of the species or is occurring at a rate too fast to allow adaptation. Small population sizes are inherently prone extinction. As population densities fall below minimum viable population (MVP), the risk from environmental stochasticity increases and is eventually sucked into extinction vortex. A meta analysis for MVP failed to detect any correlation between MVP size and intrinsic (phonological) and extrinsic (ecological) factors. The only pattern that emerges is MVP follow a log normal distribution among species and the median value is a few thousand individuals (Brooke et al. 2008).

TRAITS WITH SURVIVAL VALUE DURING MASS EXTINCTION

The attributes such as local abundance, reproductive mode, body size and inferred generation time, trophic strategy, life habit, geographic range at species level and species richness which show good correlation under "background extinction" condition tend to be a poor indicator of survival potential under mass extinction. This is possibly because the traits that show good correlation are often misinterpreted as good selectivity. The biological traits co-vary and selection on one feature drags along with it many other traits making it difficult to identify the root trait that conferred selective advantage. For example, bryozoans taxa with complex colonies are more resistant to extinction than simple colonies during background event but the reverse is true during mass extinction event as was observed in end Ordovician extinction. But, the colony complexity is also inversely proportional to genus level richness. Therefore, it is difficult to deduce the basis of selection in background as well as during mass extinction.

Jablonski (2005) has shown that mass extinction does not affect the taxa randomly and each event shows some form of selectivity. Broad geographical distribution at genus level, as shown below, promoted survival of major groups in the past.

Table 3. Extinction Events and Taxa in Which Broad Geographic range at Genus Level Enhanced Survivorship (from Jablonski, 2005)

Extinction Event	*Taxa*	*Reference*
End Ordovician	Bivalves	Bretsky, 1973
End Ordovician	Brachiopods	Harper and Rong, 2001
End Ordovician	Bryozoans	Anstey, 1986
End Ordovician	Trilobites	Robertson et al. 1991
End Ordovician	Marine invertebrates	Foote, 2003
Late Devonian	Bivalves	Bretsky, 1973
End Permian	Gastropods	Erwin, 1993
End Triassic	Bivalves	Bretsky, 1973
End Cretaceous	Bivalves and Gastropods (exception: echinoids)	Smith and Jeffery, 2000

However, extrapolation of survivorship pattern at genus level to its constituent species has to be exercised with great caution. It is normally expected that ecologically generalized and geographically widespread species of a survivor genus should also show survivorship across mass extinction events. However, fossil records show that this is seldom observed. The conclusion we may draw is that natural selection operates at higher taxonomic (>generic) level during mass extinction. Apparently, this piece of information has some practical application. Large concentration of endemics (44% of world's plants and 35% of world's animals) in 25 hotspots of the world and progressive reduction of sizes of hotspots (70% in 2000) imply that in the event of next mass extinction these taxa with limited geographical spread may go extinct.

Following generalizations about mass extinction at higher taxa level has been made by Jablonski (2005).

(a) The rate of species extinction during background extinction events correlates well with that of mass extinction for volatile taxa at class level.

(b) Subtracting the median extinction rate for each class brings the mass extinction intensities to similar magnitude.

(c) The survivorship pattern at background extinction is no indicator of mass extinction selectivity within the class.

It is fair to conclude that we do not know precisely the traits that confer selection advantage at the time of mass extinction. We know even less how these factors interact with environmental stressors to determine survival value of the organisms. Therefore, it is not possible to build, at the current level of knowledge, conservation programme targeted to minimize biodiversity loss in the face of mass extinction.

Consequences (a) First Order Effect

Under business as usual (BAU) scenario, we may lose between 1/3rd and 2/3rd of existing species in foreseeable future. The 2008 update of Red Data list shows that already 22% of the mammals, 31% of amphibians, 13.6% of birds, 27% of warm water reef building corals, 12% of groupers and 16% of fresh water crabs are either extinct or threatened with extinction. Sampled Red List Index study on plants conducted recently by Kew Botanical Garden 2010 shows that 15% of bryophytes, 14% of pteridophytes, 36% of gymnosperms and 22% of monocots are under threat to extinction. Not only the rates of species loss are very high, the rates of such loss appears to be accelerating across diverse taxonomic groups (Vié et al. 2009).

The extinction of distinct populations of same species, though seldom noticed, is going on at scale much greater than rate of species loss. This improvises the biodiversity at genetic level and compromises its ability to resist disease and pest attack, reduce ability to colonize new habitat and slow down the ability for further evolution.

Some of the biomes of earth, like tropical rainforest and coral reef etc., are already under much pressure and may decline significantly in size and productivity or disappear altogether. Overfishing and climate change has already damaged the coral ecosystem all over the world and are being replaced by fleshy macroalgae that cover the entire reef. The live coral cover in Carribean has gone down from 55% to 5% during the time period 1977 to 2001. In Indo-Pacific region, live coral coverage has halved to 22% between 1980 and 2007. Even the famous Great Barrier Reef has only 23% live coral cover. Only uninhabited protected atolls of Central pacific have 50% or more coral cover (Jackson, 2008) The threat to rainforests great significance as it is main storehouse of terrestrial biodiversity and the engine of evolution. Largest chunk of rainforest (≈ 40%) is in Amazon basin. This area is under threat particularly due to expansion of road network leading to built up of anthropogenic activity in hitherto undisturbed areas. The mean annual rate of deforestation in Brazilian Amazon between 1990 and 1994 was 1.37 million ha yr^{-1} which increased 61% to 2.20 million ha yr^{-1}, a decade later in 2000-2004 (Lawrence et al. 2004). Even by conservative estimate the Brazilian

Amazon is being lost at a rate of 0.7% per annum (Hubbell et al. 2009). In other words, we shall lose this biome within 143 years under BAU scenario. If this is not enough, Climate change may trigger even bigger extinction in next century through its effect on hydrology of the area. Hubbel et al. (2009) estimated that loss of tree species in Brazilian Amazon may of the order of 20% and 33% under optimistic and non-optimistic scenarios respectively. One may argue that past predictions of tropical deforestation (Whitmore, 1980, Myers, 1980) were not correct. But given the trend, Whitmore's dark vision prove to be accurate in long run.

Consequences (b) Evolutionary Effects

Loss of large number of species in mass extinction results in vacating many different niches. This provides opportunities for new species to evolve, even adaptive radiation to happen, in those vacant niches. However, the number of new species evolved shall not match even remotely the number of species lost. Moreover, the emerging species will mostly be generalists, often opportunists. These will replace the specialists, whose place they occupy. The evolution in post extinction scenario may throw open novel features, although they are difficult to predict. We may expect adaptive radiation of animals like small mammals and insects that live in human dominated system.

Large vertebrates like elephants, apes, rhinoceros, big cats require large areas for speciation. In the present day world, the habitats of these animals are fragmented. Even the biggest protected area is not large enough to allow further evolution of these large vertebrates. This simply means speciation has been suspended for these animals.

Tropics has served as the cradle of evolution for virtually every major groups of plants and animals. Recognizing this role, Jablonski (1993) remarked that tropics has been the engine of biodiversity for at least 250 million years. But with significant decline of tropical biomes like tropical rainforest and coral reef, the evolutionary powerhouse seems to be under threat. (Myers & Knoll, 2001).

Present Rate of Extinction

The lifetime of taxa and their extinction rates, even under conditions uninfluenced by human, varies greatly as given in Table 4.

Table 4. Lifetime and Extinction rate of various taxa

Serial	*Taxa*	*Lifetime of spp.(mya)*	*Extinction rate (extinctions/ million spp./years)*
1.	Diatoms	8	0.12
2.	Dinoflagellates	13	0.08
3.	Pteridophytes	7–15	0.14–0.07
4.	Gymnosperms	2–15	0.5-0.07
5.	Monocots	4	0.25
6.	Dicots	3	0.33
7.	All invertebrates	11	0.09
8.	Reef Corals	20	0.05
9.	Marine gastropods	10	0.1
10.	Marine bivalves	15	0.07
11.	Echinoids	6	0.17
12.	Crinoids	6.7	0.17
13.	Freshwater fish	3	0.33
14.	Birds	2.5	0.42
15.	Mammals	1.4	0.71

The rate of extinction has increased many fold over this background rate at Holocene times. A few examples are given below.

1. The current rates of amphibian extinction is 211 times the rate of their background extinction and if all the threatened species go extinct, the rate would be as high as 25000 to 45000 times greater (Wake and Vredenburg, 2008).
2. Global coral monitoring network reports that 20% of global coral reefs are degraded to a state that is beyond recovery, 24% is in imminent danger of collapse and 26% faces long-term risk (Reaka et al. 2008)
3. IUCN (2009) estimated that between 10000-20000 freshwater species are either extinct or at threatened with extinction. This number is so high that it is comparable to extinction rate observed at transitions between geological epochs like Pleistocene-to-Holocene (Vorosmarty et al. 2010)
4. *The IUCN Red List of Threatened Species (version 2010.4)* contains 55,926 species and 33% of these (18,351 species) are known to be threatened. This is a minimum estimate; as status of 8,358 Data

Deficient species are not known. If Data Deficient species are assumed to be threatened at the same level as data sufficient species, the number of known and assumed threatened species rises to 21,610 species (39% of all species assessed).

How Much Time We Have

Pimm and Brooks (1996) has compiled the global rates of extinction made by different authors, which is reproduced below.

Table 5. Estimation of global extinction raters (extinctions/million spp./year)

Serial	*Author*	*Extinction Rate*	*Method adopted*
1.	Myers (1979)	4000	Extraploration of exponentially increasing extinction
2.	Myers (1988)	7000	Destruction of hotspots
3.	Reid (1992)	1000 – 11000	Species area relationship
4.	Smith et al. (1993)	1400 – 2000	Movement of bird and mammals through red list
5.	Ehrlich (1994)	10000	Increasing human energy consumption
6.	Mace (1994)	1100 – 2200	Extinction probabilities from vertebrate list
7.	May et al. (1995)	0.1 – 1	Background rate (13 studies)
8.	Pimm and Brooks (1997)	1000	Half life of 50 years for threatened birds

Quick calculation using the above data points to a grim future for biodiversity. Assuming the lower range of extinction figures (1000 extinctions per million species per year) and assuming species extinction follows negative exponential, half of estimated number of species (9 million) will be extinct in next thousand years.

CONSERVATION INITIATIVES

Conservation initiatives are being taken all over the world for quite some time with mixed amount of success. These initiatives range from launching species specific programmes to building protected area networks. The 2008 IUCN red list reported 40 species that benefited from such programmes as indicated by improvement in their conservation status. It is further estimated that 16 bird species would have gone extinct and a further 49 critically endangered bird species would have slipped into lower conservation status, unless these conservation initiatives were undertaken. We may cite examples of Pink Pigeon, Mauritius Kestrel, Fody and Parakeet in Mauritius, Lear's Macaw in Brazil, Elephants in Africa, Mallorcan Midwife Toad in Balearic island, Mallorca, Spain as beneficiaries of successful conservation initiative in last 20 years (Vie et al. 2009). However, there is little scope of being complacent. The 2008 IUCN report shows 223 species changed their conservation status: 40 species became less threatened as quoted above whereas 183 species become more threatened. However, success story of these 40 species tells us that it is possible to design and implement species specific conservation programme in the face of intense extinction pressure.

Biotic recovery from current biodiversity crisis could occur with or without human intervention. The simplest scenario is recovery following the conclusion of current mass extinction through natural reversal of trend. This will be similar to past recoveries of five mass extinctions. This recovery will require the causative agent of current mass extinction to disappear and negative environmental effects to be naturally attenuated. Based on fossil records of past recoveries, we may conclude that this will require very long time ranging from a few million years to tens of millions years. The second scenario involves Malthusian crushing down of human population and consequent improvement of drivers of biodiversity loss like pollution, climate change, biogeochemical cycle and habitat loss. The third scenario of partial biotic recovery is possible through protection of key habitats. This kind of programme is already underway. The fourth scenario of biotic recovery goes beyond the "protected area" programme and strives to preserve key ecosystems, and mitigate the altered biogeochemical cycle.

The current rate of species loss makes it clear that conservation initiatives cannot take care of all species that are threatened with extinction. Thus, the question of prioritization arises: should we try to conserve as many species as possible or should we try to safeguard evolutionary processes at risk? In the words of Jablonski (1995) "If we are concerned with avoiding the loss of particular functional groups, or with maximizing the potential source pool for evolutionary recovery, then biodisparity measures may provide a more appropriate assessment, beyond sheer number of taxa, of how priorities should be set." Thus, areas having potential for diversity generation should be focus of our future conservation programme, rather than individual taxa based initiatives (Myers & Knoll, 2001)

ACKNOWLEDGEMENT

The author is grateful to Ms. Pratiti Chatterjee of West Bengal Biodiversity Board for help with graphics.

REFERENCES

Alvarez, L.W., Alvarez, W., Asaro, F. and Michel, H.V. (1980). Extraterrestrial cause for the Cretaceous-Tertiary extinction. *Science*, p. 208, 1095–1108.

Anstey, R.L. (1986). Bryozoan provinces and patterns of generic evolution and extinction in the late Ordovician of North America *Lethaia*, p. 19, 33–51.

Benton, M.J. (2003). *When Life Nearly Died: The Greatest Mass Extinction of All Time*. Thames and Hudson London.

Baronsky, A.D., Koch, P.L., Feranec, R.S., Wing, S.L. and Shabel, A.B. (2004). Assessing the causes of late Pleistocene extinctions on the continents. *Science*, p. 306, 70–74.

Baronsky, Anthony D. (2008). Megafauna biomass tradeoff as a driver of Quaternary and future extinctions. In: *In the Light of Evolution, Vol. II: Biodiversity and Extinction*. National Academy of Sciences, Washington, D.C., pp. 227–242.

Bayon, Ricardo and Jenkins, M. (2010). The business of biodiversity. *Nature*, Vol. 466, pp. 184–185.

Bretsky, P.W. (1973). Evolutionary patterns in the Paleozoic Bivalvia: Documentation and some theoretical considerations. *Geological Society of America Bulletin*, 84, pp. 2079–2096.

Cindy V. Looy, Richard J. Twitchett, David L. Dilcher, Johanna H.A. Van Konijnenburg-Van Cittert and Henk Visscher (2001). Life in the end Permian dead zone. *PNAS*, Vol. 98, No. 14: 7879–7883.

Diamond, Jared (1992). *The Third Chimpanzee*. HarperCollins, NY.

Douglas E. Soltis and J. Gordon Burleigh (2009). Surviving the K-T mass extinction: New perspectives of polyploidization in angiosperms. *PNAS*, Vol. 106, No. 14.

Elewa, Ashraf, M.T. and Joseph, Rhawn (2009). The history, origins and causes of mass extinctions. *Journal of Cosmology*, 2: 201–220.

Erwin, D.H. (1993). *The Great Paleozoic Crisis: Life and Death in the Permian*. Columbia University Press, New York.

Fahrig, L. (2003). Effects of habitat fragmentation on biodiversity. *Annual Review of Ecology and Systematics*, p. 34, 487–515.

Firestone, R. (2007). The case for the younger dryas extraterrestrial impact event. *Journal of Cosmology*, p. 2, 256–285.

Foote, M. (2003). Origination and extinction through the Phanerozoic: A new approach. *Journal of Geology*, p. 111, 125–148.

Gribbin, John and Gribbin, Mary (2001). *Ice Age*., Allen Lane, Penguin Press.

Harper, D.A.T. and Rong, J.Y. (2001). Palaeozoic brachiopod extinctions, survival and recovery: Patterns within the rhynchonelliformeans. *Geological Journal*, p. 36, 317–328.

Hubblell, S.P., F. He, Condit, R., Borda-de-Agua, L., Kellner, J. and Steege, H.T. (2008). How many tree species are there in Amazon and how many of them will go extinct? In: *The Light of Evolution, Volume II: Biodiversity and Extinction*. Avise, J.C. Hubbell, S.P. and Ayala, F.J. (Eds.). The National Academic Press, Washington, D.C., pp. 107–126.

Huggett, R.J. (2003). *Fundamentals of Biogeography*. Second Edition. Routledge, London and New York.

IUCN (2009). Species and climate change: More than just polar bears (http://cmsdata.iucn.org/downloads/species_and_climate_change.pdf) accessed 10th January, 2011.

International Union for Conservation of Nature and Natural Resources (2010). *IUCN Red List of Threatened Species* (version 2010.4) (*http://www.iucnredlist.org*).

Jablonski, D. (1995). Extinctions the fossil record. In: *Extinction Rates*. Lawton, J.H. and May, R.M. (Eds.). Oxford Univ. Press, Oxford, pp. 25–44.

________. (2001). Lessons from the past: Evolutionary impacts of mass extinctions. *Proceedings of National Academy of Sciences*, 98, pp. 5393–5398.

________. (2008). Species selection: Theory and data. *Annual Review of Ecology, Evolution and Systematics*, p. 39, 501–524.

________. (2005). Mass extinctions and macroevolution. *Paleobiology*, 31 (Suppl. to No. 2): 192–210.

________. (2004). Extinction: Past and present. *Nature*, Vol. 427. p. 589.

Jackson, J.B.C. (2008). Ecological extinction and evolution in the brave new ocean. In: *The Light of Evolution, Vol II: Biodiversity and Extinction.* Avise, J.C., Hubbel, S.P. and Ayala, F.J. (Eds.). The National Academies Press, Washington, D.C., pp. 5–26

Joachimski, M.J. and Buggisch, W. (2002). Conodont apatite delta 18O signatures indicate climate cooling as a trigger of the late Devonian mass extinction. *Geology*, p. 30, 711–714.

Joseph, R. (2009). Extinction, metamorphosis, evolutionary apoptosis and genetically programmed species mass death. *Journal of Cosmology*, 2, pp. 235–255.

Lane, N. (2007). Reading the book of death. *Nature*, p. 448, 122–125.

Lantzy, R., Dacey, M. and McKenzie, F. (1977). Catastrophe theory: Application to the permian mass extinction. *Geology*, p. 5, 724–728.

Lawrance, W.F., Albernaz, A.K., Fearnside, P.M., Vasconcelos, H.L. and Ferreira, L.V. (2004). Deforestation in Amazonia. *Science*, 304, 1110.

Lowe, S., Browne, M., Boudjelas, S., De Poorter, M. (2000). *100 of the World's Worst Invasive Alien Species: A Selection from the Global Invasive Species Database.* Published by The Invasive Species Specialist Group (ISSG) a specialist group of the Species Survival Commission (SSC) of the World Conservation Union (IUCN), p. 12.

Melott, A.L. et al. (2004). Did a gamma-ray burst initiate the late Ordovician mass extinction? *International Journal of Astrobiology*, 3, 55–61.

Millennium Ecosystem Assessment (2005). *Ecosystems and Human Well-being: Biodiversity Synthesis.* World Resources Institute, Washington, D.C.

Myers, N. (1980). *Conservation of Tropical Moist Forests.* National Academy of Sciences, Washington, D.C.

Myers, N. and Knoll, A.H. (2001). The biotic crisis and the future of evolution. *Proceedings of National Academy of Sciences*, Vol. 98 (10), pp. 5389–5392.

Pimm, S.L. and Brooks, T.M. (1997). The sixth mass extinction: How large, where and when? In: *Nature and Human Society: The Quest for a Sustainable Society.* Raven, P.H and Williams, T. (Eds.). National Academy Press, Washington, D.C.

Raup, D.M. (1992). *Bad Genes or Bad Luck.* Norton, New York.

Reaka, M.L., Rodgers, P.J. and Kudla, A.U. (2008). Patterns of biodiversity and endemism on Indo-West Pacific coral reefs. In: *The Light of Evolution, Volume II: Biodiversity and Extinction.* Avise, J.C, Hubbell, S.P. and Ayala, F.J. (Eds.). The National Academic Press, Washington, D.C., pp. 45–61.

Robertson, D.B.R., Brenchley, P.J. and Owen, A.W. (1991). Ecological disruption close to the Ordovician-Silurian boundary. *Historical Biology,* p. 5. 131–144.

Royal Botanical Gardens, Kew (http://www.kew.org/) accessed on 31st December, 2010.

Smith, A.B. and Jeffery, C.H. (2000). Maastrichtian and Palaeocene echinoids: A key to world faunas. Special papers in *Palaentology,* 63.

UNDP (2007). *Human Development Report 2007/2008. Fighting Climate Change: Human Solidarity in a Divided World,* Palgrave Macmillan, New York.

Vie, J.C., Hilton-Taylor, C. and Stuart, S.N. (2009). *Wildlife in a Changing World – An Analysis of the 2008 IUCN Red List of Threatened Species,* Gland, Switzerland, IUCN.

Vorosmarty, C.J. et al. (2010). Global threats to human water security and river biodiversity. *Nature,* Vol. 467, pp. 555–561.

Whitmore, T.C. (1980). The Conservation of Tropical Rain Forest. In: *Conservation Biology.* Soule, M.E. and Wilcox, B.A., Sinauer, Sunderland, MA, (Eds.). pp. 303–318.

Wilcove, D.S., McLellan, C.H. and Dobson, A.P. (1986). Habitat fragmentation in the temperate zone. In: *Conservation Biology.* M.E. Soule, (Ed.). pp. 237–56, Sunderland, MA: Sinauer.

Section – III

Indian Scenario

9

Diversity of Bacteria and Archaea: Present Status in India

Tapan Chakrabarti

INTRODUCTION

Biodiversity is projected in general biology textbooks or in public forums as diversity of plants and animals. There is hardly any mention of microorganisms in the context of biodiversity although they constitute the vast majority on the earth and play a very crucial role in sustaining all life forms and in maintaining the health of the planet. In public perception, microbes are generally associated with disease and rot, unaware of the fact that only a very few bacteria are pathogenic, majority of them are beneficial and work unseen to support diversity of life forms. A human will not be a "normal human" if he/she does not have 10 (DeLong et al. 2006) (hundred trillion) bacterial cells in and on him/her. This number is ten times higher than the number of cells a human is made up of. Lack of public knowledge regarding the enormity of microbial diversity is mainly due to their very tiny size which makes them invisible to unaided eyes.

There is hardly any doubt that major diversity of life is microbial and this diversity is distributed in all three domains of life, archaea, bacteria and eukaryota (Pace, 1997). While all members of domains archaea and bacteria are microorganisms, a substantial number of eukaryotes (for example, fungi, microalgae, protozoa etc.) are also microscopic organisms. The number of archaea and bacteria on the earth is estimated to be 5×10^{30} and the amount of carbon present in this huge biomass is between 400-500 petagram (1 petagram = 10^{15} g or one billion ton) which is almost equal to the amount of carbon present in plants (Whitman, et al. 1998).

Bacteria were the first life forms to inhabit this planet some 3.5 billion years ago and since then invented a diverse array of metabolic capabilities to establish and flourish on earth and even in environments where other life forms normally do not survive. They live almost everywhere on earth from the icy continents to high temperature environments (as high as 113°C) and in extremely acidic to alkaline habitats performing vital functions like biogeochemical cycling of C, N, S, P etc. and cleaning the environment of toxic pollutants. Bacteria are the only organisms on earth which can fix atmospheric nitrogen and therefore contribute to the fertility of soil. Although they are present in every conceivable ecological niche, majority of them live in oceanic and terrestrial sub-surfaces (35.5×10^{29} and 2.5 to 25.0×10^{29} respectively). They are an integral part of most animals including human and plants, living in association or in symbiotic relationship with them (Whitman et al. 1998). In spite of their long evolutionary history they remained small and mostly with very limited cellular-morphological features. "Their diversity is expressed in terms of physiology and metabolism while that of the larger eukaryotes is expressed in terms of structure and behavior" (Hunter-Cavera et al. 1998).

Bacteria and archaea possess enormous metabolic diversity and can utilize complex organic compounds. While many bacteria are organotrophs, also referred to as heterotrophs, (use of reduced organic compounds for carbon and energy) like us, majority of them live as autotrophs (fixation of CO_2 to reduced organic compound). Autotrophic organisms derive energy for CO_2 fixation either from light (photoautotrophs) or from oxidation of inorganic compounds such as hydrogen, elemental sulfur, ammonium ions (chemolithotrophs or lithotrophs). Photoautotrophic generation of O_2 by cyanobacteria (also by plants and algae) is of vital importance in sustaining life on earth. Non-oxygenic photoautotrophy is also a way of life for some groups of bacteria [like purple sulfur bacteria, green sulfur bacteria, nonsulfur purple bacteria (Brock, 1979)]. Lithotrophic mode of metabolism is found only in bacteria and archaea and like phototrophic bacteria they also contribute greatly in generation of biomass to support the organotrophs (Pace, 1997). No eukaryotic organism on its own is known to possess lithotrophic metabolism although some can acquire a lithotrophic way of life by harbouring sulfur oxidizing and carbon dioxide fixing bacterial symbionts.

Appearance of cyanobacteria, capable of oxygenic photosynthesis, in the early phase of evolution was directly or indirectly responsible for emergence of aerobic organisms on this planet. Cyanobacteria and proteobacteria with "specialization through coevolution with the host cells" became chloroplast and mitochondria respectively and thus contributed to photosynthetic and respiratory capacity of eukaryotic organisms. It may not be an exaggeration to say that bacteria and archaea played key roles in shaping the biodiversity landscape of our planet.

In spite of long evolutionary history, enormous diversity in metabolism and habitat, number of species of bacteria and archaea known today is about ten thousand. Why is it so? Convention has it that bacteria and archaea to be recognized as species require their cultivation in pure form followed by phenotypic and genotypic characterization. Many lines of evidence has now clearly established that majority of bacteria and archaea defy cultivation in the conventional laboratory conditions. From any sample, number of bacteria seen under a microscope is always many times more than the number of colonies formed on a culture plate [great plate count anamoly (Staley et al. 1985)]. For example, total count of bacteria in a forest soil sample was found to be 1.5×10^{10} g (dry wt) soil^{-1} which was more than two orders of magnitude compared to 4.3×10^{7} g (dry wt) soil^{-1} calculated on the basis of bacterial colonies that appeared on plates (Torsvik et al. 1990). Bacterial and archaeal cells can be detected under a microscope by staining with acridine orange, 4'6-diamino-2-phenylindole (DAP1) etc. They can also be quantified and broadly identified using fluorescently labeled rRNA targeted oligonucleotide probes by a technique called fluorescent in situ hybridization (FISH). That majority of cells in a sample are metabolically active can be shown by microautoradiography (MAR) (Staley et al. 1985) or by FISH/MAR (Schleifer et al. 2009). Less number of colonies appearing on a culture plate than the number of cells present in a sample may be due to: (i) failure to provide right nutrient and the optimum conditions, (ii) separation from the community members (Kacberlein et al. 2002) and (iii) some cells may be in a dormant state. It has been reported that cutlurable bacteria and archaea constitute only 0.3% in soil and less than 0.1% in ocean of the total bacteria and archaea present in these habitats (Schleifer et al. 2004). The fact that a forest soil sample harbours much diverse type of bacteria compared to the ones that appear on a culture plate was shown by DNA reassociation kinetics of community DNA. Number of genomes (equivalent to species) present in the sample was calculated to be about 4000 compared to much less morphotypes observed on culture plates (Torsvik et al. 1920).

Extensive use of molecular tools in the last few years has resulted in a wealth of information about microbial and archaeal diversity which would not have been possible using conventional culture dependent approach. In a culture independent molecular approach, environmental (community) DNA is extracted directly from a sample, the 16S rRNA genes are amplified by PCR (polymerase chain reaction) using bacterial or archaeal universal primers, the amplicons are then cloned in *E. coli* and sequenced followed by phylogenetic analysis. Preliminary profile or fingerprint of community structure may also be generated by techniques like amplified ribosomal DNA restricted analysis (ARDRA), denaturing gradient gel electrophoresis (DGGE), terminal restriction fragment length polymorphism (T-RFLP) and other methods. Outcome of such culture independent efforts was the discovery of many new lineages in bacteria and archaea. In fact, out of 53 phyla of bacteria, about half of them have no cultured representatives. They were classified on the basis of 16S rRNA gene sequence signatures alone. Most of the culturable bacteria belong mainly to six phyla, other phyla have only a few cultivated representatives (Handelsman, 2004). The 16S rRNA gene sequence based analysis of environmental samples is an important tool to find out phylogenetic types (phylotypes) although it throws very little light on the physiology and metabolism of so far uncultured organisms.

Many environments starting from marine sediment (Biddle et al. 2006, Lipp et al. 2008, Santelli et al. 2008) to oceanic water, (DeLong et al. 2006, Wu, 2007) fresh water bodies (Venter et al. 2004) of the tropics to the extreme of habitats of Arctic (Bano et al. 2002) and Antarctic, (Shivaji et al. 2004, DeLong et al. 1994, Niederberger et al. 2008) from termite gut (Ohkuma et al. 1996) to human gut, (Gill et al. 2006, Eckburg et al. 2005) from soil (Buckley et al. 2003) to hydrothermal vent (Kimura et al. 2003, Reysenbach et al. 2000) have been explored to ascertain the extent of bacterial and archaeal diversity using culture independent approach. In culture independent metagenomic approach, environmental DNA is cloned into a suitable vector and transformed into bacteria (without PCR amplification). Clone libraries thus constructed then can be screened using functional assays, or fishing out genes of interest using suitable probe(s) or can be sequenced to find out genomic diversity and metabolic potential of the community (Venter et al. 2004, Reisenfeld et al. 2004, Rusch et al. 2007 and Yooseph et al. 2007). Discovery of novel antibiotics, antibiotic resistance genes, degradative enzymes and a few transporters were reported by using this approach (26 for an excellent review of metagenomic approach to biotechnological applications). Complete genomes of individual components of acid mine drainage microbial community were assembled and annotated recently (Tyson et al. 2004). This type of approach not only reveals microbial diversity, it also gives information about metabolic activities of the community. Biggest use of metagenomic approach so far has been the global oceanic sampling undertaken by Craig Venter and his group (Venter et al. 2004, Rusch et al.

2007). Using shotgun genome sequencing, they assembled 7.7 billion base pairs to generate 6.6 million "sequence reads". In addition to providing a picture of the marine surface microbial diversity, the authors attempted to find out the community structure and also determine what factors are responsible in shaping such a structure. Just in Sargasso sea alone they found about 1800 species, out of which 148 were new phylotypes or species (Venter et al. 2004). Thousands of new protein families were discovered from this expedition (Yooseph et al. 2007).

Such metagenomic approach provides us with information regarding microbial diversity and potential functions of the genes but throws very little light on the metabolic activities actually taking place in the community at a given point of time. Recently attempts have been made to isolate proteins directly from the environmental samples and analyse them by a combination of sophisticated tools (LC-MS, LC-MS/MS, MALDI TOF etc.) This metaproteomics approach, although presently at an initial stage, which is complementary to metagenomics, may provide information about native metabolic activities of the microbial community and the organisms carrying out such functions (Keller et al. 2009, Wilmes et al. 2006, VerBerkmos et. al, 2009).

An organism is more than just nucleic acid sequences and proteins and biology of microorganisms from different habitats can be best studied once they become amenable to cultivation. Since majority of bacteria and archaea defy culturing, knowledge about their biology is very meager. In the last few years innovative efforts have been made to bring these so called "unculturables" into culture with some success (Kaeberlein et al. 2002). Members of phylum SAR-11 are the most predominant (more than 25%) bacteria of marine surface (Morris et al. 2002) but were known only from their sequence signatures. Some of them have recently been cultured using high throughput dilution to extinction method (Rappe et al. 2002, Vartoukian et al. 2010).

Number of bacteria and archaea on the earth is huge, they are ubiquitous, show immense diversity in genetic makeup, metabolism and biochemical functions which sustains the biosphere. What is debatable at present is the number of species of bacteria and archaea on this planet. Estimate of their number varies between 106 to 109 (Dykhuizen, 1998, Curtis et al. 2002). Even if the conservative estimate is considered, the number of species of bacteria and archaea known today represent only about 0.5% of their diversity compared to 90% of vertebrates and 84% of plants already described (Schleifer, 2004).

Very little research activities in microbial diversity and ecology are going on in India and these areas need to be strengthened and intensified not only for newer and better biotechnological products but also because the vast reservoir of unknown microbial diversity is increasingly being looked upon as possible solutions to our present day challenges of fuel, food and climate change (Tiedje and Donohue, 2008). Since bacteria and archaea drive global carbon and nitrogen cycle, understanding their diversity will enable us to develop informed strategies to address pressing global issues such as man-made climate change (Editorial, Nature Rev, 2010).

DIVERSITY OF BACTERIA AND ARCHAEA OF INDIA-PRESENT STATUS

India is one of the seventeen megadiversity countries of the world. It also hosts two of the world's biodiversity "hot spots". This unique distinction comes from its rich flora and fauna, significant portion of which is endemic. According to State of the Environment Report 2009 45,500 plant species including fungi and lower plants and 91000 animal species, representing about 7% of the world's flora and 6.5% of the world's fauna respectively have been reported from India (Biodiversity State of Environment Report, 2009). From biodiversity standpoint, India has some 59,353 insect species, 2546 fish species, 240 amphibian species, 460 reptile species, 1232 bird species and 397 mammal species of which 18.4 percent are endemic. The country is home to at least 18,664 species of vascular plants of which 26.8% are endemic (Biodiversity State of Environment Report, 2009).

The reason for quoting these numbers above is just to emphasise that the words bacteria, archaea or microorganisms are conspicuous by their absence in a serious report of the custodian of biodiversity of India. Sporadic efforts were made in the past to isolate microorganisms with specific functions from different environments for biotechnological applications. Most of the investigators were not interested to find out composition of microbial community of the environmental samples they collected. One of the reasons for lack of interest in microbial diversity research is lack of knowledge of microbial taxonomy. Identification of bacteria isolated by many investigators from different habitats, therefore, was difficult. Large scale exploration of microbial diversity using culture-dependent and culture-independent approaches has not been undertaken in India. The journal Current Science published a Special Issue (Vol. 73, 10 July, 1997) on Biodiversity of Western Ghats. Unfortunately, there was not a single paper related to microorganisms in the issue. However, the journal published another Special Issue on microbial diversity (Vol. 89, 10 July, 2005). Some records of plant growth promoting and plant associated bacteria in Indian soil, (Tilak et al. 2005) buffalo and cattle (Kamra, 2005) and Indian deserts (Bhatnagar and Bhatnagar, 2005) are available. It was also reported that more than 27000 species of fungi distributed in 2080 genera have been described from India (Manoharachary et al. 2005). Indian Journal of Microbiology also brought out a Special Issue [Vol. 48(2), 2008] on microbial diversity which provided a review of work going on around the world, but did not give a picture of microbial diversity in Indian context. In spite of being a biodiversity rich country very little information is available about diversity of bacteria and archaea of India.

Lonar Lake in Maharashtra, India is one ecological niche which has been studied by many investigators in terms of bacterial and archaeal diversity. Culture-dependent study of the lake (Joshi et al. 2008) revealed the predominance of Gram-positive *Firmicutes* represented by the genera *Bacillus, Paenibacillus, Exiguobacterium, Planococcus, Enterococcus* and *Vagococcus*. Gram-negative bacteria found in the lake belong to the genera *Halomonas, Stenotrophomonas, Providencia, Alcaligenes* and *Paracoccus*. Actinobacteria were represented by the genera *Cellulomicrobium, Dietzia, Arthrobacter* and *Micrococcus*. Cyanobacteria were represented in the the lake by species of four genera, *Arthrospira, Oscillatoria, Spirulina* and *Dactylococcopsis* (Badve et al. 1993).

Culture-independent study of the lake sediment (Wani et al. 2006) also revealed that members belonging to the phylum *Firmicutes* are the predominant (39%) bacteria in the clone library followed by *Proteobacteria* (29.5%), *Actinobacteria* (6.8%), CFB (13.3%), *Planctomycetes* (6.8%), *Deinococcus-Thermus* (4.5%), *Cyanobacteria* (4.5%) and *Spirochacte* (2.3%). Archaeal phylotypes belonging to *Crenarchaeota* and *Euryarchaeota* were also present. In another similar study of archaeal diversity in the lake many phylotypes showing affinity to the genera *Methanoculeus* and *Methanocalculus* were found to be the predominant ones (Surakashi et al. 2007). Members of these two genera were also isolated in pure form. Sequence similarity of many clones with other archaea in both the studies was very low indicating the presence of possible new taxa at various levels. *Methanosarcina* species was also reported from the lake in a previous study (Thakker and Ranade, 2002). The presence of methylotrophic organisms (*Methylomicrobium, Methylophaga* and *Bacillus*) in the lake sediment was shown recently by using SIP [stable isotope probing (Antony et al. 2010)]. Besides these studies, efforts were made to isolate from the lake bacteria with potential applications, like bioremediation of phenol (Kanekar et al. 2002) and protease (Kanekar et al. 2002). Two new genera of bacteria were discovered from this lake (Anil Kumar et al. 2010, Anil Kumar et al. Press).

Besides Lonar Lake, diversity of bacteria was reported from a few other aquatic habitats in India. Culture-independent approach was used to study bacterial diversity of a hot spring in Bakreswar, West Bengal (Ghosh et al. 2003). Predominant bacteria in this spring were found to be members of *Gammaproteobacteria* (40%) followed by *Cyanobacteria* (32%), Green non-sulfur bacteria (12%) and low G+C Gram-positive bacteria (4%). The authors also isolated a few *Shewenella* like bacteria and tried to correlate their presence and iron metabolism in the hot spring. Culture-dependent and culture-independent analysis of bacterial diversity of three spring waters (temperature 36°-39°C), two from near Puttur in Karnataka and one from Borpung region near Jorhat, Assam have been reported (Ruckmani et al. 2010, Bhattacharya, 2010, Saha, 2006). Bacteria isolated from one of them (Irde village in Puttur) were found to belong to the genera *Acinetobacter, Aeromonas, Arthrobacter, Bacillus, Cellulosimicrobium, Chromatobacterium, Chryseobacterium, Dermococcus, Enterobacter, Halomonas, Janibacter, Klebsiella, Kluyvera, Microbacterium, Micrococcus, Nocardioides, Novosphingobium, Paenibacillus, Pseudomonas, Roseomonas and Staphylococcus*. Members of the phylum *Proteobacteria* was found to be the most abundant (58%) in this habitat. Other phyla were represented by CFB (5%), *Actinobacteria* (13%) and *Firmicutes* (26%). In culture-independent analysis of the same water, the predominant sequences (>50%) in the clone library belonged to *Cyanobacteria* (affiliated to the genus *Cynacococcus* with varying degree of sequence similarity) followed by *Proteobacteria*, CFB and *Actinobacteria* (Ruckmani et al. 2010). Similar study of water and sediment samples from another spring located at Padmunja, Karnataka revealed the presence of members of the genera *Acidovorax, Acinetobacter, Bacillus, Brachybacterium, Brevibacterium, Cohnella, Enhydrobacter, Exiguobacterium, Microbacterium, Paenibacillus, Pseudomonas, Reinheimera, Staphylococcus* and *Streptomyces* in water of the spring (Bhattacharya, 2010). Like in the former spring, members of phylum *Proteobacteria* were also predominant (56%) in this spring water although generic distribution was not identical and cyanobacterial clones were not detected in the 16S rRNA gene library which showed the presence of large percentage (70%) of sequences affiliated to the genus *Thiovirga*. Bacterial species isolated from the sediment of the spring belong to the genera *Acidovorax, Altererythrobacter, Arcicella,*

Arthrobacter, Chitinimonas, Erythromicrobium, Flavobacterium, Limnobacter, Microbacterium, Reinheimera, Rhodobacter and *Undibacterium.*

In contrast to the water, sediment of the spring was dominated by *Firmicutes* (55%) followed by *Proteobacteria* (26%) as revealed by culture-dependent approach. Culture-independent analysis of the sediment on the other hand showed the predominance of *Proteobacteria* (84%).

Diversity of archaea in the sediment sample was limited and majority of the clones in the archaeal library were related to members of either *Methanosarcinales* or *Methanomicrobiales.*

Microbial diversity study of water of another spring (temperature ~ 38^0C) from Borpung, in Mikir hills, Karbi Anglong district of Assam by culture-dependent approach, revealed the presence of species belonging to the genera *Aeromonas, Aquabacterium, Azoarcus, Bacillus, Chitinomonas, Comomonas, Deinococcus, Hyderogenophaga, Ideonella, Paenibacillus, Pseudomonas, Rubrivirax, Thauera* and *Vibrio* (Saha, 2006). In addition, three new genera (*Aquimonas voraii,* (Saha et al. 2005) *Emticicia, oligotrophica* (Saha and Chakrabarti, 2006) and *Fontibacillus aquaticus* (Saha et al. 2010)) and three new species (*Paenibacillus assamensis,* (Saha et al. 2005) *Flavobacterium indicum* (Saha and Chakrabarti, 2006) and *Aeromonas sharmana* (Saha and Chakrabarti, 2006) were discovered from this spring and they have been validly published. Culture-independent molecular analysis of the spring water revealed the predominance (56%) of cyanobacterial sequences in the clone library most of which were related to *Cyanocystis* with different degree of similarity (Saha, 2006). Other sequences in the clone library were affiliated mainly to the phylum *Proteobacteria.*

From the studies mentioned above it is apparent that bacterial diversity revealed by culture-independent approach does not necessarily overlap with the one obtained by cultivation based method. These two approaches complement each other and application of both these methods give a better picture of microbial community structure of a habitat.

Municipal landfills are man-made complex ecosystems due to dumping of many varieties of organic and inorganic substances. Bacterial and archaeal community play important roles in degradation of these compounds in municipal solid wastes. Cultivation based microbial analysis of Dhapa land fill in Kolkata, West Bengal showed the presence of bacteria belonging to the genera *Aeromonas, Azotobacter, Bacillus, Paenibacillus, Pseudomonas, Rhodococcus, Sphingobacterium* (Ghosh et al. 2007). Culture independent analysis was not attempted in this study and, therefore, the presence of other bacteria and archaea remain unknown.

Culture dependent and culture independent analysis of bacterial and archaeal diversity of a municipal landfill in Chandigarh on the other hand revealed the presence of many more genera and species compared to the Dhapa land fill (Krishnamurthi, 2009). Majority (85%) of bacteria isolated from the latter land fill or sequences analysed from the 16S rRNA gene library were affiliated to Gram-positive *Firmicutes* and among them species of *Bacillus* was found to be the predominant members of this community. Very few members of *Proteobacteria* were found in this ecosystem. Besides *Bacillus* sp., other genera present in the land fill were *Actinomadura, Aerococcus, Arthrobacter, Brevibacillus, Clostridium, Cohnella, Desulfatomaculum, Dietzia, Kocuria, Lysinibacillus, Microbacterium, Micromonospora, Microvirga, Mycobacterium, Nocardia, Oceanobacillus, Ornithinibacillus, Paenibacillus, Pseudomonas, Rhodococcus, Roseomonas, Staphylococcus, Stenotrophomonas* and *Streptococcus*. One new genus, *Paenisporosarcina* was discovered from this land fill (Krishnamurthi, 2009).

Archaeal diversity in this land fill was very limited and was represented by members affiliated mainly to the order *Methanosarcinates* and *Methanomicrobiales*. It is to be noted that microbial diversity of landfills all over the world appears to be broadly similar in having members of *Firmicutes* as the predominant bacterial group and archaea are represented by the members belonging mainly to *Methanomicrobiales* and *Methanosarcinales* (Palmisano et al. 1993, Pourcher et al. 2001, Huang et al. 2005 and Huang et al. 2002). Members of Gram-positive *Firmicutes* are known to have diverse hydrolytic and fermentative capabilities. This may be one reason for their predominance in the landfill environment. Hydrolytic and fermentative activities of bacteria lead to the formation of acetate, CO_2 and H_2 which are then utilized by members of methanogenic archaea, mostly members of *Methanomicrobiales* and *Methanosarcinales* to generate methane. Methane is an important product of land fill and is increasingly being looked upon as an alternate source of energy.

Bacterial and archaeal diversity of mangrove ecosystem along the coastal regions of India is little known. Recently, bacterial community structure of mangrove sediment of Sundarban has been reported (Ghosh et al. 2010). Using culture independent method the authors found that sequences related to the members of the phylum *Proteobacteria* were the most abundant in the two clone libraries. Other sequences of the libraries were affiliated to phyla *Flexibacteria* (CFB group), *Actinobacteria, Acidobacteria, Chloroflexi, Firmicutes, Planctomycetes* and *Gammatimonadates*. Members of the phylum *Proteobacteria* were also found to be the predominant bacteria (38%) in uranium mine sites of Jaduguda (Sar et al. 2007). Other sequences in the clone libraries were affiliated to the phyla *Acidobacteria* (21%), *Firmicutes* (8%), *Bacteroidetes* (8%), *Gemmatimonadetes* (4%) and *Cyanobacteria* (4%). Like many other culture independent studies the authors also found many sequences (around 17%) of uncertain affiliation because of their low sequence similarity with known taxa and speculated that they might represent potential "new organisms adapted to this habitat."

Among extreme environments of India, archaeal and bacterial diversity of three solar salterns in Tamil Nadu was studied by both culture dependent and culture independent approaches (Manikandan et al. 2009). Predominant microbial species inhabiting this environment were found to be archaea. Several members belonging to the genera *Haloarcula, Halobacterium, Haloferax, Halgeometricum and Halorubrum* were isolated. The presence of species related to the archaeal genus *Natrinema* were indicated in culture independent analysis. Bacterial isolates belonging to the genera *Salinibacter, Cytophaga and Marinococcus* could be recovered from only one of these three salterns.

Cold regions of the Himalayas provide good habitats for diverse group of bacteria. Culture-independent and culture dependent approaches were used to study bacterial diversity of Roopkund, Hamta and Pindari Glaciers. Glacial soil of Roopkund was found to have less species richness compared to glacial lake (Pradhan et al. 2010). The authors concluded that this may be due to reduced low organic carbon and mineralizable N and P in the former. Many new species of bacteria have been described from different parts of the Himalayas (Shivaji et al. 2005, Chaturvedi et al. 2005, Reddy et al. 2008, Pindi et al. 2009, Kishore et al. 2010, Mayilraj et al. 2005, Singla et al. 2005, Mayilraj et al. 2006, Mayilarj et al. 2006, Mayilraj et al. 2006, Mayilraj et al. 2006, Ghosh et al. 2007, Anil Kumar et al. 2007 and Arunasri et al. 2008). Seven new species of bacteria were also discovered from high altitude atmosphere over Hyderabad (Shivaji et al. 2006, Shivaji et al. 2009).

An examination of websites of List of Prokaryotic Names Standing in Nomenclature, (Euzeby, www.bacterio.cict.fr) Approved list of generic cyanobacterial names (Komarek and Hauer, www.cyanodb.cz) and of International Journal of Systematic and Evolutionary Microbiology (ijs. Sgmjournals.org) revealed that 25 new genera and 317 new species (including 6 genera and 91 new species of cyanobacteria and a few species of Antarctic bacteria isolated by Indian scientists), 4 subspecies and one "candidatus species" of bacteria and three species of archaea of Indian origin have been described since 1900. It is good to see that the number of new taxa described from India increased significantly in last five years (Fig. 1) although much behind China and South Korea (A list of new taxa discovered from India is appended. Some of the taxa mentioned in the list have been reclassified since their original publications).

A well maintained database of bacteria and archaea recorded so far from India will help not only the researchers in microbial diversity, it will also be helpful to present a better and more complete picture of biodiversity of India. An effort to this effect was initiated by Department of Biotechnology, Government of India, to prepare an "Inventory of Microbial Resources of India". Data were collected from primary and secondary sources by a group of scientists from fifteen institutions. A draft database was created which included bacteria, fungi, algae, protozoa and viruses recorded from (or available in) India. Some records in the existing database need editing (mainly correction of spelling mistakes of genus and species names and removal of redundancy). Having made manual corrections of names and removing the redundant entries the number of bacterial species in the list comes to 1174 (in 271 genera) and archaeal species to 50 (distributed in 26 genera). A few species of bacteria and archaea maintained by culture collections (like MTCC, NCIM, MCM) which may not be of Indian origin were also included in this list. Present manual updating of the list with recent information available in public domain about bacteria and archaea recorded from India resulted in increase in the number of bacterial and archaeal genera to 381 containing about 1782 species.

Attempts have been made in this paper to present as much information as possible from available literature about bacterial and archaeal diversity of India. Although the picture may not be complete it gives a reasonable idea about bacteria and archaea present in different ecosystems in India and may act as a base which may be updated with additional new information.

Microbiological research in general and microbial diversity in particular requires good infrastructure support which most of the universities and research institutes in India do not have. While a lot of efforts and expenditure are committed to save a few species in India, research on microbial diversity, taxonomy and ecology do not get the attention it deserves. Conservation of this immense diversity either as an organism and/or DNA libraries is a mammoth task. It is about time that research activities in these areas and ex-situ conservation of microorganisms are taken up on priority basis.

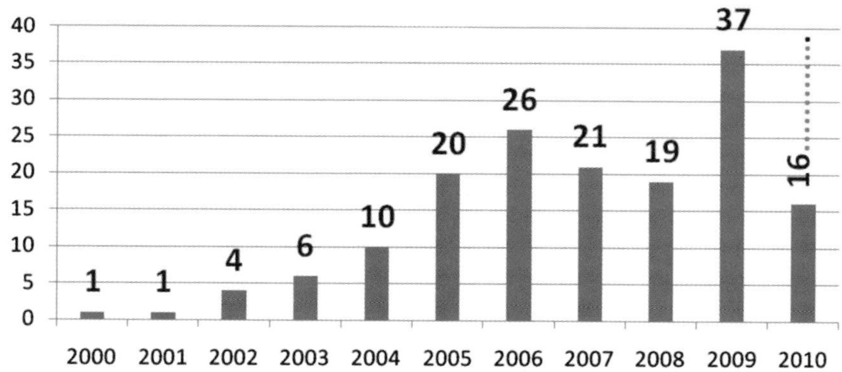

Fig. 1. Number of new bacterial taxa described from India (2000 to July 2010)

ACKNOWLEDGEMENTS

I thank Dr S Shivaji (Centre for Cellular and Molecular Biology, Hyderabad) for critically reading the manuscript and for his help in preparing the manuscript. I also sincerely thank Dr S Natesh, Senior Advisor, Department of Biotechnology, Government of India for his enthusiasm and initiative in gathering information about microbial resources of India. This part of the project was supported by DBT, Govt. of India and CSIR. Data from primary and secondary sources were collected by many project assistants under the supervision of Prof A K Paul (University of Calcutta), Prof P K Sarkar (University of North Bengal), Prof C Manoharachary (Osmania University), Prof G Subramanyam (Bharathidasan University), Prof H Shekhara Shetty (University of Mysore), Prof S C Sehgal (Regional Medical Centre, Andaman), Dr B N Ganguli and Dr P Kanekar (Agharkar Research Institute), Dr M V Deshpande (National Chemical Laboratory), Dr A Adholeya (The Energy and Resource Institute), Dr S C Nautiyal and Dr H B Singh (National Botanical Research Institute) Dr K Annapurna (Indian Agricultural Research Institute) and Dr S Shivaji (Centre for Cellular and Molecular Biology). Software for the database was developed by a succession of project assistants guided by Dr G P S Raghava and Dr. Balvinder Singh (Institute of Microbial Technology). I also thank Ms Dimple Davray for her help in preparing figure 1.

References

Anil Kumar, P., Srinivas, T.N.R., Madhu, S., et al. (2010). Indibacter alkaliphilus gen. nov., sp. nov., an alkaliphilic bacterium isolated from a haloalkaline lake. *Int. J. Syst. Evol. Microbiol*, 60: 721–726.

Anil Kumar, P., Srinivas, T.N.R., Sasikala, C. and Ramana, C.V.(2007) Rhodobacter changlensis sp. nov., a psychrotolerant, phototrophic alphaproteobacterium from the Himalayas of India. *Int. J. Syst. Evol. Microbiol.*, (2007), 57: 2568–2571.

Anil Kumar, P., T.N.R. Srinivas, P. Pavan Kumar, S. et al. *Nitritalea halalkaliphila* gen. nov., sp. nov., an alkaliphilic bacterium of the family '*Cyclobacteriaceae*', phylum *Bacteroidetes. Int. J. Syst. Evol. Microbiol.*, (in press). doi: 10.1099/ijs.0.020230-0.

Antony, C.P., Kumaresan, D., Ferrando, L., et al. (2010). Active methylotrophs in the sediments of Lonar Lake, a saline and alkaline ecosystem formed by meteor impact. *ISME Journal*, (in press). doi: 10.1038/ismej.2010.70.

Arunasri, K., Venkataramana,V., Sproer C., Sasikala,C. and Ramana, C.V. (2008). Rhodobacter megalophilus sp. nov., a phototroph from the Indian Himalayas possessing a wide temperature range for growth. *Int. J. Syst. Evol. Microbiol.*, 58: 1792–1796.

Badve, R.M., Kumaran, K.P.N. and Rajshekhar. (1993). Eutrophication of Lonar Lake. C., *Curr. Sci.*, 65: 347–354.

Bano, N. and Hollibaugh, J.T. (2002). Phylogenetic composition of bacterioplankton assemblages from the Arctic Ocean. *Appl. Environ. Microbiol.*, 68: 505–518.

Bhatnagar, A. and Bhatnagar, M. (2005). Microbial diversity in desert ecosystems. *Curr. Sci.*, 89: 91–100.

Bhattacharya, A., Ph.D. Thesis, Jawaharlal Nehru University, New Delhi (2010).

Biddle, J.F., Lipp, J.S. and Lever, M.A., et al. (2006). Heterotrophic Archaea dominate sedimentary subsurface ecosystems off Peru. *Proc.Natl.Acad. Sci.USA,* 103: 3846–3851.

Biodiversity, State of the Environment Report (2009), Ministry of Environment and Forests New Delhi, p. 50 (www.moef.org.in)

Brock, T.D. (1974). *Biology of microorganisms*, Prentice Hall, New Jersey, p. 201.

Buckley, D.H. and Schmidt, T.M. (2003). Diversity and dynamics of microbial communities in soils from agro-ecosystems. *Environ. Microbiol.*, 5: 441–452.

Chaturvedi, P., Reddy, G.S.N. and Shivaji. (2005). *Dyadobacter hamtensis* sp. nov., from Hamta glacier, located in the Himalayas, India. S. *Int. J. Syst. Evol. Microbiol.,* 55: 2113–2117.

Curtis, T.P., Sloan, W.T. and Scanell, J.W. (2002). Estimating prokaryotic diversity and its limits. *Proc. Natl. Acad. Sci. USA,* 99: 10494–10499.

DeLong, E.F., Preston, C.M., Mincer, T. et al. (2006). Community genomics among stratified microbial assemblages in the ocean's interior. *Science,* 311: 496–503.

DeLong, E.F., Wu, K.Y., Prezelin, B.B. and Jovine. R.V. (1994). High abundance of Archaea in Antarctic marine picoplankton. *Nature*, 371: 695–697.

Dykhuizen, D.E. (1998). Santa Rosalia revisited: Why are there so many species of bacteria? *Antonie Van Leeuwenhoek,* 73: 25–33.

Eckburg, P.B., Bik, E.M., Bernstein, C.N., et al. (2005). Diversity of the human intestinal microbial flora. *Science*, 308: 1635–1638.

Editorial (2010). Biodiversity 2010: The tip of the iceberg. *Nature Rev. Microbiol.,* 8S: 384.

Euzeby, J. www.bacterio.cict.fr.

Ghosh, A., Bhardwaj, M., Satyanarayana,T., et al. (2007). *Bacillus lehensis sp. nov.,* an alkalitolerant bacterium isolated from soil. *Int. J. Syst. Evol. Microbiol.,* 57: 238–242.

Ghosh, A., Dey, N. Bera, A., et al. (2010). Culture independent molecular analysis of bacterial communities in the mangrove sediment of Sundarban, India. *BMC Saline Systems,* 6(1): 1–11.

Ghosh, A., Maity, B., Chakrabarti, K. and Chattopadhyay, D.J. (2007). Bacterial diversity of East Calcutta wet land area: Possible identification of potential bacterial population for different biotechnological uses. *Microb. Ecol.*, 54: 452–459.

Ghosh, D.,Bal, B., Kashyap, V.K. and Pal, S. (2003). Molecular phylogenetic exploration of bacterial diversity in a bakreshwar (India) hot spring and culture of *Shewanella* – related thermophiles. *Appl. Env. Microbiol,* 69: 4332–4336.

Gill, S.R., Pop, M., Deboy, R.T., et al. (2006). Metagenomic analysis of the human distal gut microbiome. *Science,* 312: 1355–1359.

Handelsman, J. (2004). Metagenomics: Application of Genomics to Uncultured Microorganisms, *Microbiol. Mol. Biol. Rev.*, 68: 669–685. http://ijs.sgmjournals.org.

Huang, L.N., Zhou, H., Chen, Y.Q. et al. (2002). Diversity and structure of the archaeal community in the leachate of a full-scale recirculating landfill as examined by direct 16S rRNA gene sequence retrieval. *FEMS Microbiol Lett.,* 214: 235–240.

Huang, L.N., Zhu, S. and Qu, L.H. (2005). Molecular phylogenetic diversity of bacteria associated with the leachate of a closed municipal solid waste landfill. *FEMS Micobiol. Lett.,* 242: 297–303.

Hunter-Cavera, J.C. (1998). Evaluation of genetic diversity among *Pseudomonas citronellolis* strains isolated from oily sludge-contaminated sites. *Curr. Opinion Microbiol.,* 1: 278–285.

Joshi, A.A. , Kanekar , P.P. , Kelkar, A.S., et al. (2008). Cultivable bacterial diversity of alkaline Lonar lake, India. *Microb. Ecol.*, 55: 163–172.

Kaeberlein, T., Lewis, K. and Epstein S.S. (2002). Isolating "uncultivable" microorganisms in pure culture in a simulated natural environment. *Science,* 296: 1127–1129.

Kamra, D.N. (2005). Rumen microbial ecosystem. *Curr. Sci.*, 89: 124–135.

Kanekar, P.P., Nilegaonkar, S.S., Sarnaik, S.S., et al. (2002). Optimization of protease activity of alcaliphilic bacteria from an alkaline lake in India. *Bioresource Technology,* 85: 87–93.

Kanekar, P.P., Sarnaik S.S. and Kelkar, A.S. (1999). Bioremediation of phenol by aliphilic bacteria isolated from alkaline lake of Lonar, India. *J. Appl. Microbiol. (Suppl),* 85: 1285–1335.

Keller, M. and Hettich, R. (2009). Environmental proteomics: A paradigm shift in characterizing microbial activities at the molecular level. *Microbiol. Mol. Biol. Rev.,* 73: 62–70.

Kimura,H., Asada, R., Masta, A. and Naganuma, T. (2003). Distribution of microorganisms in the subsurface of the Manus basin hydrothermal vent field in Papua New Guinea. *Appl. Environ. Microbiol.*, 69: 644–648.

Kishore, K.H., Begum, Z., Pathan, A.A.K. et al. (2010). Paenibacillus glacialis sp. nov., isolated from the Kafni glacier of the Himalayas, India. Int. *J. Syst. Evol. Microbiol.*, 60: 1909–1913.

Komárek, J. and Hauer, T. (2011). CyanoDB. cz – On-line database of cyanobacterial genera. Word-wide electronic publication, Univ. of South Bohemia & Inst. of Botany ASCR, http:\\www.cyanodb.cz.

Krishnamurthi, S., Ph.D. Thesis, Jawaharlal Nehru University, New Delhi (2009).

Krishnamurthi, S., Bhattacharya, A., Mayiraj, S. et al. (2009). Description of *Paenisporosarcina quisquiliarum gen. nov., sp. nov.,* and reclassification of Sporosarcina macmurdoensis Reddy et al. 2003 as *Paenisporosarcina macmurdoensis comb. nov. Int. J. Syst. Evol. Microbiol.,* 59: 1364–1370.

Lipp, J.S., Morono, Y., Inagaki F. and Hinrichs, K.U. (2008) Significant contribution of Archaea to extant biomass in marine subsurface sediments. *Nature,* 454: 991–994.

Manikandan, M., Kannan, V. and Pasˇic, L. (2009). Optimization of growth medium for protease production by Haloferax Lucentensis VKMM 007 by response surface methodology. *World J. Microbiol. Biotechnol.,* 25: 1007–1117.

Manoharachary, C., Sridhar, K., Singh, R., et al. (2005). Fungal biodiversity: Distribution, conservation and prospecting of fungi from India. *Curr. Sci.,* 89: 58–71.

Mayilraj, S., Kroppenstedt, R.M., Suresh, K., et al. (2006). *Kocuria himachalensis sp. nov.,* an actinobacterium isolated from the Indian Himalayas. *Int. J. Syst. Evol. Microbiol.,* 56: 1971–1975.

Mayilraj, S., Prasad, G.S., Suresh, K., et al. (2005). *Planococcus stackebrandtii sp. nov.,* isolated from a cold desert of the Himalayas, India. *Int. J. Syst. Evol. Microbiol.* 55: 91–94.

Mayilraj, S., Saha, P., Suresh, K., et al. (2006). *Ornithinimicrobium kibberense sp. nov.,* isolated from the Indian Himalayas. *Int. J. Syst. Evol. Microbiol.,* 56: 1657–1661.

Mayilraj, S., Suresh, K., Kroppenstedt, R.M., et al. (2006). *Dietzia kunjamensis sp. nov.,* isolated from the Indian Himalayas. *Int. J. Syst. Evol. Microbiol.,* 56: 1667–1671.

Mayilraj, S., Suresh, K., Schumann, P., et al. (2006). *Agrococcus lahaulensis sp. nov.,* isolated from a cold desert of the Indian Himalayas. *Int. J. Syst. Evol. Microbiol.,* 56: 1807–1810.

Morris, R.M., Rappé, M.S., Connon, S.A., et al. (2002). SAR11 clade dominates ocean surface bacterioplankton communities. *Nature,* 420: 806–810.

Niederberger, T.D., McDonald, I.R., Hacker, A.L., et al. (2008). Microbial community composition in soils of Northern Victoria Land, Antarctica. *Environ. Microbiol.,* 10: 1713–1724.

Ohkuma, M. and Kudo, T. (1996). Phylogenetic diversity of the intestinal bacterial community in the termite Reticulitermes speratus. *Appl. Environ. Microbiol.* 62: 461–468.

Pace, N.R. (1997). A molecular view of microbial diversity and the biosphere. *Science,* 276: 734–740.

Palmisano, A.C., Maruscik, D.A. and Schwab, B.S. (1993). Enumeration of fermentative and hydrolytic micro-organisms from three sanitary landfills. *J. Gen. Microbiol.,* 139: 387–391.

Pindi, P.K., Kishore, K.H., Reddy, G.S.N., et al. (2009). Description of *Leifsonia kafniensis sp. nov.* and *Leifsonia antarctica sp. nov. Int. J. Syst. Evol. Microbiol.,* 59: 1348–1352.

Pourcher, A.M., Sutra, L., Hebe, I., Moguedet, G., et al. (2001). Enumeration and characterization of cellulolytic bacteria from refuse of a landfill. *FEMS Microbiol. Ecol.,* 34: 229–241.

Pradhan, S., Srinivas, T. N. R., Pindi, P.K., et al. (2010). Bacterial biodiversity from Roopkund Glacier, Himalayan mountain ranges, India. *Extremophiles,* 14: 377–395.

Rappe, M.S., Connon, S. Avergin, K.L. and Giovannoni. S.J. (2002). Cultivation of the ubiquitous SAR11 marine bacterioplankton clade. *Nature,* 418: 630–633.

Reddy, G.S.N., Prabagaran, S.R. and Shivaji S. (2008). Leifsonia pindariensis sp. nov., isolated from the Pindari glacier of the Indian Himalayas, and emended description of the genus Leifsonia. *Int. J. Syst. Evol. Microbiol.,* 58: 2229–2234.

Reddy, G.S.N., Pradhan, S., Manoram, R., et al. (2010). Cryobacterium psychrotolerans sp. nov., a novel psychrotolerant bacterium isolated from the China No. 1 glacier. *Int. J. Syst. Evol. Microbiol.,* 60: 866–869.

Reysenbach, A.L., Longnecker, K. and Kirshtein (2000). Novel bacterial and archaeal lineages from an in-situ growth chamber deployed at a Mid-Atlantic Ridge hydrothermal vent. J., *Appl. Environ. Microbiol.,* 66: 3798–3806.

Riesenfeld, C.S., Schloss, P.D. and Handelsman. (2004). Metagenomics: Genomic analysis of microbial communities. J., *Annu. Rev. Genet.,* 38: 525–552

Ruckmani, A. and Chakrabarti, T. (2010). Analysis of bacterial community composition of a spring water from the Western Ghats, India using culture dependent and molecular approaches. *Curr. Microbiol.,* (in press). doi: 10.1007/s00284-010-9663-9.

Rusch, D.B., Halpern, A.L., Sutton, G., et al. (2007) The sorcerer II global ocean sampling expedition: Northwest atlantic through eastern tropical pacific. *PLoS. Biol.,* 5: e77.

Saha, P. and Chakrabarti, T. (2006). *Emticicia oligotrophica gen. nov., sp. nov.,* a new member of the family 'Flexibacteraceae', phylum Bacteroidetes. *Int. J. Syst. Evol. Microbiol.,* 56: 991–995.

Saha, P. and Chakrabarti, T. (2006). A replacement name of the specific epithet aurantiaca in Micromonospora aurantiaca Sveshnikova et al. 1969 (Approved Lists 1980) and a proposal to treat the combination Micromonospora aurantiaca Sveshnikova et al. 1969 as a rejected name. Request for an opinion. *Int. J. Syst. Evol. Microbiol.,* 56: 1905–1906.

Saha, P. and Chakrabarti, T. (2006). *Flavobacterium indicum sp. nov.,* isolated from warm spring water in Assam, India. *Int. J. Syst. Evol. Microbiol.,* 56: 2617–2621.

Saha, P., Ph.D.Thesis, Jawaharlal Nehru University, New Delhi (2006).

Saha, P., Krishnamurthi, S., Bhattacharya, A. et al. (2010). *Fontibacillus aquaticus gen. nov., sp. nov.,* isolated from a warm spring, *Int. J. Syst. Evol. Microbiol,* 60: 422–428.

Saha, P., Krishnamurthi, S., Mayilraj, S., et al. (2005). *Aquimonas voraii gen. nov., sp. nov.,* a novel gammaproteobacterium isolated from a warm spring of Assam, India. *Int. J. Syst. Evol. Microbiol.,* 55: 1491–1495.

Saha, P., Mondal, A.K., Krishnamurthi, S., et al. (2005). *Paenibacillus assamensis sp. nov.,* a novel bacterium isolated from a warm spring in Assam, India. *Int. J. Syst. Evol. Microbiol.,* 55: 2577–2581.

Santelli, C.M., Orcutt, B.N., Banning, E., et al. (2008). Abundance and diversity of microbial life in ocean crust. *Nature,* 453: 653–657.

Sar, P., Dhal, P.K., Islam, E. and Kazy, S.K. (2007). Molecular assessment of microbial diversity and community structure at uranium mines of jaduguda, India. *Advanced Materials Research,* 20–21: 413.

Schleifer, K.H. (2004). Microbial diversity: Facts, problems and prospects., *System. Appl. Microbiol.,* 27: 3–9.

Shivaji, S., Chaturvedi, P., Begum, Z., et al. (2009). *Janibacter hoylei sp. nov., Bacillus isronensis* sp. nov. and Bacillus aryabhattai *sp. nov.,* isolated from cryotubes used for collecting air from the upper atmosphere. *Int. J. Syst. Evol. Microbiol.,* 59: 2977–2986.

Shivaji, S., Chaturvedi, P., Reddy, G.S.N., et al. (2005). *Pedobacter* himalayensis *sp. nov.,* from the Hamta glacier located in the Himalayan mountain ranges of India. *Int. J. Syst. Evol. Microbiol.,* 55: 1083–1088.

Shivaji, S., Chaturvedi, P., Suresh, K., et al. (2006). *Bacillus aerius* sp. nov., *Bacillus aerophilus* sp. nov., *Bacillus stratosphericus* sp. nov. and

Bacillus altitudinis sp. nov., isolated from cryogenic tubes used for collecting air samples from high altitudes. *Int. J. Syst. Evol. Microbiol.*, 56: 1465–1473.

Shivaji, S., Reddy, G.S., Aduri, R.P., Kutty, R. and Ravenschlaq, K. (2004). Bacterial diversity of a soil sample from Schirmacher Oasis, Antarctica. *Cell Mol. Biol. (Noisy-le-qrand).* 50: 525–535.

Singla, A.K., Mayilraj, S., Kudo, T., et al. (2005). *Actinoalloteichus spitiensis sp. nov.*, a novel actinobacterium isolated from a cold desert of the Indian Himalayas. *Int. J. Syst. Evol. Microbiol.*, 55: 2561–2564.

Staley, J.T. and Konopka, A. (1985). Measurement of in-situ activities of nonphotosynthetic microorganisms in aquatic and terrestrial habitats. *Annu. Rev. Microbiol.*, 39: 321–346.

Surakashi, V.P., Wani, A.A., Shouche, Y.S. and Ranade, D.R. (2007). Phylogenetic analysis of methanogenic enrichment cultures obtained from Lonar lake in India: Isolation of *Methanocalculus sp.* and *Methanoculleus* sp. *Microbial. Ecol.*, 54: 697–704.

Thakker, C.D. and Ranade, D.R. (2002). An alkalophilic methanosarcina isolated from Lonar crater, *Curr. Sci.*, 82: 455–458.

Tiedje, J. and Donohue, T. (2008). Microbes in the energy grid. *Science*, 320: 985–987.

Tilak, K.V.B.R., Ranganayaki, N., Pal, K.K. and De, R., et al. (2005). Diversity of plant growth and soil health supporting bacteria. *Curr. Sci.*, 89: 136–150.

Torsvik, V., Salte, K., Sørheim, R., et al. (1990). Comparison of phenotypic diversity and DNA heterogeneity in a population of soil bacteria. *Appl. Environ. Microbiol.*, 56: 776–781.

Tyson, G.W., Chapman, J., Hugenholtz, P., et al. (2004). Community structure and metabolism through reconstruction of microbial genomes from the environment. *Nature,* 428: 37–43.

Vartoukian, S.R., Palmer, R.M. and Wade, W.G. (2010). Strategies for culture of 'unculturable' bacteria. *FEMS Microbiol. Lett.*, 309: 1–7.

Venter, J.C., Karin Remington, K. and Heidelberg, J.F., et al. (2004). Environmental genome shotgun sequencing of the Sargasso Sea. *Science,* 304: 66–74.

VerBerkmos, C.V., Denef, V.J., Hettich, R.L., Banfield, J.F. et al. (2009). Systems biology: Functional analysis of natural microbial consortia using community proteomics. *Nat. Rev. Microbiol.,* 7: 196–205.

Wani, A.A., Surakasi V.P., Siddharth, J. et al. (2006). Molecular analyses of microbial diversity associated with the Lonar soda Lake in India: An impact crater in a basalt area. *Res. Microbiol.*, 157: 928–937.

Whitman, W.B., Coleman, D.C. and Weibe, W.J. (1998). Prokaryotes: The unseen majority. *Proc. Natl. Acad. Sci. USA*, 95: 6578–6583.

Wilmes, P. and Bond, P.L. (2006). Metaproteomics: Studying functional gene expression in microbial ecosystems. *Trends Microbiol.*, 14: 92–97.

Xin, Wu., Wanyan, Xi., Wenjin, Ye. And Hong Yang. (2007). Bacterial community composition of a shallow hypertrophic fresh water lake in China, revealed by 16S rRNA sequences. *FEMS Microbiology Ecology.* (2007); 61: 85–96.

Yooseph, S., Sutton, G., Rusch, D.B., et al. (2007). The sorcerer II global ocean sampling expedition: Expanding the universe of protein families. *PLoS Biol.*, 5: e16.

Novel taxa of Archaea described from India

1. *Methanobrevibacter acididurans* sp. nov. Savant et al. 2002
2. *Methanocaldococcus indicus* sp. nov. L'Haridon et al. 2003
3. *Methanolobus bombayensis* sp. nov. Kadam et al. 1994

References

Kadam, P.C., Ranade, D.R., Mandelco, L. and Boone, D.R. (1994). Isolation and characterization of *Methanolobus bombayensis* sp. nov., a methylotrophic methanogen that requires high concentrations of divalent cations. *Int. J. Syst. Bacteriol.*, 44: 603–607.

L'haridon, S., Reysenbach, A.L., Banta, A., Messner, P., Schumann, P., Stackebrandt, E. and Jeanthon, C. (2003). *Methanocaldococcus indicus* sp. nov., a novel hyperthermophilic methanogen isolated from the Central Indian Ridge. *Int. J. Syst. Evol. Microbiol.*, 53: 1931–1935.

Savant, D.V., Shouche, Y.S., Prakash, S. and Ranade, D.R. (2002). *Methanobrevibacter acididurans* sp. nov., a novel methanogen from a sour anaerobic digester. *Int. J. Syst. Evol. Microbiol.*, 52: 1081–1087.

Archaea recorded from or available in India:

	Genera	Number of species
1.	*Acidianus*	1
2.	*Haloarcula*	3
3.	*Halobacillus*	1
4.	*Halobacterium*	4
5.	*Halococcus*	4
6.	*Haloferax*	2
7.	*Halogeometricum*	1
8.	*Halorubrum*	1
9.	*Methanobacterium*	6
10.	*Methanobrevibacterium*	2
11.	*Methanocalculus*	1
12.	*Methanococcoides*	2
13.	*Methanococcus*	2
14.	*Methanoculeus*	1
15.	*Methanogenium*	1
16.	*Methanohalobium*	1
17.	*Methanohalophilus*	1
18.	*Methanolobus*	2
19.	*Methanomicrobium*	1
20.	*Methanoplanus*	1
21.	*Methanosarcina*	4
22.	*Methanospirilum*	1
23.	*Methanothermobacter*	1
24.	*Methanothrix*	1
25.	*Natranobacterium*	2
26.	*Sulfolobus*	3

Novel Taxa of Bacteria described from India.

(references arranged at the end according to numerical order of the taxa)

1. *Acetobacter nitrogrifigens* sp. nov. Dutta and Gachhui 2006
2. *Actinoalleteichus spitiensis* sp. nov. Singla et al. 2005
3. *Actinomadura oligospora* sp. nov. Mertz and Yao 1986
4. *Aeromonas culicicola* sp. nov. Pidiyar et al. 2002
5. *Aeromonas sharman* sp. nov. Saha and Chakrabarti 2006

6. *Agrococcus carbonis* sp. nov. Dhanjal et al. 2010
7. *Agrococcus lahaulensis* sp. nov. Mayilraj et al. 2006
8. *Allochromatium phaeobacterium* sp. nov. Srinivas et al. 2009
9. *Allochromatium renukae* sp. nov. Anil Kumar et al. 2008
10. *Altererythobacter indicus* sp. nov. Ramesh Kumar et al. 2008
11. *Amycolatopsis decaplanina* sp. nov. Wink et al. 2004
12. *Amycolatopsis eurytherma* sp. nov. Kim et al. 2002
13. *Amycolatopsis sacchari* sp. nov. Goodfellow et al.2001
14. *Aquaspirillum bengal* sp. nov. Kumar et al. 1974
15. *Aquimonas voraii* gen. nov., sp. nov. Saha et al. 2005
16. *Arthrobacter antarcticus* sp. nov. Pindi et al. 2009
17. *Arthrobacter flavus* sp. nov. Reddy et al. 2000
18. *Arthrobacter gangotriensis* sp. nov. Gupta et al. 2004
19. *Arthrobacter kerguelensis* sp. nov. Gupta et al. 2004
20. *Arthrobacter mysorens* sp. nov. Nand and Rao 1972
21. *Arthrobacter roseus* sp. nov. Reddy et al. 2002
22. *Bacillus aerius* sp. nov. Shivaji et al. 2006
23. *Bacillus aerophilus* sp. nov. Shivaji et al. 2006
24. *Bacillus altitudinis* sp. nov. Shivaji et al. 2006
25. *Bacillus arsenicus* sp. nov. Shivaji et al. 2005
26. *Bacillus aryabhattai* sp. nov. Shivaji et al. 2009
27. *Bacillus cecembensis* sp. nov. Reddy et al. 2008
28. *Bacillus indicus* sp. nov. Suresh et al. 2004
29. *Bacillus isroensis* sp. nov. Shivaji et al. 2009
30. *Bacillus lehensis* sp. nov. Ghosh et al. 2007
31. *Bacillus methylotrophicus* sp. nov. Madhaiyan et al. 2009
32. *Bacillus okhensis* sp. nov. Nowlan et al. 2006
33. *Bacillus stratosphaericus* sp. nov. Shivaji et al. 2006
34. *Bhargavaea cecembiensis* gen. nov., sp. nov. Manorama et al. 2009
35. *Blastochloris gulmargensis* sp. nov. Venkata Ramana et al. 2010
36. *Bosea thiooxidans* gen. nov. sp. nov. Das et al. 1996
37. *Brevibacterium oceani* sp. nov. Bhadra et al. 2008
38. *Caloramator indicus* sp. nov. Chrisostomos et al. 1996
39. *Catenuloplanes atrovinosus* sp. nov. Tamura et al. 995

40. *Catenuloplanes indicus* sp. nov. Tamura et al. 1995
41. *Catenuloplanes niger* sp. nov. Tamura et al. 1995
42. *Cesiribacter andamanensis* gen. nov., sp. nov. Srinivas et al. 2010
43. *Chainia antibiotica* gen. nov., sp. nov. Thirumalachar 1955
44. *Chainia flava* sp. nov. Thirumalachar and Sukapure 1964
45. *Chainia fumigata* sp. nov. Thirumalachar et al. 1966
46. *Chainia minutisclerotica* sp. nov. Thirumalachar et al. 1965
47. *Chainia nigra* sp. nov. Thirumalachar 1955
48. *Chainia olivacea* sp. nov. Thirumalachar and Sukapure 1964
49. *Chainia* poonensis sp. nov. Thirumalachar 1960
50. *Chainia purpurogena* sp. nov. Thirumalachar and Sukapure 1964
51. *Chainia rosea* sp. nov. Thirumalachar 1966
52. *Chainina rubra* sp. nov. Thirumalachar 1955
53. *Chelatococcus shambhunathii* sp. nov. Panday and Das 2010
54. *Chryseobacterium arthrosphaerae* sp. nov. Kaempfer et al. 2009
55. *Chryseobacterium culicis* sp. nov. Kaempfer et al. 2009
56. *Chryseomicrobium imtechense* gen. nov., sp. nov. Arora et al. 2010
57. *Clostridium nitrophenolicum* sp. nov. Suresh et al. 2007
58. *Clostridium schirmacherense* sp. nov. Alam et al. 2006
59. *Comamonas thiooxidans* sp. nov. Kanwar et al. 2010
60. *Cryobacterium rupkundense* sp. nov. Reddy et al. 2010
61. *Cyclobacterium qasimii* sp. nov. Shivaji et al. 2010
62. *Deinococcus indicus* sp. nov. Suresh et al. 2004
63. *Deinococcus mumbaensis* sp. nov. Shashidhar and Bandekar 2006
64. *Deinococcus piscis* sp. nov. Shashidhar and Bandekar 2009
65. *Desulfovibrio psychrotolerans* sp. nov. Sasi Jyothsna et al. 2008
66. *Desulfurispirillum indicum* sp. nov. Rauschenbach et al. 2010
67. *Devosia albogilva* sp. nov. Verma et al. 2009
68. *Devosia chinhatensis* sp. nov. Kumar et al. 2008
69. *Devosia croci*na sp. nov. Verma et al. 2009
70. *Devosia neptuniae* sp. nov. Rivas et al. 2003
71. *Dietzia kunjamensis* sp. nov. Mayilraj et al. 2006
72. *Dyadobacter hamtensis* sp. nov. Chaturvedi et al. 2005
73. *Effluvibacter roseus* gen. nov., sp. nov. Suresh et al. 2006

74. *Emticicia oligotrophica* gen. nov., sp. nov. Saha and Chakrabarti 2006
75. *Exiguobacterium indicum* sp. nov. Chaturvedi and Shivaji 2006
76. *Exiguobacterium soli* sp. nov. Chaturvedi et al. 2008
77. *Flavobacterium indicum* sp. nov. Saha and Chakrabarti. 2006
78. *Flavobacterium lindanitolerans* sp. nov. Jit et al. 2008
79. *Flavobaterium glycines* sp. nov. Madhaiyan et al. 2009
80. *Fontibacillus aquaticus* gen. nov., sp. nov. Saha et al. 2010
81. *Georgenia soli* sp. nov. Kämpfer et al. 2010
82. *Gluconoacetobacter combuchae* sp. nov. Datta and Gachhui 2007
83. *Gulbenkiana indiensis* sp. nov. Jyoti et al. 2009
84. *Halochromatium roseum* sp. nov. Anil Kumar et al. 2007
85. *Halomonas glaciei* sp. nov. Reddy et al. 2003 ep nvp
86. *Ignatzschineria indica* sp. nov. Gupta et al. 2010
87. *Ignatzschineria ureaclastica* sp. nov. Gupta et al. 2010
88. *Indibacter alkaliphilus* gen. nov., sp. nov. Anil Kumar et al. 2010
89. *Janibaacter hoylei* sp. nov. Shivaji et al. 2009
90. *Kineococcus aurentiacus* gen. nov., sp. nov. Yokota et al. 1993
91. *Kitasatospora sampliensis* sp. nov. Mayilraj et al. 2005
92. *Kocuria himachalensis* sp. nov. Mayilraj et al. 2006
93. *Kocuria polaris* sp. nov. Reddy et al. 2003
94. *Lactobacillus delbrueckii* subsp. *indicus* subsp. nov. Dellaglio et al. 2005
95. *Leifsonia antarctica* sp. nov. Pindi et al. 2009
96. *Leifsonia aurea* sp. nov. Reddy et al. 2003
97. *Leifsonia kafniensis* sp. nov. Pindi et al. 2009
98. *Leifsonia pindariensis* sp. nov. Reddy et al. 2008
99. *Leifsonia rubra* sp. nov. Reddy et al. 2003
100. *Leifsonia soli* sp. nov. Madhaiyan et al. 2010
101. *Leucobacter iarius* sp. nov. Somvanshi et al. 2007
102. *Lysobacter defluvii* sp. nov. Yassin et al. 2007
103. *Mangrovibacter plantisponsor* gen. nov., sp. nov. Rameshkumar et al. 2010
104. *Maricaulis parjimensis* sp. nov. Abraham et al. 2002
105. *Marichromatium bheemlicum* sp. nov. Anil Kumar et al. 2007
106. *Marichromatium fluminis* sp. nov. Sucharita et al. 2010

107. *Marichromatium indicum* sp. nov. Arunasri et al. 2005
108. *Marinobacter maritimus* sp. nov. Shivaji et al. 2005
109. *Marinomonas polaris* sp. nov. Gupta et al. 2006
110. *Marinomonas ushuaiensis* sp. nov. Prabagaran et al. 2005
111. *Mesorhizobium thiogangeticum* sp. nov. Ghosh and Roy 2006
112. *Microbacterium arthrosphaerae* sp. nov. Kaempfer et al. 2010
113. *Microbacterium azadirachtae* sp. nov. Madhaiyan et al. 2009
114. *Microbacterium indicum* sp. nov. Shivaji et al. 2007
115. *Microbacterium linditolerans* sp. nov. Lal et al. 2009
116. *Microbispora indica* sp. nov. Rao et al. 1987
117. *Microbispora karnakatensis* sp. nov. Rao et al. 1987
118. *Micrococcus niistensis* sp. nov. Dastager et al. 2010
119. *Mucilaginibacter gossypii* sp. nov. Madhaiyan et al. 2010
120. *Mucilaginibacter gossypiicola* sp. nov. Madhaiyan et al. 2010
121. *Mycoplasma indiense* sp. nov. Hill 1993
122. *Nitritalea halalkaliphila* gen. nov., sp. nov. Anil Kumar et al. 2009
123. *Novosphingobium mathurense* sp. nov. Gupta et al. 2009
124. *Novosphingobium panipatense* sp. nov. Gupta et al. 2009
125. *Oceanisphaera arctica* sp. nov. 2010 Srinivas et al.
126. *Oceanotoga teriensis* sp. nov. Jayasinghearachchi and Lal 2010
127. *Ochrobactrum oryzaea* sp. nov. Tripathi et al. 2006
128. *Ornithinimicrobium kibberense* sp. nov. Mayiraj et al. 2006
129. *Paenibacillus assamensis* sp. nov. Saha et al. 2005
130. *Paenibacillus glacialis* sp. nov. Kishore et al. 2009
131. *Paenibacillus tezpurensis* sp. nov. Rai et al. 2010
132. *Paenisporosarcina quisquiliarum* gen. nov., sp. nov. Krishnamurthi et al. 2009
133. *Paracoccus bengalensis* sp. nov. Ghosh et al. 2006
134. *Pedobacter himalayensis* sp. nov. Shivaji et al. 2005
135. *Phaeospirillum chandramohanii* sp. nov. Anil Kumar et al. 2009
136. *Phaeospirillum oryzae* sp. nov. Lakshmi et al. 2010
137. *Phaeovibrio sulfidiphilus* gen. nov., sp. nov. Lakshmi et al. 2010
138. *Planococcus antarcticus* sp. nov. Reddy et al. 2002
139. *Planococcus columbae* sp. nov. Suresh et al. 2007

140. *Planococcus maitriensis* sp. nov. Alam et al. 2003
141. *Planococcus psychrophilus* sp. nov. Reddy et al. 2002
142. *Planococcus stackebrandtii* sp. nov. Mayilraj et al. 2005
143. *Planomonospora sphaerica* sp. nov. Mertz 1994
144. *Pontibacter niistensis* sp. nov. Dastager et al. 2010
145. *Prosthecochloris indica* sp. nov. Anil Kumar et al. 2009
146. *Providencia vermicola* sp. nov. Somvanshi et al. 2006
147. *Pseudomonas antarctica* sp. nov. Reddy et al. 2004
148. *Pseudomonas delhiensis* sp. nov. Prakash et al. 2007
149. *Pseudomonas indica* sp. nov. Pandey et al. 2002
150. *Pseudomonas indoloxydans* sp. nov. Manickam et al. 2008
151. *Pseudomonas meridiana* sp. nov. Reddy et al. 2004
152. *Pseudomonas panipatensis* sp. nov. Gupta et al. 2008
153. *Pseudomonas proteolytica* sp. nov. Reddy et al. 2004
154. *Pseudonocardia antarctica* sp. nov. Prabahar et al. 2004
155. *Psychrobacter adeliensis* sp. nov. Shivaji et al. 2005
156. *Psychrobacter aquaticus* sp. nov. Shivaji et al. 2005
157. *Psychrobacter salsus* sp. nov. Shivaji et al. 2005
158. *Psychrobacter vallis* sp. nov. Shivaji et al. 2005
159. *Rhizobium rosettiformans* sp. nov. Kaur et al. 2010
160. *Rhodobacter aestuarii* sp. nov. Venkata Ramana et al. 2009
161. *Rhodobacter changlensis* sp. nov. Anil Kumar et al. 2007
162. *Rhodobacter johrii* sp. nov. Girija et al. 2009
163. *Rhodobacter maris* sp. nov. Venkata Ramana et al. 2008
164. *Rhodobacter megalophilus* sp. nov. Arunasri et al. 2008
165. *Rhodobacter ovatus* sp. nov. Srinivas et al. 2008
166. *Rhodobacter vinaykumarii* sp. nov. Srinivas et al. 2007
167. *Rhodobium gokarnense* sp. nov. Srinivas et al. 2007
168. *Rhodococcus imtechensis* sp.nov. Ghosh et al. 2006
169. *Rhodococcus kroppenstedtii* sp. nov. Mayilraj et al. 2005
170. *Rhodoplanes pokkaliisoli* sp. nov. Lakshmi et al. 2009
171. *Rhodospirillum sulfurexigens* sp. nov. Anil Kumar et al. 2008
172. *Rhodovulum imhoffii* sp. nov. Srinivas et al. 2007
173. *Rhodovulum kholense* sp. nov. Anil Kumar et al. 2008

174. *Rhodovulum lacipunicei* sp. nov. Chakravarthy et al. 2009
175. *Rhodovulum marinum* sp. nov. Srinivas et al. 2006
176. "Candidatus *Rickettsia kellyi*," Rolain et al. 2006
177. *Roseospira goensis* sp. nov. Chakravarthy et al. 2007
178. *Roseospira vishakhapatnemensis* sp. nov. Chakravarthy et al. 2007
179. *Rubrivivax benzoatilyticus* sp. nov. Raman et al. 2006
180. *Saccharomonospora saliphila* sp. nov. Syed et al. 2008
181. *Salmonella enterica* subsp. *indica* (Le Minor et al. 1987) Le Minor and Popoff 1987,
182. *Selenomonas lipolytica* sp. nov. Dighe et al. 1998
183. *Serratia ureilytica* sp. nov. Bhadra et al. 2005
184. *Shewanella chilikensis* sp. nov. Sucharita et al. 2009
185. *Shewanella fodinae* sp. nov. Sravan Kumar et al. 2009
186. *Sphingobacterium antarcticus* sp. nov. Shivaji et al. 1992 (corrig. *S.antarcticum* 1998)
187. *Sphingobium abikonense* sp. nov. Kumari et al. 2009
188. *Sphingobium chinhatense* sp. nov. Dadhwal et al. 2009
189. *Sphingobium fuliginis* sp. nov. Prakash and Lal 2006
190. *Sphingobium indicum* sp. nov. Pal et al. 2005
191. *Sphingobium lactosutens* sp. nov. Kumari et al. 2009
192. *Sphingobium quisquiliarum* sp. nov. Bala et al. 2010
193. *Sphingobium ummariense* sp.nov. Singh and Lal 2009
194. *Sphingomonas histidinilytica* sp. nov. Nigam et al. 2010
195. *Sphingopyxis ummariensis* sp. nov. Sharma et al. 2010
196. *Sporosarcina macmurdoensis* sp. nov. Reddy et al. 2003 (*Paenisporosarcina macmurdoensis* comb. nov. Krishnamurthi et al. 2009)
197. *Stappia indica* sp. nov. Lai et al. 2010
198. *Streptoalloteichus hindustanus* gen. nov., sp. nov. Tomita et al. 1987
199. *Streptomyces albofaciens* sp. nov. Thirumalachar and Bhatt 1963
200. *Streptomyces anandii* sp. nov. Batra and Bajaj 1965
201. *Streptomyces champavathii* sp. nov. Uma and Narashimha Rao 1957
202. *Streptomyces deccanensis* sp. nov. Dastager et al. 2008
203. *Streptomyces gulbargensis* sp. nov. Dastager et al. 2007
204. *Streptomyces himastatinicus* sp. nov. Kumar and Goodfellow 2008

205. *Streptomyces hyderabadensis* sp. nov Reddy et al. 2010
206. *Streptomyces kashmirensis* sp. nov. (Gupta and Chopra 1963) Witt and Stackebrandt 1991, comb. nov.
207. *Streptomyces katrae* sp. nov. Gupta and Chopra 1963
208. *Streptomyces kavutarensis* sp. nov. Ellaiah et al. 2003-2004
209. *Streptomyces osmaniensis* sp. nov. Reddy et al. 2009
210. *Streptomyces prasinosporus* sp. nov. Tresner et al. 1966
211. *Streptomyces regensis* sp. nov. Gupta et al. 1963
212. *Streptomyces thermonitrificans* sp. nov. Desai and Dhala 1967
213. *Streptomyces tritolerans* sp. nov. Dastager et al. 2009
214. *Streptosporangium indianense* sp. nov. corrig. Gupta 1965 (*Streptomyces indiaensis* Kudo and Seino 1987, comb. nov.)
215. *Streptoverticillium acido carbonicum* sp. nov. Lakshmi 1993
216. *Streptoverticillium cinnamoneum* subsp. *albosporum* subsp. nov. Thirumalachar 1968
217. *Streptoverticillium cinnamoneum* subsp. *lanosum* subsp. nov. Thirumalachar 1968
218. *Streptoverticillium cinnamoneum* subsp. *sparsum* subsp. nov. Thirumalachar 1968
219. *Swaminathania salitolerans* gen. nov., sp. nov. Loganathan and Nair 2004
220. *Tetrathiobacter kashmirensis* gen. nov., sp. nov. Ghosh et al. 2005
221. *Thiohalocapsa marina* sp. nov. Anil Kumar et al. 2009
222. *Thiomonas bhubaneswarensis* sp. nov. Panda et al. 2009
223. *Thiophaeococcus mangrovi* gen. nov., sp. nov. Anil Kumar et al. 2008
224. *Thiorhodococcus bheemlicus* sp. nov. Anil Kumar et al. 2007
225. *Thiorhodococcus kakinadensis* sp. nov. Anil Kumar et al. 2007
226. *Thiorhodococcus modestalkaliphilus* sp. nov. Sucharita et al. 2010
227. *Vibrio portersiae* sp. nov. Rameshkumar et al. 2008
228. *Vibrio rhizosphaerae* sp. nov. Rameshkumar and Nair 2007
229. *Vibrio mangrovi* sp. nov. Ramesh Kumar et al. 2010
230. *Xenorhabdus indica* sp. nov. Somvanshi et al. 2009
231. *Yaniella fodinae* sp. nov. Dhanjal et al. 2010

References

Abraham, W.R., Strömpl, C., Bennasar, A., Vancanneyt, M., Snauwaert, C., Swings, J., Smit, J. and Moore, E.R.B. (2002). Phylogeny of *Maricaulis* Abraham et al. 1999 and proposal of *Maricaulis virginensis* sp. nov., *M. parjimensis* sp. nov., *M. washingtonensis* sp. nov., and *M. salignorans* sp. nov. *Int. J. Syst. Evol. Microbiol.*, 52: 2191–2201.

Alam, S.I., Dixit, A., Reddy, G.S.N., Dube, S., Palit, M., Shivaji, S. and Singh, L. (2006). *Clostridium schirmacherense* sp. nov., an obligately anaerobic, proteolytic, psychrophilic bacterium isolated from lake sediment of Schirmacher Oasis, Antarctica. *Int. J. Syst. Evol. Microbiol.*, 56: 715–720.

Alam, S.I., Singh, L., Dube, S., Reddy, G.S.N. and Shivaji, S. (2003). Psychrophilic *Planococcus maitriensis* sp. nov. from Antarctica. *Syst. Appl. Microbiol.*, 26: 505–510.

Anil Kumar, P., Aparna, P., Srinivas, T.N.R., Sasikala, C. and Ramana, C.V. (2008). *Rhodospirillum sulfurexigens* sp. nov., a phototrophic alphaproteobacterium requiring a reduced sulfur source for growth. *Int. J. Syst. Evol. Microbiol.*, 58: 2917–2920.

________. (2008). *Rhodovulum kholense* sp. nov., *Int. J. Syst. Evol. Microbiol.*, 58: 1723–1726.

Anil Kumar, P., Sasi Jyothsna, T.S., Srinivas, T.N.R., Sasikala, C., Ramana, C.V. and Imhoff, J.F. (2007). *Marichromatium bheemlicum* sp. nov., a non-diazotrophic, photosynthetic gammaproteobacterium from a marine aquaculture pond. *Int. J. Syst. Evol. Microbiol.*, 57: 1261–1265.

________. (2007). Two novel species of marine phototrophic *Gammaproteobacteria*: *Thiorhodococcus bheemlicus* sp. nov. and *Thiorhodococcus kakinadensis* sp. nov., *Int. J. Syst. Evol. Microbiol.*, 57: 2458–2461.

Anil Kumar, P., Srinivas, T.N.R., Madhu, S., Manorama, R. and Shivaji, S. (2010). *Indibacter alkaliphilus* gen. nov., sp. nov., an alkaliphilic bacterium isolated from a haloalkaline lake. *Int. J. Syst. Evol. Microbiol.*, 60: 721–726.

Anil Kumar, P., Srinivas, T.N.R., Sasikala, C. and Ramana, C.V. (2008). *Allochromatium renukae* sp. nov. *Int. J. Syst. Evol. Microbiol.*, 58: 404–407.

________. (2007). *Halochromatium roseum* sp. nov., a non-motile phototrophic gammaproteobacterium with gas vesicles, and emended description of the genus *Halochromatium*. *Int. J. Syst. Evol. Microbiol.*, 57: 2110–2113.

________. (2007). *Rhodobacter changlensis* sp. nov., a psychrotolerant, photo-trophic alphaproteobacterium from the Himalayas of India. *Int. J. Syst. Evol. Microbiol.*, 57: 2568–2571.

Anil Kumar, P., Srinivas, T.N.R., Sasikala, C., Ramana, C.V. and Imhoff, J.F. (2008). *Thiophaeococcus mangrovi* gen. nov., sp. nov., a photosynthetic, marine gammaproteobacterium isolated from the Bhitarkanika mangrove forest of India. *Int. J. Syst. Evol. Microbiol.*, 58: 2660–2664.

Anil Kumar, P., Srinivas, T.N.R., Takaichi, S., Maoka, T., Sasikala, C. and Ramana, C.V. (2009). *Phaeospirillum chandramohanii* sp. nov., a phototrophic alphaproteobacterium with carotenoid glycosides. *Int. J. Syst. Evol. Microbiol.*, 59: 2089–2093.

Anil Kumar, P., Srinivas, T.N.R., Takaichi, Thiel, V., Tank, M., Sasikala, C., Ramana, C.V. and Imhoff, J.F. (2009). *Thiohalocapsa marina* sp. nov., from an Indian marine aquaculture pond. *Int. J. Syst. Evol. Microbiol.*, 59: 2333–2338.

Anil Kumar, P., Naga Radha Srinivas, T., Sasikala, C., Venkata Ramana, C., Süling, J, and Imhoff, J. (2009). *Prosthecochloris indica* sp. nov., a novel green sulfur bacterium from a marine aquaculture pond, Kakinada, India. *J. Gen. Appl. Microbiol.*, 55(2): 163–169.

Anil Kumar, A., Srinivas, T. N. R., Pavan Kumar, P. , Madhu, S. and Shivaji, S. (2009). *Nitritalea halalkaliphila* gen. nov., sp. nov., a novel alkaliphilic bacterium of the family '*Cyclobacteriaceae*', phylum '*Bacteroidetes*'. *Int. J. Syst. Evol. Microbiol.*, (in press). doi: 10.1099/ijs.0.020230-0.

Arora, P.K., Chauhan, A., Pant, B., Suresh, K., Mayilraj, S., and Jain, R.K. (2010). Description of *Chryseomicrobium imtechense* gen. nov., sp. nov., a new member of the family *Planococcaceae Int. J. Syst. Evol. Microbiol.*, (in press). doi: 10.1099/ijs.0.023184-0.

Arunasri, K., Sasikala, C., Ramana, C.V., Süling, J. and Imhoff, J.F. (2005). *Marichromatium indicum* sp. nov., a novel purple sulfur gammaproteobacterium from mangrove soil of Goa, India. *Int. J. Syst. Evol. Microbiol.*, 55: 673–679.

Arunasri, K., Venkata, Ramana, V., Spröer, C., Sasikala, C. and Ramana, C.V. (2008). *Rhodobacter megalophilus* sp. nov., a phototroph from the Indian Himalayas possessing a wide temperature range for growth. *Int. J. Syst. Evol. Microbiol.*, 58: 1792–1796.

Bala, K., Sharma, P. and Lal, R. (2010). *Sphingobium quisquiliarum* sp. nov., a hexachlorocyclohexane (HCH)-degrading bacterium isolated from an HCH-contaminated soil. *Int. J. Syst. Evol. Microbiol.*, 60: 429–433.

Batra, S.K. and Bajaj, B.S. (1965) *Streptomyces anandii* – a new species of *Streptomyces* isolated from soil. *Indian Journal of Experimental Biology*, 3: 240–242.

Bhadra, B., Raghukumar, C., Pindi, P.K. and Shivaji, S. (2008). *Brevibacterium oceani* sp. nov., isolated from deep-sea sediment of the Chagos Trench, Indian Ocean. *Int. J. Syst. Evol. Microbiol.*, 58: 57–60.

Bhadra, B., Roy, P. and Chakraborty, R. (2005). *Serratia ureilytica* sp. nov., a novel urea-utilizing species. *Int. J. Syst. Evol. Microbiol.*, 55: 2155–2158.

Chakravarthy, S.K., Sucharitha, K., Sasikala, C. and Ramana, C.V. (2009). *Rhodovulum lacipunicei* sp. nov., an obligate sulfide-demanding phototrophic alphaproteobacterium isolated from a purple pond in India. *Int. J. Syst. Evol. Microbiol.*, 59: 1615–1619.

Chaturvedi, P. and Shivaji, S. (2006). *Exiguobacterium indicum* sp. nov., a psychrophilic bacterium from the Hamta glacier of the Himalayan mountain ranges of India. *Int. J. Syst. Evol. Microbiol.*, 56: 2765–2770.

Chaturvedi, P., Prabahar, V., Manorama, R., Kumar Pindi, P., Bhadra, B., Begum, Z. and Shivaji, S. (2008). *Exiguobacterium soli* sp. nov., a psychrophilic bacterium from the McMurdo Dry Valleys, Antarctica. *Int. J. Syst. Evol. Microbiol.*, 58: 2447–2453.

Chaturvedi, P., Reddy, G.S.N. and Shivaji, S. (2005). *Dyadobacter hamtensis* sp. nov., from Hamta glacier, located in the Himalayas, India. *Int. J. Syst. Evol. Microbiol.*, 55: 2113–2117.

Chrisostomos, S., Patel, B.K.C., Dwivedi, P.P. and Denman, S.E. (1996). *Caloramator indicus* sp. nov., a new thermophilic anaerobic bacterium isolated from the deep-seated nonvolcanically heated waters of an Indian artesian aquifer. *Int. J. Syst. Bacteriol.*, 46: 497–501.

Dadhwal, M., Jit, S., Kumari, H. and Lal, R. (2009). *Sphingobium chinhatense* sp. nov., a hexachlorocyclohexane (HCH)-degrading bacterium isolated from an HCH dumpsite. *Int. J. Syst. Evol. Microbiol.*, 59: 3140–3144.

Das, S.K., Mishra, A.K., Tindall, B.J., Rainey, F.A. and Stackebrandt, E. (1996). Oxidation of thiosulfate by a new bacterium, *Bosea thiooxidans* (strain BI-42) gen. nov., sp. nov., analysis of phylogeny based on chemotaxonomy and 16S ribosomal DNA sequencing. *Int. J. Syst. Bacteriol.*, 46: 981–987.

Dastager, G.S., Agasar, D., Kim, C.J., Li, W.J., Lee, J.C., Park, D.J., Xu, L.H., Tian, X.P. and Jiang, C.L. (2007). *Streptomyces tritolerans* sp. nov., a novel actinomycete isolated from soil in Karnataka, India. *Antonie Van Leeuwenhoek*, 92: 391–397.

Dastager, S.G., Kim, C.J., Lee, J.C., Agasar, D., Park, D.J. and Li, W.J. (2008). *Streptomyces deccanensis* sp. nov., an alkaliphilic species isolated from soil. *Int. J. Syst. Evol. Microbiol.*, 58: 1089–1093.

Dastager, S.G., Li, W.J., Agasar, D., Sulochana, M.B., Tang, S.K., Tian, X.P. and Zhi, X.Y. (2007). *Streptomyces gulbargensis* sp. nov., isolated from soil in Karnataka, India. *Antonie Van Leeuwenhoek*, 91: 99–104.

Dastager, S.G., Raziuddin, Q.S., Deepa, C.K., Li, W.J. and Pandey, A. (2010). *Pontibacter niistensis* sp. nov., isolated from forest soil in India. *Int. J. Syst. Evol. Microbiol.*, (in press). doi: 10.1099/ijs.0.019935-0.

Dastager, S.G., Raziuddin, S., Deepa, C.K., Tang, S.K., Li, W.J. and Pandey, A. (2010). *Micrococcus niistensis* sp. nov., isolated from forest soil, India. *Int J. Syst. Evol. Microbiol.*, (in press). doi: 10.1099/ijs.0.023382-0.

Dellaglio, F., Felis, G.E., Castioni, A., Torriani, S. and Germond, J.E. (2005). *Lactobacillus delbrueckii* subsp. *indicus* subsp. nov., isolated from Indian dairy products. *Int. J. Syst. Evol. Microbiol.*, 55: 401–404.

Desai, A.J. and Dhala, S.A. (1967). *Streptomyces thermonitrificans* sp. nov., a thermophilic streptomycete. *Antonie van Leeuwenhoek Journal of Microbiology and Serology*, 33: 137–144.

Dhanjal, S., Kaur, I., Suresh, K., Schuman, P., Cameotra, S.S., Pukall, R., Klenk, H.P. and Mayilraj, S. (2010). *Agrococcus carbonis* sp. nov., a novel species of the genus *Agrococcus* isolated from coal mine soil sample. *Int. J. Syst. Evol. Microbiol.*, (in press). doi: 10.1099/ijs.0.024745-0.

Dhanjal, S., Ruckmani, A., Cameotra, S.S., Pukall, R., Klenk, H.P. and Mayilraj, S. (2010). *Yaniella fodinae* sp. nov., a novel member of the genus *Yaniella* isolated from coal mine sample. *Int. J. Syst. Evol. Microbiol.*, (in press). doi: 10.1099/ijs.0.020636-0.

Dighe, A.S., Shouche, Y.S. and Ranade, D.R. (1998). *Selenomonas lipolytica* sp. nov., an obligately anaerobic bacterium possessing lipolytic activity. *Int. J. Syst. Bacteriol.*, 48: 783–791.

Dutta, D. and Gachhui, R. (2007). Nitrogen-fixing and cellulose-producing *Gluconacetobacter kombuchae* sp. nov., isolated from Kombucha tea. *Int. J. Syst. Evol. Microbiol.*, 57: 353–357.

Dutta, D. and Gachhui, R. (2006). Novel nitrogen-fixing *Acetobacter nitrogenifigens* sp. nov., isolated from Kombucha tea. *Int. J. Syst. Evol. Microbiol.*, 56: 1899–1903.

Ellaiah, P., Adinarayana, G., Kumar, J.P., Saisha, V. and Vasu, P. Characterisation of a new bioactive actinomycete AUB N5/8 from marine sediments. *Hindustan Antibiot. Bull.*, 2003–2004, 45–46, 16–21.

Ghosh, A., Bhardwaj, M., Satyanarayana, T., Khurana, M., Mayilraj, S. and Jain, R.K. (2007). *Bacillus lehensis* sp. nov., an alkalitolerant bacterium isolated from soil. *Int. J. Syst. Evol. Microbiol.*, 57: 238–242.

Ghosh, A., Paul, D., Prakash, D., Mayilraj, S. and Jain, R.K. (2006). *Rhodococcus imtechensis* sp. nov., a nitrophenol-degrading actinomycete. *Int. J. Syst. Evol. Microbiol.*, 56: 1965–1969.

Ghosh, W. and Roy, P. (2006). *Mesorhizobium thiogangeticum* sp. nov., a novel sulfur-oxidizing chemolithoautotroph from rhizosphere soil of an Indian tropical leguminous plant. *Int. J. Syst. Evol. Microbiol.*, 56: 91–97.

Ghosh, W., Bagchi, A., Mandal, S., Dam, B. and Roy, P. (2005). *Tetrathiobacter kashmirensis* gen. nov., sp. nov., a novel mesophilic, neutrophilic, tetrathionateoxidizing, facultatively chemolithotrophic betaproteobacterium isolated from soil from a temperate orchard in Jammu and Kashmir, India. *Int. J. Syst. Evol. Microbiol.*, 55: 1779–1787.

Ghosh, W., Mandal, S. and Roy, P. (2006). *Paracoccus bengalensis* sp. nov., a novel sulfur-oxidizing chemolithoautotroph from the rhizospheric soil of an Indian tropical leguminous plant. *Syst. Appl. Microbiol.*, 29: 396–403.

Girija, K.R., Sasikala, C., Ramana, C.V., Spröer, C., Takaichi, S., Thiel, V. and Imhoff, J.F. (2010). *Rhodobacter johrii* sp. nov., an endospore-producing cryptic species isolated from semi-arid tropical soils. *Int. J. Syst. Evol. Microbiol.*, 60: 2099–2107.

Goodfellow, M., Kim, S.B., Minnikin, D.E., Whitehead, D., Zhou, Z.H. and Mattinson-Rose, A.D. (2001). *Amycolatopsis sacchari* sp. nov., a moderately thermophilic actinomycete isolated from vegetable matter. *Int. J. Syst. Evol. Microbiol.*, 51: 187–193.

Gupta, A.K., Dharne, M.S., Rangrez, A.Y., Verma, P., Ghate, H.V., Rohde, M., Patole, M.S., and Shouche, Y.S. (2010). *Ignatzschineria indica* sp. nov. and *Ignatzschineria ureaclastica* sp. nov., isolated from adult flesh fly (Diptera: *Sarcophagidae*). *Int. J. Syst. Evol. Microbiol.*, (in press). doi: 10.1099/ijs.0.018622-0.

Gupta, K.C. and Chopra, I.C. (1963). A new whorl-forming species of *Streptomyces*. *Hindustan Antibiotics Bulletin*, 5: 110–112.

________. (1963). *Streptomyces Katrae* – a new species of *Streptomyces* isolated from soil, *Indian Journal of Microbiology*, 3: 1–4.

Gupta, K.C., Sobti, R.R. and Chopra, I.C. (1963). Actinomycin produced by a new species of *Streptomyces*. *Hindustan Antibiotics Bulletin*, 6: 12–16.

Gupta, K.C. (1965). A new species of the genus *Streptosporangium* isolated from Indian soil, *Journal of Antibiotics (Tokyo) Series A*, 18: 125–127.

Gupta, P., Chaturvedi, P., Pradhan, S., Delille, D. and Shivaji, S. (2006). *Marinomonas polaris* sp. nov., a psychrohalotolerant strain isolated from coastal sea water off the subantarctic Kerguelen islands. *Int. J. Syst. Evol. Microbiol.*, 56: 361–364.

Gupta, P., Reddy, G.S.N., Delille, D. and Shivaji, S. (2004). *Arthrobacter gangotriensis* sp. nov. and *Arthrobacter kerguelensis* sp. nov. from Antarctica. *Int. J. Syst. Evol. Microbiol.*, 54: 2375–2378.

Gupta, S.K., Kumari, R., Prakash, O. and Lal R. (2008). *Pseudomonas panipatensis* sp. nov., isolated from an oil-contaminated site. *Int. J. Syst. Evol. Microbiol.*, 58: 1339–1345.

Gupta, S.K., Lal, D. and Lal, R. (2009). *Novosphingobium panipatense* sp. nov. and *Novosphingobium mathurense* sp. nov., from oil-contaminated soil. *Int. J. Syst. Evol. Microbiol.*, 59: 156–161.

Hill, A.C. (1993). *Mycoplasma indiense* sp. nov., isolated from the throats of nonhuman primates. *Int. J. Syst. Bacteriol.*, 43: 36–40.

Jayasinghearachchi, H. and Lal, B. (2010). *Oceanotoga teriensis* gen. nov. sp. nov., a thermophilic, bacterium isolated from off-shore oil-producing wells of Bombay High in India. *Int. J. Syst. Evol. Microbiol.*, (in press). doi: 10.1099/ijs.0.018036-0.

Jit, S., Dadhwal, M., Prakash, O. and Lal R. (2008). *Flavobacterium lindanitolerans* sp. nov., isolated from hexachlorocyclohexane-contaminated soil. *Int. J. Syst. Evol. Microbiol.*, 58: 1665–1669.

Jyoti, V., Narayan, K.D. and Das S.K. (2010). *Gulbenkiania indica* sp. nov., isolated from a sulfur spring. *Int. J. Syst. Evol. Microbiol.*, 60: 1052–1055.

Kalyan Chakravarthy, S., Srinivas, T.N.R., Anil Kumar, P., Sasikala, C.H. and Ramana, C.V. (2007). *Roseospira visakhapatnamensis* sp. nov. and *Roseospira goensis* sp. nov. *Int. J. Syst. Evol. Microbiol.*, 57: 2453–2457.

Kämpfer, P., Arun, A.B., Busse, H.J., Langer, S., Young, C.C., Chen, W.M., Schumann, P., Syed, A.A. and Rekha, P.D. (2010). *Georgenia soli* sp. nov., isolated from iron-ore-contaminated soil in India. *Int. J. Syst. Evol. Microbiol.*, 60: 1027–1030.

Kämpfer, P., Arun, A.B., Young, C.C., Chen, W.M., Sridhar, K.R. and Rekha, P.D. (2010). *Chryseobacterium arthrosphaerae* sp. nov., isolated from the faeces of the pill millipede *Arthrosphaera magna* Attems. *Int. J. Syst. Evol. Microbiol.*, 60: 1765–1769.

Kämpfer, P., Rekha, P.D., Schumann, P., Arun, A.B., Young, C.C., Chen, W.M. and Sridhar, K.R. (2010). *Microbacterium arthrosphaerae* sp. nov., isolated from the faeces of the pill millipede *Arthrosphaera magna* Attems from India. *Int. J. Syst. Evol. Microbiol.*, doi: 10.1099/ijs.0.026401-0.

Kämpfer, Peter, Chandel, Kshitij, Prasad, G.P.K.S., Shouche, Y.S. and Veer, Vijay, (2009). *Chryseobacterium culicis* sp. nov., isolated from the midgut of *Culex quinquefasciatus* say, a mosquito from India. *Int. J. Syst. Evol. Microbiol.*, (in press). doi: 10.1099/ijs.0.019794-0.

Kaur, J., Verma, M., and Lal, R. (2010). *Rhizobium rosettiformans* sp. nov., isolated from hexachlorocyclohexane (HCH) dump site in India, and reclassification of *Blastobacter aggregatus* Hirsch and Müller (1985) as *Rhizobium aggregatum* comb. nov., *Int. J. Syst. Evol. Microbiol.*, doi: 10.1099/ijs.0.017491-0.

Kim, B., Sahin, N., Tan, G.Y.A., Zakrzewska-Czerwinska, J. and Goodfellow, M. (2002). *Amycolatopsis eurytherma* sp. nov., a thermophilic actinomycete isolated from soil. *Int. J. Syst. Evol. Microbiol.*, 52: 889–894.

Kishore, K.H., Begum, Z., Pathan, A.A.K. and Shivaji, S. (2010). *Paenibacillus glacialis* sp. nov., isolated from the Kafni glacier of the Himalayas, India. *Int. J. Syst. Evol. Microbiol.*, 60: 1909–1913.

Krishnamurthi, S., Bhattacharya, A., Mayilraj, S., Saha, P., Schumann, P. and Chakrabarti, T. (2009). Description of *Paenisporosarcina quisquiliarum* gen. nov., sp. nov., and reclassification of *Sporosarcina macmurdoensis* Reddy et al. 2003 as *Paenisporosarcina macmurdoensis* comb. nov., *Int. J. Syst. Evol. Microbiol.*, 59: 1364–1370.

Kudo, T. and Seino, A. (1987). Transfer of *Streptosporangium indianense* Gupta 1965 to the genus *Streptomyces* as *Streptomyces indiaensis* (Gupta 1965) comb. nov. *Int. J. Syst. Bacteriol.*, 37: 241–244.

Kumar, M., Verma, M. and Lal, R. (2008). *Devosia chinhatensis* sp. nov., isolated from a hexachlorocyclohexane (HCH) dump site in India. *Int. J. Syst. Evol. Microbiol.*, 58: 861–865.

Kumar, R., Banerjee, A.K., Bowdre, J.H., Mcelroy, L.J. and Krieg, N.R. (1974). Isolation, Characterization, and Taxonomy of *Aquaspirillum bengal* sp. nov., *Int. J. Syst. Evol. Microbiol.*, 24: 453–458.

Kumar, Y. and Goodfellow, M. (2008). Five new members of the *Streptomyces violaceusniger* 16S rRNA gene clade: *Streptomyces castelarensis* sp. nov., comb. nov., *Streptomyces himastatinicus* sp. nov., *Streptomyces mordarskii* sp. nov., *Streptomyces rapamycinicus* sp. nov. and *Streptomyces ruanii* sp. nov., *Int. J. Syst. Evol. Microbiol.*, 58: 1369–1378.

Kumari, H., Gupta, S.K., Jindal, S., Katoch, P. and Lal, R. (2009). *Sphingobium lactosutens* sp. nov., isolated from a hexachlorocyclohexane dump site and *Sphingobium abikonense* sp. nov., isolated from oil-contaminated soil. *Int. J. Syst. Evol. Microbiol.*, 59: 2291–2296.

Kumari, H., Gupta, S.K., Jindal, S., Katoch, P. and Lal, R. (2009). *Sphingobium lactosutens* sp. nov., isolated from a hexachlorocyclohexane dump site and *Sphingobium abikonense* sp. nov., isolated from oil-contaminated soil. *Int. J. Syst. Evol. Microbiol.*, 59: 2291–2296.

Kunwar, D.N., Pandey, S.K. and Das, S.K. (2010) Characterization of *Comamonas thiooxidans* sp. nov. and Comparison of Thiosulfate Oxidation with *Comamonas testosteroni* and *Comamonas composti*. *Curr. Microbiol.* (in press). doi: 10.1007/s00284-010-9602-9.

Lai, Q., Qiao, N., Wu, C., Sun, F., Yuan, J. and Shao, Z. (2010). *Stappia indica* sp. nov., isolated from deep seawater of the Indian Ocean. *Int. J. Syst. Evol. Microbiol.*, 60: 733–736.

Lakshmi, B.V. (1993). *Streptoverticillium acido-carbonicum* sp. nov. *Hindustan Antibiot. Bull.*, 35: 191–194.

Lakshmi, K.V.N.S., Sasikala, C.H., Kumar Ashok, G.V., Chandrasekaran, R. and Ramana, C.H.V. (2010). *Phaeovibrio sulfidiphilus* gen. nov., sp. nov., phototrophic alphaproteobacterium isolated from brackish water. *Int. J. Syst. Evol. Microbiol.*, (in press). doi: 10.1099/ijs.0.018911-0.

Lakshmi, K.V.N.S., Sasikala, C.H., Takaichi, S., and Ramana, C.H.V. (2010). *Phaeospirillum oryzae* sp. nov., a spheroplast forming phototrophic alphaproteobacterium from a paddy soil. *Int. J. Syst. Evol. Microbiol.*, (in press). doi: 10.1099/ijs.0.025544-0.

Lakshmi, K.V.N.S., Sasikala, C. and Ramana, C.V. (2009). *Rhodoplanes pokkaliisoli* sp. nov., a phototrophic alphaproteobacterium isolated from a waterlogged brackish paddy soil. *Int. J. Syst. Evol. Microbiol.*, 59: 2153–2157.

Lal, D., Gupta, S.K., Schumann, P. and Lal, R. (2009). *Microbacterium lindanitolerans* sp. nov., isolated from hexachlorocyclohexane (HCH) contaminated soil *Int. J. Syst. Evol. Microbiol.*, (in press). doi: 10.1099/ijs.0.017699-0.

Loganathan, P. and Nair, S. (2004). *Swaminathania salitolerans* gen. nov., sp. nov., a salt-tolerant, nitrogen-fixing and phosphate-solubilizing bacterium from wild rice (*Porteresia coarctata* Tateoka). *Int. J. Syst. Evol. Microbiol.*, 54: 1185-1190.

Madhaiyan, M., Poonguzhali, S., Kwon, S.W., and Sa, T.M. (2009). *Bacillus methylotrophicus* sp. nov. a novel species of methanol utilizing, plant-growth promoting bacterium isolated from rice. *Int. J Syst. Evol. Microbiol.*, (in press). doi: 10.1099/ijs.0.015487-0.

Madhaiyan, M., Poonguzhali, S., Lee, J.S., Lee, K.C. and Sundaram, S. (2010). *Flavobacterium glycines* sp. nov., a facultative methylotroph isolated from the rhizosphere of soybean. *Int. J. Syst. Evol. Microbiol.*, 60: 2187–2192.

Madhaiyan, M., Poonguzhali, S., Lee, J.S., Lee, K.C., Saravanan, V.S. and Santhanakrishnan, P. (2010). *Microbacterium azadirachtae* sp. nov., a plant-growth-promoting actinobacterium isolated from the rhizoplane of neem seedlings. *Int. J. Syst. Evol. Microbiol.*, 60: 1687–1692.

Madhaiyan, M., Poonguzhali, S., Lee, J.S., Senthilkumar, M., Lee, K.C. and Sundaram, S. (2010). *Leifsonia soli* sp. nov., a yellow-pigmented actinobacterium isolated from teak rhizosphere soil. *Int. J. Syst. Evol. Microbiol.*, 60: 1322–1327.

________. (2009). *Mucilaginibacter gossypii* sp. nov. and *Mucilaginibacter gossypiicola* sp. nov., plant-growth promoting bacteria isolated from cotton rhizosphere soils. *Int. J. Syst. Evol. Microbiol.*, doi: 10.1099/ijs.0.018713-0.

Manickam, N., Ghosh, A., Jain, R.K. and Mayilraj, S. (2008). *Syst. Appl. Microbiol.*, 31: 101–107.

Manorama, R., Pindi, P.K., Reddy, G.S.N. and Shivaji, S. (2009). *Bhargavaea cecembensis* gen. nov., sp. nov., isolated from the Chagos-Laccadive ridge system in the Indian Ocean. *Int. J. Syst. Evol. Microbiol.*, 59: 2618–2623.

Mayilraj, S., Krishnamurthi, S., Saha, P. and Saini, H.S. (2006). *Kitasatospora sampliensis* sp. nov., a novel actinobacterium isolated from soil of a sugar-cane field in India. *Int. J. Syst. Evol. Microbiol.*, 56: 519–522.

________. (2006). *Rhodococcus kroppenstedtii* sp. nov., a novel actinobacterium isolated from a cold desert of the Himalayas, India. *Int. J. Syst. Evol. Microbiol.*, 56: 979–982.

Mayilraj, S., Kroppenstedt, R.M., Suresh, K. and Saini, H.S. (2006). *Kocuria himachalensis* sp. nov., an actinobacterium isolated from the Indian Himalayas. *Int. J. Syst. Evol. Microbiol.*, 56: 1971–1975.

Mayilraj, S., Prasad, G.S., Suresh, K., Saini, H.S., Shivaji, S. and Chakrabarti, T. (2005). *Planococcus stackebrandtii* sp. nov., isolated from a cold desert of the Himalayas, India. *Int. J. Syst. Evol. Microbiol.*, 55: 91–94.

Mayilraj, S., Saha, P., Suresh, K. and Saini, H.S. (2006). *Ornithinimicrobium kibberense* sp. nov., isolated from the Indian Himalayas. *Int. J. Syst. Evol. Microbiol.*, 56: 1657–1661.

Mayilraj, S., Suresh, K., Kroppenstedt, R.M. and Saini, H.S. (2006). *Dietzia kunjamensis* sp. nov., isolated from the Indian Himalayas. *Int. J. Syst. Evol. Microbiol.*, 56: 1667–1671.

Mayilraj, S., Suresh, K., Schumann, P., Kroppenstedt, R.M. and Saini, H.S. (2006). *Agrococcus lahaulensis* sp. nov., isolated from a cold desert of the Indian Himalayas. *Int. J. Syst. Evol. Microbiol.*, 56: 1807–1810.

Mertz, F.P. and Yao, R.C. (1986). *Actinomadura oligospora* sp. nov., the producer of a new polyether antibiotic. *Int. J. Syst. Bacteriol.*, 36: 179–182.

Mertz, F.P. (1994). *Planomonospora alba* sp. nov. and *Planomonospora sphaerica* sp. nov., two new species isolated from soil by baiting techniques. *Int. J. Syst. Bacteriol.*, 44: 274–281.

Nand, K. and Rao, D.V. (1972). *Arthrobacter mysorens* – a new species excreting L-glutamic acid. *Zentralblatt für Bakteriologie, Parasitenkunde, Infektionskrankheiten und Hygiene. Abteilung* II, 127: 324–331.

Nigam, A., Jit, S. and Lal, R. (2010). *Sphingomonas histidinilytica* sp. nov., isolated from a hexachlorocyclohexane dump site. *Int. J. Syst. Evol. Microbiol.*, 60: 1038–1043.

Nowlan, B., Dodia, M.S., Singh, S.P. and Patel, B.K.C. (2006). *Bacillus okhensis* sp. nov., a halotolerant and alkalitolerant bacterium from an Indian saltpan. *Int. J. Syst. Evol. Microbiol.*, 56: 1073–1077.

Pal, R., Bala, S., Dadhwal, M., Kumar, M., Dhingra, G., Prakash, O., Prabagaran, S.R., Shivaji, S., Cullum, J., Holliger, C. and Lal, R. (2005). Hexachlorocyclohexane-degrading bacterial strains *Sphingomonas paucimobilis* B90A, UT26 and Sp+, having similar *lin* genes, represent three distinct species, *Sphingobium indicum* sp. nov., *Sphingobium japonicum* sp. nov. and *Sphingobium francense* sp. nov. and reclassification of [*Sphingomonas*] *chungbukensis* as *Sphingobium chungbukense* comb. nov. *Int. J. Syst. Evol. Microbiol.*, 55: 1965–1972.

Panda, S.K., Jyoti, V., Bhadra, B., Nayak, K.C., Shivaji, S., Rainey, F.A. and Das, S.K. (2009). *Thiomonas bhubaneswarensis* sp. nov., an obligately mixotrophic, moderately thermophilic, thiosulfate-oxidizing bacterium. *Int. J. Syst. Evol. Microbiol.*, 59: 2171–2175.

Panday, D. and Das, S.K. (2010). *Chelatococcus sambhunathii* sp. nov., a moderately thermophilic alphaproteobacterium isolated from hot spring sediment. *Int. J. Syst. Evol. Microbiol.*, 60: 861–865.

Pandey, K.K., Mayilraj, S. and Chakrabarti, T. (2002). *Pseudomonas indica* sp. nov., a novel butane-utilizing species. *Int. J. Syst. Evol. Microbiol.*, 52: 1559–1567.

Pidiyar, V., Kaznowski, A., Narayan, N.B., Patole, M. and Shouche, Y.S. (2002). *Aeromonas culicicola* sp. nov., from the midgut of *Culex quinquefasciatus*. *Int. J. Syst. Evol. Microbiol.*, 52: 1723–1728.

Pindi, P.K., Kishore, K.H., Reddy, G.S.N. and Shivaji, S. (2009). Description of *Leifsonia kafniensis* sp. nov. and *Leifsonia antarctica* sp. nov. *Int. J. Syst. Evol. Microbiol.*, 59: 1348–1352.

Pindi, P.K., Manorama, R., Begum, Z. and Shivaji, S. (2009). *Arthrobacter antarcticus* sp. nov., isolated from a sediment from the Antarctic sea. *Int. J. Syst. Evol. Microbiol.*, doi: 10.1099/ijs.0.012989-0.

Prabagaran, S.R., Suresh, K., Manorama, R., Delille, D. and Shivaji, S. (2005). *Marinomonas ushuaiensis* sp. nov., isolated from coastal sea water in Ushuaia, Argentina, sub-Antarctica. *Int. J. Syst. Evol. Microbiol.*, 55: 309–313.

Prabahar, V., Dube, S., Reddy, G.S.N. and Shivaji, S. (2004). *Pseudonocardia antarctica* sp. nov. an *Actinomycetes* from McMurdo Dry Valleys, Antarctica. *Syst. Appl. Microbiol.*, 27: 66–71.

Prakash, O. and Lal, R. (2006). Description of *Sphingobium fuliginis* sp. nov., a phenanthrene-degrading bacterium from a fly ash dumping site, and reclassification of *Sphingomonas cloacae* as *Sphingobium cloacae* comb. nov. *Int. J. Syst. Evol. Microbiol.*, 56: 2147–2152.

Prakash, O., Kumari, K. and Lal, R. (2007). *Pseudomonas delhiensis* sp. nov., from a fly ash dumping site of a thermal power plant. *Int. J. Syst. Evol. Microbiol.*, 57: 527–531.

Rai, S.K., Roy, J.K. and Mukherjee, A.K. (2010). Characterisation of a detergent-stable alkaline protease from a novel thermophilic strain Paenibacillus tezpurensis sp. nov. AS-S24-II. *Appl. Microbiol. Biotechnol.*, 85: 1437–1450

Ramana, C.V., Sasikala, C., Arunasri, K., Anil Kumar, P., Srinivas, T.N.R., Shivaji, S., Gupta, P., Süling, J. and Imhoff, J.F. (2006). *Rubrivivax benzoatilyticus* sp. nov., an aromatic, hydrocarbon-degrading purple betaproteobacterium. *Int. J. Syst. Evol. Microbiol.*, 56: 2157–2164.

Ramesh Kumar, N. and Nair, S. (2007). *Vibrio rhizosphaerae* sp. nov., a red-pigmented bacterium that antagonizes phytopathogenic bacteria. *Int. J. Syst. Evol. Microbiol.*, 57: 2241–2246.

Ramesh Kumar, N., Nair, S., Langer, S., Busse, H.J. and Kämpfer, P. (2008). *Altererythrobacter indicus* sp. nov., isolated from wild rice (*Porteresia coarctata* Tateoka). *Int. J. Syst. Evol. Microbiol.*, 58: 839–844.

Ramesh Kumar, N., Sproer, C., Lang, E. and Nair, S. (2010). *Vibrio mangrovi* sp. nov., a diazotrophic bacterium isolated from mangrove-associated wild rice (*Poteresia coarctata* Tateoka), FEMS Microbiology Letters, 30: 35–40.

Ramesh Kumar, N., Fukui, Y., Sawabe, T. and Nair, S. (2008). *Vibrio porteresiae* sp. nov., a diazotrophic bacterium isolated from a mangrove-associated wild rice (*Porteresia coarctata* Tateoka). *Int. J. Syst. Evol. Microbiol.*, 58: 1608–1615.

Ramesh Kumar, N., Lang, E. and Nair, S. (2010). *Mangrovibacter plantisponsor* gen. nov., sp. nov., a nitrogen-fixing bacterium isolated from a mangrove-associated wild rice (*Porteresia coarctata* Tateoka). *Int. J. Syst. Evol. Microbiol.*, 60: 179–186.

Rao, V.A., Prabhu, K.K., Sridhar, B.P., Venkateswarlu, A. and Actor, P. (1987). Two new species of *Microbispora* from Indian soils: *Microbispora karnatakensis* sp. nov. and *Microbispora indica* sp. nov. *Int. J. Syst. Bacteriol.*, 37: 181–185.

Rauschenbach, I., Narasingarao, P. and Haggblom, M. (2010). *Desulfurispirillum indicum* sp. nov., a selenate and selenite respiring bacterium isolated from an estuarine canal in southern India. *Int. J. Syst. Evol. Microbiol.*, (in press). doi: 10.1099/ijs.0.022392-0.

Reddy, G.N.S., Raghavan, P.U.M., Sarita, N.B. Prakash, J.S.S., Narayan, Nagesh, Delille, D. and Shivaji, S. (2003). *Halomonas glacei* sp. nov. isolated from fast ice of Adelie Land, Antarctica. *Extremophiles*, 7: 55–61.

Reddy, G.S.N., Aggarwal, R.K., Matsumoto, G.I. and Shivaji, S. (2000). *Arthrobacter flavus* sp. nov., a psychrophilic bacterium isolated from a pond in McMurdo Dry Valley, Antarctica. *Int. J. Syst. Evol. Microbiol.*, 50: 1553–1561.

Reddy, G.S.N., Matsumoto, G.I. and Shivaji, S. (2003). *Sporosarcina macmurdoensis* sp. nov., from a cyanobacterial mat sample from a pond in the McMurdo Dry Valleys, Antarctica. *Int. J. Syst. Evol. Microbiol.*, 53: 1363–1367.

Reddy, G.S.N. Matsumoto, G.I., Schumann, P., Stackebrandt, E. and Shivaji, S. (2004). Psychrophilic pseudomonads from Antarctica: *Pseudomonas antarctica* sp. nov., *Pseudomonas meridiana* sp. nov. and *Pseudomonas proteolytica* sp. nov. *Int. J. Syst. Evol. Microbiol.*, 54: 713–719.

Reddy, G.S.N., Prabagaran, S.R. and Shivaji, S. (2008). *Leifsonia pindariensis* sp. nov., isolated from the Pindari glacier of the Indian Himalayas, and emended description of the genus *Leifsonia*. *Int. J. Syst. Evol. Microbiol.*, 58: 2229–2234.

Reddy, G.S.N., Pradhan, S., Manorama, R. and Shivaji, S. (2010). *Cryobacterium roopkundense* sp. nov., a psychrophilic bacterium isolated from glacial soil. *Int. J. Syst. Evol. Microbiol.*, 60: 866–870.

Reddy, G.S.N., Prakash, J.S.S., Matsumoto, G.I., Stackebrandt, E. and Shivaji, S. (2002). *Arthrobacter roseus* sp. nov., a psychrophilic bacterium isolated from an Antarctic cyanobacterial mat sample. *Int. J. Syst. Evol. Microbiol.*, 52: 1017–1021.

Reddy, G.S.N., Prakash, J.S.S., Prabahar, V., Matsumoto, G.I., Stackebrandt, E. and Shivaji, S. (2003). *Kocuria polaris* sp. nov., an orange-pigmented psychrophilic bacterium isolated from an Antarctic cyanobacterial mat sample. *Int. J. Syst. Evol. Microbiol.*, 53: 183–187.

Reddy, G.S.N., Prakash, J.S.S., Srinivas, R., Matsumoto, G.I. and Shivaji, S. (2003). *Leifsonia rubra* sp. nov. and *Leifsonia aurea* sp. nov., psychrophiles from a pond in Antarctica. *Int. J. Syst. Evol. Microbiol.*, 53: 977–984.

Reddy, G.S.N., Prakash, J.S.S., Vairamani, M. Prabhakar, S., Matsumoto, G.I. and Shivaji, S. (2002). *Planococcus antarcticus* and *Planococcus psychrophilus* sp. nov. isolated from cyanobacterial mat samples collected from ponds in Antarctica. *Extremophiles*, 6: 253–261.

Reddy, G.S.N., Uttam, A. and Shivaji, S. (2008). *Bacillus cecembensis* sp. nov., isolated from the Pindari glacier of the Indian Himalayas. *Int. J. Syst. Evol. Microbiol.*, 58: 2330–2335.

Reddy, T.V.K., Mahmood, S., Paris, L., Reddy, Y. H.K., Wellington, E.M.H. and Idris, M.M. (2010). *Streptomyces hyderabadensis* sp. nov., a novel actinomycete isolated from soil in Andhra Pradesh, India. *Int. J. Syst. Evol. Microbiol.*, (in press). doi: 10.1099/ijs.0.020446-0,

Reddy, T.V.K., Mahmood, S., Idris, M.M. and Ciesielski, S. (2010). *Streptomyces osmaniensis* sp. nov., isolated from soil. *Int. J. Syst. Evol. Microbiol.*, 60: 1755–1759.

Rivas, R., Willems, A., Subba-Rao, N.S., Mateos, P.F., Dazzo, F.B., Kroppenstedt, R.M., Martínez-Molina, E., Gillis, M. and Velázquez, E. (2003). Description of *Devosia neptuniae* sp. nov., that nodulates and fixes nitrogen in symbiosis with *Neptunia natans*, an aquatic legume from India. *Syst. Appl. Microbiol.*, 26: 47–53.

Rolain, J.M., Mathai, E., Lepidi, H., Somashekar, H.R., Mathew, L.G., Prakash, J.A. and Raoult, D. (2006). "Candidatus Rickettsia kellyi," India. Emerg Infect Dis., 12(3): 483–5.

Saha, P. and Chakrabarti, T. (2006). *Aeromonas sharmana* sp. nov., isolated from a warm spring. *Int. J. Syst. Evol. Microbiol.*, 56: 1905–1909.

________. (2006). *Emticicia oligotrophica* gen. nov., sp. nov., a new member of the family '*Flexibacteraceae*', phylum *Bacteroidetes. Int. J. Syst. Evol. Microbiol.*, 56: 991–995.

________. (2006). *Flavobacterium indicum* sp. nov., isolated from warm spring water in Assam, India. *Int. J. Syst. Evol. Microbiol.*, 56: 2617–2621.

Saha, P., Krishnamurthi, S., Bhattacharya, A., Sharma, R. and Chakrabarti, T. (2010). *Fontibacillus aquaticus* gen. nov., sp. nov., isolated from a warm spring. *Int. J. Syst. Evol. Microbiol.*, 60: 422–428.

Saha, P., Krishnamurthi, S., Mayilraj, S., Prasad, G.S., Bora, T.C. and Chakrabarti, T. (2005). *Aquimonas voraii* gen. nov., sp. nov., a novel gammaproteobacterium isolated from a warm spring of Assam, India. *Int. J. Syst. Evol. Microbiol.*, 55: 1491–1495.

Saha, P., Mondal, A.K., Mayilraj, S., Krishnamurthi, S., Bhattacharya, A. and Chakrabarti, T. (2005). *Paenibacillus assamensis* sp. nov., a novel bacterium isolated from a warm spring in Assam, India. *Int. J. Syst. Evol. Microbiol.*, 55: 2577–2581.

Sasi, Jyothsna, T.S., Sasikala, C. and Ramana, C.V. (2008). *Desulfovibrio psychrotolerans* sp. nov., a psychrotolerant and moderately alkaliphilic sulfate-reducing deltaproteobacterium from the Himalayas. *Int. J. Syst. Evol. Microbiol.*, 58: 821–825.

Sharma, P., Verma, M., Bala, K., Nigam, A. and Lal, R. (2010). *Sphingopyxis ummariensis* sp. nov., isolated from a hexachlorocyclohexane dump site. *Int. J. Syst. Evol. Microbiol.*, 60: 780–784.

Shashidhar, R. and Bandekar, J.R. (2006). *Deinococcus mumbaiensis* sp. nov., a radiation-resistant pleomorphic bacterium isolated from Mumbai, India. *FEMS Microbiol. Lett.*, 254: 275–280.

_______. (2009). *Deinococcus piscis* sp. nov., a radiation-resistant bacterium isolated from a marine fish. *Int. J. Syst. Evol. Microbiol.*, 59: 2714–2717.

Shivaji, S., Bhadra, B., Rao, R.S., Chaturvedi, P., Pindi, P.K. and Raghukumar, C. (2007). *Microbacterium indicum* sp. nov., isolated from a deep-sea sediment sample from the Chagos Trench, Indian Ocean. *Int. J. Syst. Evol. Microbiol.*, 57: 1819–1822.

Shivaji, S., Chaturvedi, P., Begum, Z., Pindi, P.K., Manorama, R., Padmanaban, D.A., Shouche, Y.S., Pawar, S., Vaishampayan, P., Dutt, C.B.S., Datta, G.N., Manchanda, R.K., Rao, U.R., Bhargava, P.M. and Narlikar, J.V. (2009). *Janibacter hoylei* sp. nov., *Bacillus isronensis* sp. nov. and *Bacillus aryabhattai* sp. nov., isolated from cryotubes used for collecting air from the upper atmosphere. *Int. J. Syst. Evol. Microbiol.*, 59: 2977–2986.

Shivaji, S., Chaturvedi, P., Reddy, G.S.N. and Suresh, K. (2005). *Pedobacter himalayensis* sp. nov., from the Hamta glacier located in the Himalayan mountain ranges of India. *Int. J. Syst. Evol. Microbiol.*, 55: 1083–1088.

Shivaji, S., Chaturvedi, P., Suresh, K., Reddy, G.S.N., Dutt, C.B.S., Wainwright, M., Narlikar, J.V. and Bhargava, P.M. (2006). *Bacillus aerius* sp. nov., *Bacillus aerophilus* sp. nov., *Bacillus stratosphericus* sp. nov. and *Bacillus altitudinis* sp. nov., isolated from cryogenic tubes used for collecting air samples from high altitudes. *Int. J. Syst. Evol. Microbiol.*, 56: 1465–1473.

Shivaji, S., Gupta, P., Chaturvedi, P., Suresh, K. and Delille, D. (2005). *Marinobacter maritimus* sp. nov., a psychrotolerant strain isolated from sea water off the subantarctic Kerguelen islands. *Int. J. Syst. Evol. Microbiol.*, 55: 1453–1456.

Shivaji, S., Ray, M.K., Rao, N.S., Saisree, L., Jagannadham, M.V., Kumar, G.S., Reddy, G.S.N. and Bhargava, P.M. (1992). *Sphingobacterium antarcticus* sp. nov., a psychrotrophic bacterium from the soils of Schirmacher Oasis, Antarctica. *Int. J. Syst. Bacteriol.*, 42: 102–106.

Shivaji, S., Reddy, G.S.N., Raghavan, P.U.M., Sarita, N.B. and Delille, D. (2004). *Psychrobacter salsus* sp. nov. and *Psychrobacter adeliensis* sp. nov. isolated from fast ice from Adelie Land, Antarctica. *Syst. Appl. Microbiol.*, 27: 628–635.

Shivaji, S., Reddy, G.S.N., Suresh, K., Gupta, P., Chintalapati, S., Schumann, P., Stackebrandt, E. and Matsumoto, G.I. (2005). *Psychrobacter vallis* sp. nov. and *Psychrobacter aquaticus* sp. nov., from Antarctica. *Int. J. Syst. Evol. Microbiol.*, 55: 757–762.

Shivaji, S., Reddy, P.V.V., Nageshwara, Rao, S.S.S., Begum, Z. and Srinivas, T.N.R. (2010). *Cyclobacterium qasimii* sp. nov., a psychrotolerant bacterium isolated from a marine sediment of Kongsfjorden, Svalbard. *Int. J. Syst. Evol. Microbiol.*, (in press). doi: 10.1099/ijs.0.024901-0.

Shivaji, S., Suresh, K., Chaturvedi, P., Dube, S. and Sengupta, S. (2005). *Bacillus arsenicus* sp. nov., an arsenic-resistant bacterium isolated from a siderite concretion in West Bengal, India. *Int. J. Syst. Evol. Microbiol.*, 55: 1123–1127.

Singh, A. and Lal, R. (2009). *Sphingobium ummariense* sp. nov., a hexachlorocyclohexane (HCH)-degrading bacterium, isolated from HCH-contaminated soil. *Int. J. Syst. Evol. Microbiol.*, 59: 162–166.

Singla, A.K., Mayilraj, S., Kudo, T., Krishnamurthi, S., Prasad, G.S. and Vohra, R.M. (2005). *Actinoalloteichus spitiensis* sp. nov., a novel actinobacterium isolated from a cold desert of the Indian Himalayas. *Int. J. Syst. Evol. Microbiol.*, 55: 2561–2564.

Somvanshi, V.S., Lang, E., Ganguly, S., Swiderski, J., Saxena, A.K. and Stackebrandt, E. (2006). A novel species of *Xenorhabdus*, family Enterobacteriaceae: *Xenorhabdus indica* sp. nov., symbiotically associated with entomopathogenic nematode *Steinernema thermophilum* Ganguly and Singh, 2000. *Syst. Appl. Microbiol.*, 29: 519–525.

Somvanshi, V.S., Lang, E., Schumann, P., Pukall, R., Kroppenstedt, R.M., Ganguly, S. and Stackebrandt, E. (2007). *Leucobacter iarius* sp. nov., in the family *Microbacteriaceae*. *Int. J. Syst. Evol. Microbiol.*, 57: 682–686.

Somvanshi, V.S., Lang, E., Sträubler, B., Spröer, C., Schumann, P., Ganguly, S., Saxena, A.K. and Stackebrandt, E. (2006). *Providencia vermicola* sp. nov., isolated from infective juveniles of the entomopathogenic nematode *Steinernema thermophilum*. *Int. J. Syst. Evol. Microbiol.*, 56: 629–633.

Sravan Kumar, R., Sasi Jyothsna, T.S., Sasikala, C., Seong, C.N., Lim, C.H., Park, S.C. and Ramana, C.V. (2010). *Shewanella fodinae* sp. nov., isolated from a coal mine and from a marine lagoon. *Int. J. Syst. Evol. Microbiol.*, 60: 1649–1654.

Srinivas, T.N.R., Anil Kumar, P., Madhu, S., Sunil, B., Sharma, T.V.R.S, and Shivaji, S. (2010). *Cesiribacter andamanensis* gen. nov., sp. nov., a novel bacterium isolated from a soil sample of a mud volcano, Andaman Islands, India. *Int. J. Syst. Evol. Microbiol.* first published on July 23, as doi: 10.1099/ijs.0.025429-0.

Srinivas, T.N.R., Anil Kumar, P., Sasikala, C., Ramana, C.V. and Imhoff, J.F. (2007). *Rhodobacter vinaykumarii* sp. nov., a marine phototrophic alphaproteobacterium from tidal waters, and emended description of the genus *Rhodobacter*. *Int. J. Syst. Evol. Microbiol.*, 57: 1984–1987.

Srinivas, T.N.R., Anil Kumar, P., Sasikala, C., Spröer, C. and Ramana, C.V. (2008). *Rhodobacter ovatus* sp. nov., a phototrophic alphaproteobacterium isolated from a polluted pond. *Int. J. Syst. Evol. Microbiol.*, 58: 1379–1383.

Srinivas, T.N.R., Anil Kumar, P., Sucharitha, K., Sasikala, C. and Ramana, C.V. (2009). *Allochromatium phaeobacterium* sp. nov. *Int. J. Syst. Evol. Microbiol.*, 59: 750–753.

Srinivas, T.N.R., Kumar, P.A., Sasikala, C. and Ramana, C.V. (2007). *Rhodovulum imhoffii* sp. nov. *Int. J. Syst. Evol. Microbiol.*, 57: 228–232.

Srinivas, T.N.R., Kumar, P.A., Sasikala, C., Ramana, C.V., Süling, J. and Imhoff, J.F. (2006). *Rhodovulum marinum* sp. nov., a novel phototrophic purple non-sulfur alphaproteobacterium from marine tides of Visakhapatnam, India. *Int. J. Syst. Evol. Microbiol.*, 56: 1651–1656.

Srinivas, T.N.R., Kumar, P.A., Sasikala, C., Ramana, C.V. and Imhoff, J.F. (2007). *Rhodobium gokarnense* sp. nov., a novel phototrophic alphaproteo-bacterium from a saltern. *Int. J. Syst. Evol. Microbiol.*, 57: 932–935.

Srinivas, T.N.R., Reddy, P.V.V., Begum, Z., and Shivaji, S. (2010). *Oceanisphaera arctica* sp. nov., isolated from a marine sediment of Kongsfjorden, Svalbard, Arctic. *Int. J. Syst. Evol. Microbiol.*, (in press). doi: 10.1099/ijs.0.024539-0.

Sucharita, K., Sasikala, C. and Venkata Ramana, C. (2010). *Thiorhodococcus modestalkaliphilus* sp. nov. a phototrophic gammaproteobacterium from Chilika salt water lagoon, India. *J. Gen. Appl. Microbiol.*, 56: 93–99.

Sucharita, K., Sasikala, C., Park, S.C., Baik, K.S., Seong, C.N. and Ramana, C.V. (2009). *Shewanella chilikensis* sp. nov., a moderately alkaliphilic gammaproteobacterium isolated from a lagoon. *Int. J. Syst. Evol. Microbiol.*, 59: 3111–3115.

Sucharita, K., Shiva Kumar, E., Sasikala, C., Panda, B.B., Takaichi, S. and Ramana, C.V. (2010). *Marichromatium fluminis* sp. nov., a slightly alkaliphilic, phototrophic gammaproteobacterium isolated from river sediment. *Int. J. Syst. Evol. Microbiol.*, 60: 1103–1107.

Suresh, K., Mayilraj, S. and Chakrabarti, T. (2006). *Effluviibacter roseus* gen. nov., sp. nov., isolated from muddy water, belonging to the family '*Flexibacteraceae*'. *Int. J. Syst. Evol. Microbiol.*, 56: 1703–1707.

Suresh, K., Mayilraj, S., Bhattacharya, A. and Chakrabarti, T. (2007). *Planococcus columbae* sp. nov., isolated from pigeon faeces. *Int. J. Syst. Evol. Microbiol.*, 57: 1266–1271.

Suresh, K., Prabagaran, S.R., Sengupta, S. and Shivaji, S. (2004). *Bacillus indicus* sp. nov., an arsenic-resistant bacterium isolated from an aquifer in West Bengal, India. *Int. J. Syst. Evol. Microbiol.*, 54: 1369–1375.

Suresh, K., Prakash, D., Rastogi, N. and Jain, R.K. (2007). *Clostridium nitrophenolicum* sp. nov., a novel anaerobic *p*-nitrophenol-degrading bacterium, isolated from a subsurface soil sample. *Int. J. Syst. Evol. Microbiol.*, 57: 1886–1890.

Suresh, K., Reddy, G.S.N., Sengupta, S. and Shivaji, S. (2004). *Deinococcus indicus* sp. nov., an arsenic-resistant bacterium from an aquifer in West Bengal, India. *Int. J. Syst. Evol. Microbiol.*, 54: 457–461.

Syed, D.G., Tang, S.K., Cai, M., Zhi, X.Y., Agasar, D., Lee, J.C., Kim, C.J., Jiang, C.L., Xu, L.H. and Li, W.J. (2008). *Saccharomonospora saliphila* sp. nov., a halophilic actinomycete from an Indian soil. *Int. J. Syst. Evol. Microbiol.*, 58: 570–573.

Tamura, T., Yokota, A., Huang, L.H., Hasegawa, T. and Hatano, K. (1995). Five new species of the genus *Catenuloplanes*: *Catenuloplanes niger* sp. nov., *Catenuloplanes indicus* sp. nov., *Catenuloplanes atrovinosus* sp. nov., *Catenuloplanes castaneus* sp. nov. and *Catenuloplanes nepalensis* sp. nov. *Int. J. Syst. Bacteriol.*, 45: 858–860.

Thirumalachar, M.J. and Bhatt, V.V. (1960). Some *Streptomyces* species producing oxytetracycline. *Hindustan Antibiotics Bulletin*, 3: 61–63.

Thirumalachar, M.J. and Sukapure, R.S. (1964). Studies on species of the genus *Chainia* from India. *Hindustan Antibiotics Bulletin*, 6: 157–166.

Thirumalachar, M.J., Rahalkar, P.W., Deshmukh, P.V. and Sukapure, R.S. (1965). Production of aburamycin by *Chainia minutisclerotica*, a new species of actinomycete. *Hindustan Antibiotics Bulletin*, 8: 6–9.

Thirumalachar, M.J., Sukapure, R.S., Rahalkar, P.W. and Gopalkrishnan, K.S. (1966). Studies on species of the genus *Chainia* from India II. *Hindustan Antibiotics Bulletin*, 9: 10–14.

Thirumalachar, M.J. and Gopalkrishnan, K.S. (1966). Thirumalachar, M.J., Sukapure, R.S., Rahalkar, P.W. (Eds.). In: Studies on species of the genus *Chainia* from India II. *Hindustan Antibiotics Bulletin*, 9: 10–14.

Thirumalachar, M.J. (1960). Kalakoutskii, L.V. and Krasil'nikov, N.A. (Eds.). On formation of sclerotia by actinomycetes and on systematic status of the genus *Chainia*. *In*: Rautenshtein, Y.I. Biology of special groups of actinomycetes. Producers of antibiotics. *Transactions of the Institute of Microbiology, Academy of Sciences USSR*, 8: 1–344.

Thirumalachar, M.J. (1968). Rahalkar, P.W. and Thirumalachar, M.J. (Eds.). In: Cultural characteristics and identity of some *Streptoverticillium* species producing polyene antibiotics. *Hindustan Antibiotics Bulletin*, 11: 90–96.

Thirumalachar, M.J. (1955). *Chainia*, a new genus of the *Actinomycetales*. *Nature*, London, 176: 934–935.

Tindall, B.J. Grimont, P.A.D., Garrity, G.M. and Euzeby, J.P. (2005). Nomenclature and taxonomy of the genus *Salmonella*. *Int. J. Syst. Evol. Microbiol.*, 55: 521–524.

Tomita, K., Nakakita, Y., Hoshino, Y., Numata, K. and Kawaguchi, H. (1987). New genus of the *Actinomycetales*: *Streptoalloteichus hindustanus* gen. nov., nom. rev., sp. nov., nom. rev. *Int. J. Syst. Bacteriol.*, 37: 211–213.

Tresner, H., Hayes, J.A. and Backus, E.J. (1966). *Streptomyces prasinosporus* sp. nov. a new green-spored species. *Int. J. Syst. Bacteriol.*, 16: 161–169.

Tripathi, A.K., Verma, S.C., Chowdhury, S.P., Lebuhn, M., Gattinger, A. and Schloter, M. (2006). *Ochrobactrum oryzae* sp. nov., an endophytic bacterial species isolated from deep-water rice in India. *Int. J. Syst. Evol. Microbiol.*, 56: 1677–1680.

Uma, B.N. and Narasimha Rao, P.L. Actinomycetes. I. (1959). Distribution of streptomycetes in Indian soils. Formation of antifungal antibiotics by *Streptomyces champavati* nov. sp. Indian Institute of Sciences Golden Jubilee Research Vol. 1909–1959, Bangalore, India, pp. 130–141.

Venkata, Ramana, V., Anil Kumar, P., Srinivas, T.N.R., Sasikala, C. and Ramana, C.V. (2009). *Rhodobacter aestuarii* sp. nov., a phototrophic alphaproteobacterium isolated from an estuarine environment. *Int. J. Syst. Evol. Microbiol.*, 59: 1133–1136.

Venkata Ramana, V., Kapoor, S., Shobha, E., Ramprasad, E.V.V., Sasikala, C.H. and Ramana, C.H.V. (2010). *Blastochloris gulmargensis* sp. nov., isolated from an epilithic phototrophic biofilm. *Int. J. Syst. Evol. Microbiol.*, (in press). doi: 10.1099/ijs.0.025965-0.

Venkata Ramana, V., Sasikala, C. and Ramana, C.V. (2008). *Rhodobacter maris* sp. nov., a phototrophic alphaproteobacterium isolated from a marine habitat of India. *Int. J. Syst. Evol. Microbiol.*, 58: 1719–1722.

Verma, M., Kumar, M., Dadhwal, M., Kaur, J. and Lal, R. (2009). *Devosia albogilva* sp. nov. and *Devosia crocina* sp. nov., isolated from a hexachlorocyclohexane dump site. *Int. J. Syst. Evol. Microbiol.*, 59: 795–799.

Wink, J., Gandhi, J., Kroppenstedt, R.M., Seibert, G., Sträubler, B., Schumann, P. and Stackebrandt, E. (2004). *Amycolatopsis decaplanina* sp. nov., a novel member of the genus with unusual morphology. *Int. J. Syst. Evol. Microbiol.*, 54: 235–239.

Witt, D. and Stackebrandt, E. (1990). Unification of the genera *Streptoverticillium* and *Streptomyces*, and amendation of *Streptomyces* Waksman and Henrici 1943, 339AL. *Syst. Appl. Microbiol.*, 13: 361–371.

Yassin, A.F., Chen, W.M., Hupfer, H., Siering, C., Kroppenstedt, R.M., Arun, A.B., Lai, W.A., Shen, F.T., Rekha, P.D. and Young, C.C. (2007). *Lysobacter defluvii* sp. nov., isolated from municipal solid waste. *Int. J. Syst. Evol. Microbiol.*, 57: 1131–1136.

Yokota, A., Tamura, T., Nishii, T. and Hasegawa, T. (1993). *Kineococcus aurantiacus* gen. nov., sp. nov., a new aerobic, gram-positive, motile coccus with *meso*-diaminopimelic acid and arabinogalactan in the cell wall. *Int. J. Syst. Bacteriol.*, 43: 52–57.

Novel taxa of Cyanobacteria described from India.

(Source: www.cyanodb.cz)

1. *Anabaena iyengari* Bharadwaya 1935
2. *Anabaena orientalis* Dixit 1936
3. *Anabaena sankargarhii* Varma 1964
4. *Anabaena indica* Tiwari & Pandey 1976
5. *Anabaenopsis gangetica* Nair 1967
6. *Aphanothece benaresensis* Bhardwaya 1935
7. *Aphanothece crassa* Ghose 1927
8. *Arthrospira balkrishnii* Kamat 1963
9. *Arthrospira desikacharyensis* Vasishta 1962
10. *Arthrospira joshii* Vasishta 1963
11. *Aulosira bombayensis* Gonzalves 1946
12. *Aulosira fertilissima* Ghose 1923
13. *Aulosira fritschii* Bhardwaya 1933
14. *Aulosira manipurensis* Bhardwaya 1963

15. *Aulosira prolifica* Bhardwaya 1933
16. *Aulosira pseudoramosa* Bhardwaya 1933
17. *Aulosira terrestre* Subba Raju 1964
18. *Calothrix bharadwajae* De Toni 1939
19. *Calothrix clavatoides* Ghose 1927
20. *Calothrix fritschii* Bharadwaja 1935
21. *Calothrix ghosei* Bharadwaja 1935
22. *Camptylonemopsis boldii* Dutt, Datta & Gupta 1973
23. *Camptylonemopsis danilovi* (Hollerbach) Desikachary 1948
24. *Camptylonemopsis iyengari* Desikachary 1948
25. *Camptylonemopsis lahorensis* (Ghose) Desikachary 1948 gen. nov.
26. *Camptylonemopsis minor* Desikachary 1948
27. *Camptylonemopsis paperai* Subramanyam 1973
28. *Camptylonemopsis pulneyensis* Desikachary 1948
29. *Camptylonemopsis semiaquatica* Subramanyam 1973
30. *Chlorogloeopsis fritschii* Mitra & Pandey 1967 gen. nov.
31. *Chroococcidiopsis doonensis* R.B.Singh 1968
32. *Chroococcidiopsis indica* Desikachary 1959
33. *Chroococcidiopsis mysorensis* Tiwari 1972
34. *Cylindrospermum bengalense* Biswas 1925
35. *Cylindrospermum gloeotrichoides* Sinha & Pandey 1972
36. *Cylindrospermum gracile* Gonsalves & Kamat 1960
37. *Cylindrospermum sphaericum* Prasad 1935
38. *Cylindrospermum echinulatum* Gonsalves & Kamat 1960
39. *Gloeotrichia atra* Biswas 1934
40. *Gloeotrichia ghosei* R.N. Singh 1939
41. *Gloeotrichia indica* Schmidle 1900
42. *Iyengariella tirupaiensis* Desikachary 1953 gen. nov.
43. *Lyngbya birgei* Tiwari 1975
44. *Lyngbya ceylinica* Pandey & Mitra 1965
45. *Lyngbya maharashtrensis* Kamat 1963
46. *Lyngbya majuscule* Tiwari 1975
47. *Lyngbya oryzae* Gupta 1957
48. *Lyngbya stagninum* Rao 1938
49. *Lyngbya truncicola* Ghose 1924

50. *Mastigocladopsis jogensis* Iyengar & Desikachary 1946 gen. nov.
51. *Mastigocladopsis fremyi* Iyengar & Desikachary 1946
52. *Microcoleus sikkimensis* Santra 1984
53. *Microcoleus amplus* Kamat 1963
54. *Nodularia rajkotii* Gupte 1964
55. *Nostoc citrisporum* Prasad & Malhotra 1976
56. *Nostoc hatei* Dixit 1936
57. *Nostoc wartsporum* Prasad et al 1977
58. *Oscillatoria bhardwajii* Kamat 1963
59. *Oscillatoria brevis* Tiwari 1975
60. *Oscillatoria maharashtrensis* Kamat 1961
61. *Oscillatoria mitrae* Kamat 1963
62. *Oscillatoria salina* Biswas 1922
63. *Oscillatoria vizagaptensis* Rao 1938
64. *Oscillatoria yamadae* Kamat 1963
65. *Parthasarathiella prolifica1* Subba Raju 1962 gen. nov.
66. *Phormidium biswasii* Kamat 1963
67. *Phormidium croasdaleae* Kamat 1964
68. *Phormidium dixitii* Kamat 1963
69. *Phormidium schmidlei* Kamat 1963
70. *Phormidium mahabaleswarensis* Kamat 1962
71. *Phormidium spiralis* Vasishta 1962
72. *Phormidium truncicola* Ghose 1924
73. *Rivularia joshii* Vasishta 1963
74. *Rivularia maharashtrensis* Kamat 1961
75. *Scytonema dilatatum* Bharadwaya 1934
76. *Scytonema fremyi* Desikachary 1959
77. *Scytonema fritschii* Ghose 1924
78. *Scytonema iyengari* Bharadwaya 1934
79. *Scytonema malaviyanense* Bharadwaya 1930
80. *Scytonema pascherii* Bharadwaya 1934
81. *Scytonema pseudo guyanense* Bharadwaya 1934
82. *Scytonema simplex* Bharadwaya 1934
83. *Scytonematopsis ghazipurensis* Pandey & Mitra 1973
84. *Symphonemopsis katriensis* Tiwari & Mitra 1969 gen. nov.

85. *Tolypothrix nodosa* Bharadwaya 1934
86. *Wollea ambigua* R.N. Singh 1942
87. *Wollea bharadwayae* R.N. Singh 1942
88. *Wollea cylindrica* R.N. Singh 1942
89. *Wollea epiphytica* R.N. Singh 1942
90. *Wollea udaipurensis* Gupta & Kumar 1968
91. *Wollea vaginicola* R.N. Singh 1942

10

Spectrum of Diversity in Indian Flora: An Assessment

Neera Sen Sarkar and Anirban Roy

INTRODUCTION

The vastness of the seventh largest country in the world – India – lies not only in its expanse in terms of geographic area but more so in terms of the numerous life forms that it supports owing primarily to its variety in climate and habitat. India is one of the twelve identified centres of origin of a number of taxa which represent a wide array of biodiversity. It is also a megadiversity area with four global biodiversity hotspots (Eastern Himalayas, Indo-Burma, Western Ghats and Sundaland) and several gene centres. Such gene centres or centres of origin of different plants of a taxon represent maximum number of species, maximum number of individuals in a population and maximum variability (Sharma, 2005). This rich and varied assemblage of useful genes and traits in the plant wealth of India has been utilized for crop improvement since ages. About 166 cultivated species are reported to be native to this region along with 320 wild relatives distributed in different agro-ecological zones in India (Arora, 1991). India also has a bounty of complex and diverse ecosystems with three major realms – Palaeoarctic, Indomalayan and Afro-tropical, five biomes – tropical humid forests, tropical dry deciduous forests (including monsoon forests), warm and semi arid deserts, coniferous forests and alpine meadows, ten biogeographic zones – Trans Himalaya, Himalaya, Desert, Semi arid, Western Ghats, Deccan Peninsula, Gangetic plains, Coasts, North-east and Islands and twenty seven biogeographic provinces detailed in Table – 1 (MoEF, 2009; Rodgers & Panwar, 1988; Swarup, 2008; Venu, 1998). Owing to the great variety in physiographic and phytoclimatic conditions, India as a landmass fosters immense variety of vegetation from humid tropical evergreen to moist mixed deciduous sal forests, marshes, swamps, mixed deciduous forests, subtropical pine forests, broad leaved temperate forests, moist temperate broad leaved deciduous forests, temperate conifer forests, subalpine and alpine vegetations, alpine meadows, alpine scrubs and many more types. This is reflected in the fact that with just 2.4% of the global area in terms of landmass, India accounts for more than 11% of the world's known floral diversity. This rich floristic wealth of the nation makes it occupy positions of pride in being the tenth species rich nation in the world and the fourth species rich nation in Asia in terms of floral wealth.

Table 1. Biogeographic zones of India

S. No.	*Biogeographic Zones*	*Biogeographic Provinces*	*% geographical area of India*
1.	Trans Himalaya	1A: Himalaya – Ladakh Mountains 1B: Himalaya – Tibetan Plateau 1C: Trans – Himalaya Sikkim	3.3 2.2 < 0.1
2.	The Himalaya	2A: Himalaya – North West Himalaya 2B: Himalaya – West Himalaya 2C: Himalaya – Central Himalaya 2D: Himalaya – East Himalaya	2.1 1.6 0.2 2.5
3.	The Indian Desert	3A: Desert – Thar 3B: Desert - Katchchh	5.4 1.1
4.	The Semi Arid	4A: Semi – Arid – Punjab Plains 4B: Semi – Arid – Gujarat Rajputana	3.7 12.9
5.	The Western Ghats	5A: Western Ghats – Malabar Plains 5B: Western Ghats – Western Ghat Mountains	2.0 2.0
6.	The Deccan Peninsula	6A: Deccan Peninsular – Central Highlands 6B: Deccan Peninsular – Chotta Nagpur 6C: Deccan Peninsular – Eastern Highlands 6D: Deccan Peninsular – Central Plateau 6E: Deccan Peninsular – Deccan South	7.3 5.4 6.3 12.5 10.4
7.	The Gangetic Plains	7A: Gangetic Plain – Upper Gangetic Plains 7B: Gangetic Plain – Lower Gangetic Plains	6.3 4.5
8.	The Coasts	8A: Coasts – West Coast 8B: Coasts – East Coast 8C: Coasts - Lakshadweep	0.6 1.9 < 0.1
9.	North East India	9A: North – East – Brahmaputra Valley 9B: North – East – North East Hills	2.0 3.2
10.	The Islands	10A: Islands – Andaman 10B: Islands - Nicobars	0.2 0.1

Source: Wildlife Institute of India, 2009

Surveys and explorations have added new discoveries – since 1954, the scientists of BSI have discovered 1 new family, 30 new genera and 859 new species, subspecies and varieties. During 2008, scientists of BSI and other institutes have discovered 3 genera, 108 species and nothospecies, 16 subspecies, 12 varieties and 1 forma as new to science from India and 26 species, 1 subspecies and 3 varieties as new records for Indian flora (BSI, 2008). During 2009, the new discoveries related to plant diversity include among seed plants 1new genus, 53 new species, 1 nothospecies, 7 varieties and 13 new distributional records; among ferns 2 new distributional records; among bryophytes 2 new generic distribution record, 13 new species distributional record; among lichens 30 new species, 1 new generic distributional record, 8 new species distributional record; among fungi 5 new genera and 58 new species and 1 new variety, 2 new generic distribution records, 19 new species distribution record; among algae 4 new species, 1 new distributional record of species; among microbes 4 new genera, 39 new species (BSI, 2009).

The unique features of the plant diversity, among others, include 60 monotypic families and over 6000 endemic species. Recent estimates indicate the presence of over 256 globally threatened plant species in India. India's Fourth National Report to CBD records its impressive contribution to crop biodiversity with repositories of over 50,000 varieties of rice, 5,000 of sorghum, 1,000 varieties of mango, and the National Genebank, primarily responsible for *ex-situ* conservation of unique germplasm on long-term basis, as holding 3,66,933 unique accessions of plant genetic resources. The floristic diversity of the nation is summarized in Table 2.

Table 2. Floristic diversity of India as species number in major groups of plants and microorganisms.

Plant groups	*No. of species described*		*% of India to the world*
	India	*World*	
Virus/Bacteria	850	8,050	10.6
Algae	7,175	40,000	17.9
Fungi	14,500	72,000	20.1
Lichens	2,223	13,500	16.4
Bryophytes	2,500	14,500	17.2
Pteridophytes	1,200	10,000	12.0
Gymnosperms	67	650	10.3
Angiosperms	17,527	2,50,000	7.0
Total described forms	46,042	4,08,700	11.26

Source: BSI, 2009

HIGHLIGHTS OF THE INDIAN FLORISTIC DIVERSITY

A. Algae

The algae as a group are represented by a multitude of phylogenetically unrelated forms. They are ubiquitous and occupy as many and diverse habitats as they are in form and structure, although, they are most commonly found to inhabit aquatic environments. With about two third of the earth surface covered by oceans and seas and another considerable terrestrial area under inland aquatic systems it is easy to fathom the enormous available habitats for the algae to occupy and flourish. India has around 14,500 km of inland navigable waterways along with an enormous coastline with the Bay of Bengal, Indian Ocean and the Arabian Sea washing its coasts. There are twelve major rivers apart from the numerous other small and big rivers, with the total catchment area exceeding 2,528,000 km^2, and a platter of other inland water bodies in the form of lakes, springs, falls, dams, ponds and others. India also has a variety of wetland ecosystems ranging from high altitude cold desert wetlands to hot and humid wetlands in coastal zones. According to recent estimates by the National Wetland Conservation Programme, 115 wetlands have been identified out of which 25 wetlands are of international importance under Ramsar Convention. Added to these, about 4,445 sq. km area of the country is under mangroves. These and others provide an immense diversity and range of habitats within the sub-continent for the algae to inhabit.

Species Diversity

The Fourth National Report submitted to The Convention on Biological Diversity (2009) by the Ministry of Environment and Forests, Government of India accounts for 7,175 algal species from India which is 17.9% of the known world species. These species falling within 666 genera include 4,495 species belonging to Chlorophyceae, 1,453 species belonging to the Cyanophyceae and 516 species belonging to Bacillariophyceae, the rest 711 species belong to the other algal families taken together. The report mentions about the endemecity of 1,925 species to India, which is 26.8% of the described Indian algal species. The report also highlights the well documented account for marine algae from India and has reported the occurrence of 844 marine algal species belonging to 217 genera. Rao & Gupta (1997) have outlined the habitat - wise algal diversity alongwith some dominant genera occurring in India based on reports by various workers on this particular aspect (Table 3).

Table 3. Algal diversity in India based on specific habitats

Algal habitats	*Families and diversity*	*Important genera*
Fresh water algae in rivers, ponds and lakes	390 genera, 4500 species; all families except brown and red algae	*Anacystis, Microcystis, Chara, Nitella, Spirogyra, Vaucheria, Volvox, Chlamydomonas*
Marine algae	169 genera, 680 species; almost all divisions	*Ulva, Halimeda, Sargassum, Padina, Gelidium, Gracilaria, Hypnea*
Mangrove habitat & Brackish-water algae (Sundarbans)	133 genera, 694 species; almost all divisions	*Catenella, Polysiphonia, Rhizoclonium, Enteromorpha, Chaetomorpha, Ulva, Navicula, Coscinodiscus, Lyngbya, Oscillatoria*
Terrestrial algae	125 genera, 615 species; Chlorophyceae, Bacillariophyceae, Xanthophyceae and Cyanophyceae	*Aphanothece, Oscillatoria, Nostoc*
Soil algae	80 genera, 1500 species; Cyanophyceae and Chlorophyceae	*Anabaena, Anabaenopsis, Aulosira, Nostoc, Calothrix, Phormidium, Scytonema, Scenedesmus, Navicula, Pinnularia*
'Usar' soil algae	15 species; Cyanophyceae	*Aulosira, Nostoc, Tolypothrix*
Alpine algae	Chlorophyceae and Cyanophyceae	*Chlamydomonas, Schizothrix, Microcoleus*
Thermal algae	Cyanophyceae	*Mastigocladius, Scytonema*
Parasitic algae	Chlorophyceae	*Chlamydomonas, Chlorella, Cephaleuras, Zoochlorella*
Symbiotic algae	Chlorophyceae, Cyanophyceae	*Zoochlorella, Anabaena, Nostoc*
Pollution-indicator algae	Cyanophyceae, Chlorophyceae, Bacillariophyceae	*Microcystis, Spirulina, Oscillatoria, Nitella*
Sub-aerial algae in bio-films on old monuments	32 genera, 89 species; Cyanophyceae and Bacillariophyceae	*Chroococcus, Tolypothrix, Plectonema, Cyanosarcina, Aphanothece, Scytonema, Nostoc, Lyngbya, Phormidium, Calothrix, Nitzschia, Fischerella*

Source: Modified from Rao & Gupta (1997); additional inputs from Adhikary (2006), Rath (2010) and Sen (2011)

The fresh water system, which amounts to the major portion of Indian algae, Rao and Gupta (1997) have recorded the presence of 4,500 species under 300 genera. An analysis of the described algal diversity from India reflects a greater variety at the generic level than at the species level. The Indian algal flora is also found to be better represented by the classes – Cyanophyta, Dinophyta, Chlorophyta and the subclass Bacillariophyceae (comprising of the diatoms) both at the generic and species levels. Fresh water algal diversity has been a much researched area in India with noteworthy contributions made for Cyanophyta, Volvocales, Chlorococcales, Ulotrichales, Cladophorales, Oedogoniales, Chaetophorales, Zygnemataceae and Desmidiaceae.

The long Indian coastline of above 6000 km undoubtedly harbours a rich marine algal diversity but is definitely low when compared to the world totals. The lower representation can be attributed to the fact that the climate in temperate and arctic regions is more conducive to the growth, diversity and abundance of many representatives of this group. A floristic analysis of the described Indian marine algae by Rao and Mudgal (1997) shows that maximum diversity both at generic and species level is noticed among Rhodophyceae with over 100 genera and over 350 species. This is followed by Chlorophyceae with 36 genera and more than 159 species, with next in line being Phaeophyceae with 33 genera and about 150 species; Xanthophyceae and Cyanophyceae are very poorly represented with just 1 genus (*Vaucheria*) and 3 species of the former and 3 genera and 17 species of the later class. *Caulerpa* – a green algae is the most dominant species among the Chlorophyceae with about 19 species found in India. Other than these marine forms, India is also home to a number of crustose coralline algae, Jayagopal (1985) has reported as many as 60 species belonging to 13 genera of such algae from the coast of India.

With regards to endemism, the fact remains that the aquatic species are especially unique on account of their characteristic forms and adaptation to the specific environments and also due to the absence of factors like isolation and barriers, endemicity is less pronounced (Rao & Mudgal, 1997).

Uses of this group for the benefit of mankind are manifold in direct as well as indirect terms. The algae as a group have a considerable role in maintaining environmental equilibrium; they are also the most important contributors as the primary producer group for sustaining the aquatic ecosystem that they inhabit at large and take care of the diverse consumer groups which eventually form the backbone of the economics related to the fisheries in such systems. People throughout the globe use algae for purposes of food, fodder and fish food, medicine, cosmetics and also in rituals. The algae that are put into maximum use for meeting these requirements include the seaweeds and to some extent the blue greens. Following their use as source of food, the next most valuable commercial use of seaweeds is as raw material for extraction of phycocolloids (agar, alginate and carrageenan), which are used in several industries. In the health food sector, poly-unsaturated fatty acids (PUFA) from algal and fungal sources are quite in demand, because of their value in prevention and treatment of heart and circulatory diseases, inflammation, asthma, arthritis, atherosclerosis, thrombosis, diabetes, ageing and certain viral infections and also because of growing world vegan preferences. An avenue of research and development that has gained momentum throughout the world in the recent past relates to using algae for the purpose of meeting the ever increasing fuel needs. In the scenario of meeting the challenges of climate change and fast dwindling sources of fuel, this is an area to look forward to. Since the unutilized and obnoxious algal scums and blooms that meet our eyes everyday are nothing but waste unless and until properly utilized, simple rural technologies need to be developed for taking up the challenge of converting waste to wealth.

Status

With the current state-of-art of algal research in India pertaining to status of Indian algae, it would be difficult to designate rare, threatened and endangered species for this group. At the same time, it is easy to comprehend that such resources, as is with all other bio-resources, are definitely not unlimited or secure from the vagaries of climate change and other anthropogenic pressures. So, it will be imprudent to neglect their immense role per se, immaterial of whether they are freshwater or marine, aquatic or terrestrial. It is of utmost importance to explore, survey and inventorise all habitats of occurrence of this group, so as to provide exhaustive information regarding algal diversity for future sustainable use and conservation of the same.

B. Fungi

The fungi constitute a large group of organisms found throughout all ecological niches. The organisms traditionally referred to as fungi are not closely related but in general comprise of eukaryotic, spore-producing, achlorophyllous organisms with absorptive nutrition that generally reproduce both sexually and asexually and whose usually filamentous, branched somatic structures – hyphae, are surrounded by cell walls. These non-flowering organisms subsist as parasites or saprophytes and are as numerous and varied as their flowering counterparts, ranging from microscopic to huge solid bodies in size to a multitude of colours and forms. India has 14,500 species of fungi within 2,300 genera and 250 families (Sharma, 1997a; Bilgrammi *et al.* 1991). The maximum diversity of the group is recorded in the Western Ghats followed by the eastern Himalaya and the western Himalaya. Deuteromycetes with 900 genera and 6000 species (40%) is the largest group of Indian mycoflora, followed by Ascomycetes [680 genera / 3500 species (25 %)] and Basidiomycetes [520 genera/3400 species (23 %)]. *Cercospora* with 707 species is the largest genus of Indian fungi followed by *Puccinia* (328 species) and *Phyllosticta* (280 species). Sharma (1997a) has reported that out of the total species known from India about 3,500 are endemic. Sharma (1997b) has also accounted the aphyllophoraceous or the wood rotting fungi from India and has documented the presence of 388 wood rotting fungi from India, out of which about 50 species (about 12%) grow on ground probably forming mycorrhizic associations with forest trees, while the rest (about 88%) are lignicolous and one species grows exclusively on fungi.

Species Diversity

The Indian fungal diversity is found to comprise of ten dominant groups – True slime molds (Myxomycetes), Fungi with motile cells (Chytridiomycetes, Hypochytidiomycetes and Oomycetes), Bread molds (Zygomycetes), Sac fungi (Ascomycetes), Smuts (Ustiliginales), Rusts (Uredinales), Mushrooms (Agaricales), Wood rotting fungi (Aphyllophorales), Stomach fungi (Gasteromycetes) and the Fungi Imperfecti (Deuteromycotina) (Sharma, 1997a). The Deuteromycotina alongwith Ascomycetes and Basidiomycetes (including the rusts, smuts, mushrooms, wood rotting fungi and the stomach fungi) account for more than 88% of the Indian fungal diversity. Deuteromycotina alone represents about 40% followed by Ascomycotina (about 25%) and Basidiomycotina (about 23%). An interesting observation by Sharma (1997) is that about 46% of the Indian fungal genera have only one species, while 48% genera have two to ten species.

The true slime molds are represented by about 350 species under about 45 genera, out of which about 30 species are known to be endemic. The endoparasitic slime molds, found to be parasitic on algae, fungi and vascular plants have not more than 5 genera and 15 species reported from India. The dominant genera within the Indian representatives of true slime molds include *Physarum* (about 65 species), *Didymium* (about 32 species) and *Diderma* (about 30 species). These organisms have also gained importance as ideal tools for experimental studies on the mitotic cycle, morphogenesis, chemical changes that govern reproduction and the structure and physiology of protoplasm. The Mastigomygotina or the fungi with motile spores are represented by 710 species under 87 genera of which the Hypochytridiomycetes do not have more than 2 genera and 10 species in India. The endemic species under this group amounts to 140 species. Within this group, *Synchytrium* and *Physoderma* are the largest representative genera with 80 and 45 species respectively, which inhabit soil and parasitize vascular plants. The Zygomycotina with a majority of representatives as saprobes, have more than 57 genera and 360 species in India, of which 52 are endemic. The dominant and familiar genera include *Mucor, Rhizopus, Cunninghamella, Conidiobolus* and *Entomophthora.* The sac fungi in India are represented by 680 genera and about 3,500 species of which 900 species are endemic. These include the yeasts and yeast like organisms, cup-fungi, earth-tongues, morels, truffles and many parasitic forms which cause serious plant diseases like powdery mildews, anthracnose, die back disease, scabs and blights. The Basidiomycetes are known in India by about 520 genera and about 3,400 species of which 500 are endemic. These include among many others the smut and rust causing organisms for which they are organisms of immense interest and research as they are responsible for incurring heavy damage to cereal crops, ornamentals and conifers. These also comprise of the mushrooms whose fruiting bodies are fleshy of which many are of commercial value for being excellent edible species and many more find importance in not necessarily being market commodities but means of sustenance for the ethnic groups and rural communities in India. The group also includes the wood rotting fungi growing from tropical to timberline zone in the Himalayas on both hardwoods and softwoods causing either brown rot or white rot infections in the hosts. The group is also represented by the stomach fungi including the puffball, earth star, stink horn and bird's nest fungi. The Deuteromycotina are represented in India by over 900 genera and 6000 species of which about 1,850 are endemic. These organisms have septate mycelium and reproduce only by means of conidia. Many of these are parasites and cause serious diseases of plants, animals and human beings. The group is also known for a number of antibiotic producing species. Other than the uses mentioned above, the fungi as a group are considered important for other aspects as in their role in mycorrhizal associations which results in increased plant nutrient supply, in commercial extraction of antibiotics and antioxidants for medicinal purposes.

Status

As is with lower plant groups, there is extreme paucity of information on Indian fungal diversity with regards to information pertaining to distribution, habitat, ecology and status of the same. Added to these is the non existence of a uniformly accepted agreement on how to define threatened, endangered fungi. A significant observation about this group is that the habitat preferences and species distribution are so interdependent that it makes it a prudent choice to define and conserve habitats in the first place to conserve fungi. Also documentation is necessary regarding wild populations and their conservation requirements before the fungal flora can be subdivided into rare or endangered species.

C. Lichens

Lichens representing symbiotic association of fungi and algae, constitute a dominant component of epiphytic and saxicolous vegetation, and comprise 2,223 species in 283 genera and 72 families. Lichens grow on various substrata wherever conditions conducive to their growth are available. The major categories of lichens belong to saxicolous, corticolous, terricolous, ramulicolous and muscicolous types. The nature of lichens growing in an area is dependent on the variations in climate and altitude of the area. Indian lichen flora can be broadly classified into three major types – tropical and subtropical, temperate and alpine (Singh & Sinha, 1997). Singh and Sinha (1997) have also divided the Indian region into 8 lichenogeographical regions. Western Ghats are the richest region with 800 species (38%) followed by Eastern Himalaya with 759 species (37%) and Western Himalaya with 550 species (27%). Families such as Parmeliaceae, Graphidaceae, Physciaceae, Usneaceae, Cladoniaceae, and genera like *Parmelia, Graphina, Usnea, Graphis and Lecanora* are among the dominant families and genera of Indian lichens. About 23% species, mainly belonging to genera *Graphina, Trypethelium, Graphis* and *Porina*, are endemic to India. Andaman & Nicobar Islands (24%), Western Ghats (20%) and Eastern Himalaya (18%) show high percentage of endemic species. A region wise distribution of lichen flora and endemic taxa in each lichenogeographical region as enlisted by Singh and Sinha (1997) is detailed in Table 4.

Table 4. Lichen diversity and endemism in the lichenogeographical regions of India.

Lichenogeographical region	*Total No. of lichen species*	*No. of Endemic species*
Western Himalayan region	550	22
Eastern Himalayan region	759	133
Western Dry region	39	6
Gangetic Plain	224	14
Central India	48	4
Western Ghats	800	161
Eastern Ghats & Deccan Plateau	31	4
Andaman & Nicobar Islands	307	73

Source: Singh & Sinha (1997)

Species Diversity

The tropical and subtropical lichen vegetation is represented by crustose, foliose, fruticose and foliicolous species. The temperate climate is best suited for the growth of foliose and fruticose forms. The generic composition of alpine lichen vegetation is very similar to the ones found in higher temperate areas but the species components are different. Singh and Sinha (2010) in their exhaustive work on the Indian lichens have consolidated information regarding diversity of lichens and reported the presence of 2303 species under 305 genera and 74 families occurring on various substrata in tropical, subtropical, temperate and alpine regions of India. Out of these about 520 species are endemic, many of these being neoendemics belonging to crustose forms, described mostly from Western Ghats and Andaman and Nicobar Islands. The foliose family Parmeliaceaea shows maximum species diversity of about 345 species and is found to be widely distributed in the subtropical, temperate and alpine regions of India. This is followed by the crustose family Graphidaceae with 279 species with distributional records from the tropical and subtropical regions. The 10 dominant families reported from India include Parmeliaceae, Graphidaceae, Thelotremataceae, Pyrenulaceae, Caliciaceae, Lecanoraceae, Physciaceae, Trepetheliaceae, Telochistaceae and Collemataceae. At the generic level, the genus *Graphis* exhibits maximum diversity represented by 111 species distributed in the tropical and subtropical regions of India. This is followed by *Pyrenula* with 90 described species distributed in the tropical regions of the country. The dominant lichen genera described from India other than *Graphis* and *Pyrenula* include *Lecanora, Caloplaca, Usnea, Porina, Cladonia, Parmotrema, Pertusaria* and *Thelotrema*.

Status

Lichens do not differ from other plant groups from being under threat from different sources. Lichens have very limited ability to adapt themselves to abrupt environmental changes. Consequently they exhibit abundance and diversity in any particular region only when the conditions have remained stable over a long period of time. Among other factors, pollution takes it toll on the lichen diversity and abundance. The commercial value of lichens is well known and is varied as food, antibiotics, UV absorbers, antioxidants, anti-cancer agent, dye stuff, fodder and perfume. As such, overexploitation of many commercially important species for various industrial purposes is another major threat to these species. According to an estimate, about 1000 tonnes of lichen material is collected every year from the western and central Himalayas (Singh & Sinha, 1997). In India, they are mainly used in dyeing, flavouring, smoking tobacco and cigars, in sambar masala, scenting soaps and cosmetics and manufacture of incense sticks. They are also used as spices and flavouring agents by santhal tribal populations in many states.

D. Bryophytes

Bryophytes represented by 2500 species are the second largest group of green plants in India distributed largely in Eastern Himalaya, North-eastern India, Western Himalaya and the Western Ghats. Based on the structure and other characteristics, the bryophytes are broadly grouped as liverworts, mosses and hornworts. Mosses constitute the major component of Indian bryoflora with 1576 species followed by liverworts and hornworts (924 species). The bryophytes are more adapted and found to flourish in the moist tropical - tropical mountain forests to arctic tundra. The Indian forms are found to reach an altitude of over 5000 m above the mean sea level in the Himalayas (Singh, 1997). They grow in groups and patches, but their small size renders them unable to compete with higher order vascular plants for light and space and are totally outcompeted in certain habitats like in grasslands. Interestingly they grow on unusual surfaces like animal carcasses, abandoned articles, humified fecal material, etc., and majority of representatives of this group are found to be shade and moisture loving predominantly terrestrial forms, while some are epiphytic and epiphyllous, a few others absolutely aquatic and still fewer xeromorphic in nature, capable of enduring drought and may even grow under arid conditions. Pande (1958) recognized seven Bryogeographical regions in India based on the studies of bryophytic flora of India – Eastern Himalayas, Western Himalayas, Panjab and Western Rajasthan, Gangetic Plains, Central India, Western Ghats, and the Eastern Ghats and Deccan Plateau, and Singh and Semwal (1995) proposed the eighth region of Andaman and Nicobar Islands to be included within these regions. The liverwort vegetation of the country is primarily found in three broad regions – tropical-subtropical, temperate and alpine. They usually prefer higher elevations, showing best manifestation between the altitudinal range of 1000-3500 m above mean sea level. In the context of mosses, Himalayas form the largest repository in the country for moss diversity. With its wide altitudinal ranges and varied climatic conditions from humid to cold deserts, the area offers suitable habitats for a wide spectrum of species diversity.

Species Diversity

The three major liverwort vegetation types in India, support a wide range of species diversity comprising about 850 species, including infraspecific categories, belonging to over 140 genera and 52 families (Singh, 1997). Jungermanniales with about 99 genera and 630 species is the largest liverwort Order in India. The order Treubiales is represented by a single genus and single species only. The dominant families of liverworts in India include Lejeuneaceae with 36 genera and 155 species representing the largest family amongst the liverworts followed by Plagiochilaceae (119 species), Jungermanniaceae (76 species), Jubulaceae (57 species), Rebouliaceae (40 species), Geocalycaceae (37 species), Ricciaceae (37 species), Lepidoziaceae (36 species), Metzgeriaceae (25 species) and Scapaniaceae (24 species). The dominant genera include *Plagiochila* (114 species), *Frullania* (54 species), *Jungermannia* (38 species), *Riccia* (36 species), *Cololejeunea* (30 species), *Porella* (26 species), *Asterella* (24 species), *Metzgeria* (23 species), *Scapania* (22 species) and *Lejeunea* (21 species). The India liverwort diversity is also very important in the world context since family Haplomitriaceae [with 6 out of 14 species in the world (Chandra, et al. 1987)], genera *Cyathodium* [with 9 out of 12 species in the world (Singh, 1997)], and *Notothylas* [with 8 out of 18 species in the world (Singh, 1997)] have maximum representation in India, which is also highest for any geographical region of the world. This apart from the discovery of certain fossil liverworts considered together with considerably high endemism is indicative of the uniqueness of this group in the country. The diversity of mosses in India includes 2000 species belonging to 342 genera under 54 families, of these 1030 occur in the Eastern Himalayas, 751 in Western Himalayas and 540 in Western Ghats (Vohra & Aziz, 1997). The dominant families of mosses as consolidated and reported by Vohra and Aziz (1997) are represented by Pottiaceae (37 genera, 190 species), Bryaceae (12 genera, 150 species), Dicranaceae (29 genera, 146 species), Hypnaceae (18 genera, 92 species), Sematophyllaceae (25 genera, 84 species), Brachytheciaceae (9 genera, 78 species), Neckeraceae (11 genera, 73 species), Fissidentaceae (1 genera, 67 species), Orthotrichaceae (12 genera, 67 species), Thuidiaceae (10 genera, 53 species), Polytrichaceae (6 genera, 47 species) and Meteroriceae (11 genera, 46 species). The dominant genera include *Fissidens* (67 species), *Bryum* (59 species), *Campylopus* (41 species) and *Brachythecium* (39 species). Out of the reported moss diversity of India, 678 species are reported to be endemics.

Status

The bryophytes have a substantial and distinctive influence on the ecosystems, primarily by means of habitat modification, nutrient cycling and maintenance of nutrient status of the soil, primary production and by being bio-accumulators and bio-monitors. The group contributes significantly towards primary succession on bare rocks. They also influence the composition of other communities by creating a moist environment, available round the year and associated with equitable temperature regimes providing suitable conditions for a variety of organisms like orchids, ferns, microorganisms and invertebrates to colonise such areas. This apart, the organisms are themselves adapted to narrow niches and have succeeded within the restrictions of their evolutionary limits. Though no immediate data is available on the rarity or non-availability of these entities from respective localities in the country, the group with such a delicate choice for microhabitats, is highly susceptible to mass habitat destruction. Biological imparity also contributes towards the vulnerability of this group at large, rarity of male plants in some cases, reliance on asexual reproduction by a large number of species and the tendency of many to adapt themselves somatically rather than genetically to various environmental changes makes these non-adaptive species more prone to elimination than any other group in case of habitat destruction. In terms of commercial exploitation, the mosses are facing a new threat from household gardening enthusiasts. Truck loads of mosses are transported for meeting the needs of such demands, invariably this total extraction is directly from the wild.

E. Pteridophytes

Pteridophytes differ from their phanerogamic counterparts in not forming dominant vegetations anywhere, instead inhabiting areas wherever suitable conditions area are available. The different pteridophytic flora has been categorized into the broad groups – terrestrial, epiphytic, hill epiphytes (upper hill, and lower and mid hill), lithophytic, xerophytic, sun-loving, dark shady forest loving, mangrove, climbing, and aquatic and humid pteridophytes (Ghosh & Ghosh, 1997). The north-eastern region (including Eastern Himalaya) is rich in pteridophytic diversity with about 845 species, followed by south India (including Eastern and Western Ghats) and north India (including Western Himalaya).

Species Diversity

India has about 1200 species under 204 genera with about 17% of the species endemic to India (MoEF, 2010). Ghosh and Ghosh (1997) have broadly divided the country into 4 pteridophytic zones – Eastern India (including Eastern Himalayas, North Eastern states and Eastern India), Southern India (including Central India, Eastern and Western Ghats), Western India (including Garhwal and Kumaon Himalaya, Jammu and Kashmir state, parts of U.P., states of North Western India) and Andaman & Nicobar Islands (including all the islands in the Bay of Bengal under Indian jurisdiction). A floristic analysis of the described pteridophytes of India reveals that maximum number of species occurs in Eastern India (810 species), followed by Southern India (336 spp.), Western India (321 species) and Andaman & Nicobar Islands (125 species). The families such as Polypodiaceae (137 species), Dryopteridaceae (125 species), Athyriaceae (97 species), Thelypteridaceae (83 species), Selaginellaceae (62 species), and Pteridaceae (61 species) are some of the dominant families and genera of the pteridophytic flora of Indian region. The genera like *Selaginella* (62 species), *Pteris* (53 species) *Dryopteris* (60 species), *Asplenium* (45 species) and *Polystichum* (45 species) show maximum species diversity amongst the described Indian genera. Occurrence of monotypic genera in the flora imparts element of interest to the pteridophytic floral diversity of India. Endemism is reported in about 193 species of pteridophytes in different parts of India. Maximum endemism is reported to occur in Eastern India particularly in the Himalayan region. Pteridophytes have limited uses in commercial terms, though there exist a number of species which have great ornamental and medicinal values. Several species of *Pteris, Adiantum, Asplenium* and a few other are popular for their beautiful and variable frond forms and have gained importance in horticulture. A few are eaten as vegetables, specially the circinnately coiled fronds are favourites of hill people.

Status

Ghosh and Ghosh (1997) have classified 254 species (21.58%) as falling under various threat categories based on herbarium and field observations. The main causes of such threat perceptions are attributed to habitat destruction, availability of thin population and over exploitation. The endangered species include *Isoetes coromandelina* ssp. *brachyglossa* (distributional record from Serampur, West Bengal), *Emodiopteris elwesii* (distributional record from Sikkim), *Christiopteris tricuspis* (distributional record from Sikkim and West Bengal) and *Drynaria meeboldii* (distributional record from Manipur). The rare and vulnerable species include a number of species of *Isoetes, Selaginella, Ophioglossum, Osmunda, Adiantum, Pteris, Onychium, Pyrrosia, Lepisorus, Phymatopsis, Asplenium, Loxogramma, Woodsia, Dryopteris, Lunathyrium, Polystichum, Christella* and many more. The pteridophytes suffer from the same problems of lack of intensive documentation and research as all of its lower plant group counterparts. Change in habitat and in extreme cases total habitat destruction pose serious challenge to such habitat specific plants. Protection of the habitats will go on to ensure the conservation of this important plant group. Pteridophyte rich habitats need to be identified to facilitate such endeavours. Moreover, species of *Selaginella*, *Lycopodium* and other such lower group plants are becoming vulnerable in nature on account of indiscriminate collection by students. College and University teachers can play very important role in this direction for the purpose of conservation.

F. Gymnosperms

Gymnosperms are the most primitive type of seed plants and are of great importance due to the presence of both fossil and living forms. They are represented by woody perennials as either shrubs or trees. The extant gymnosperms are commonly divided into five orders – the Cycadales, Ginkgoales, Taxales, Coniferales and Gnetales. The Cycadales and Ginkgoales include living members that have long fossil records and represent the living fossils. Among the living Gymnosperms, the Coniferales form the most conspicuous order that presently are responsible for the formation of a typical forest type in the temperate regions of the Northern and Southern Hemisphere. The genera which build up such forests include *Pinus, Cedrus, Abies, Juniperus, Cupressus, Thuja,* etc. This group is represented by about 67 species in India (MoEF, 2010).

Species Diversity

In India, the living gymnosperms grow luxuriantly and form extensive forests in the Western and Eastern Himalayas. All the five orders known in gymnosperms - Cycadales, Coniferales, Taxales, Ginkgoales, Gnetales are represented in the Indian flora. Cycadales, represented by a single family Cycadaceae and single genus *Cycas* with 4 species is widely distributed. The Taxales comprise single family Taxaceae with two genera – *Taxus* and *Amentotaxus* having one species each. *Amentotaxus assamica* is extremely rare and confined to certain places only in Delai valley of Arunachal Pradesh (Singh & Mudgal, 1997). The Coniferales represented by 10 genera and 25 species in India, are well represented in the tropics but are abundant in temperate areas of both the Hemispheres. The genera that represent the Indian Coniferales include *Pinus, Abies, Picea, Tsuga, Cedrus, Larix, Cephalotaxus, Podocarpus, Juniperus, Cupressus, Araucaria, Agathis* and *Podocarpus*. Ginkgoales is represented by a monotypic family Ginkgoaceae, containing both extinct and only one living species – *Ginkgo biloba*, which still exists in the wild state in restricted areas of South-eastern China. It has been introduced in India and has survived in the parks and gardens of Mussorie, Dehradun, Darjeeling, Shimla and elsewhere with moist and cool climate. Among the known gymnosperm families found in India amounting to 11, Ginkgoaceae, Araucariaceae and Taxodiaceae are exotics. Within the remaining 8 families (Pinaceae, Cupressaceae, Ephedraceae, Gnetaceae, Cycadaceae, Taxaceae, Cephalotaxaceae and Podocarpaceae), Pinaceae with 6 genera and 15 species is the largest family. This is followed by Cupressaceae (13 genera and 13 species), Taxaceae (2 genera and 2 species), Ephedraceae (1 genus, 8 species), Gnetaceae (1 genus and 5 species), Cycadaceae (1 genera and 4 species), Cephalotaxaceae (1 genera and 2 species) and Podocarpaceae (1 genera and 2 species). The species of *Gnetum* and *Cycas* are mostly confined to North Eastern region, Eastern and Western Ghats, and Andaman & Nicobar Islands. It is also of interest to note that among the Indian gymnosperm genera, maximum number of genera belong to single species category. The Indian gymnosperms are represented by 7 endemic species and 26 exotic species (Singh & Mudgal, 1997). The endemic taxa are distributed in restricted areas and demand attention from the conservation point of view.

Status

About 8 species of the gymnosperm representatives in India have been found to be rare with two of them (*Cycas beddomei* and *Amentotaxus assamica*) being rare (Singh & Mudgal, 1997). Habitat destruction and other threats to biodiversity pose similar problems to the gymnosperm flora in India also. The major threat to the members comes in the form of overexploitation because of the highly valued timber of the representative species. The extraction of resins from very young trees of *Pinus* species is also a matter of concern, since it affects the growth and reduces the life span of the individual trees. Components of such economic importance need to be included within the scope of afforestation programmes to ensure sustainable use of the same. There exist many rare, endangered and endemic species within the group (*Cycas circinalis* var. *orixensis, Cephalotaxus mannii, Cephalotaxus griffithii, Pinus wallichiana* var. *parva, Pinus gerardiana, Pinus bhutanica, Pinus merkusii, Abies delavayi, Gnetum montanum* var. *megalocarpum, Gnetum contractum* and *Gnetum latifolium* var. *macrocarpum*) which need special attention and immediate conservation efforts.

G. Angiosperms

Angiosperms are by far the most numerous, diverse and 'successful' extant plant group, containing well over 95% of all land plant species alive today (Simpson, 2006). This monophyletic group is presently thought to be the sister group of the Gymnosperms. The members of this large group have well habitated almost all ecosystems of the globe. The diversity and distribution of this flowering plant group varies across countries through the continents depending upon different environmental conditions. India, being one of the megadiversity countries of the world possesses a richness of angiospermic floral species mainly due to it's varied topography, climate and geology. The country has 2.4% of the global area and harbours 48,000 species of plants which is about 11% of the world's flora.

Species Diversity

Out of these 48,000 species, angiosperms constitute 17,527 species under 2,984 genera and 247 families (more than 7% of the world's known flowering plants) (BSI, 2009). Among these the Dicotyledonous group has 12,750 species under 2,282 genera and 203 families whereas monocots are with 4,250 species under 702 genera and 44 families (BSI, 2001). The species composition of India reflects that within the angiosperm group dicot flora constitutes 75% and monocots the rest 25% both at generic and species level. The dominant families with more than 500 species are Poaceae, Orchidaceae, Fabaceae, Asteraceae, Rubiaceae, Cyperaceae, Euphorbiaceae and Acanthaceae. But maximum species diversity is found in Poaceae having 1,291 number under

263 genera followed by Orchidaceae with 1,229 species under 184 genera. Closely following these two monocot families are Fabaceae (Leguminasae) which occupies the third position in terms of species diversity (1,225 species) and Asteraceae with 892 species. The ten top families (Table 5) of dicots cover 40% of the genera and 41.6% of species recorded so far. The remaining 193 families of dicots constitute 60% genera and 58.4% species. The dicot families show 61 families with 1 genus, 87 families with 2-10 genera and 27 families with 36 genera. In monocots 18 families are with a single genus, 19 families with 2-10 genera and 6 families with more than 100 genera. The numerical data reflects low generic diversity among most of the angiosperm families prevailing in India. The 7 top monocot families cover 87% genera of the monocots and remaining 37 families constitute only 95 genera. About 2,863 (16.4%) are trees, which include some of the highly valued timber species of the world. India is also a storehouse of primitive flowering plants, confined mainly to the North Eastern region of the country. Diversity of such plants led Takhtajan (1969) to designate this region as the "Cradle of Flowering Plants. The Indian flora also shows a rich diversity in aquatic flowering plants. The insectivorous plant families, yet another group of unique plants, are represented by Lentibulariaceae (36 species), Droseraceae (3 species), and Nepenthaceae (1 species). About 11,058 species are endemic to Indian region, 6,200 of which belong to flowering plants alone. Eastern Himalaya and north-eastern region (about 2,500 species), peninsular India including western and Eastern Ghats (about 2,600 species), north-western Himalaya (about 800 species) and Andaman & Nicobar Islands (about 250 species) are the areas rich in endemic plants.

Table 5. Angiospermic families with maximum diversity in India

DICOT			*MONOCOT*		
Name of Families	*No. of Genera*	*No. of Species*	*Name of Families*	*No. of Genera*	*No. of Species*
Fabaceae	133	1225	Poaceae	263	1291
Asteraceae	166	892	Orchidaceae	184	1229
Rubiaceae	113	616	Cyperaceae	38	545
Euphorbiaceae	84	527	Liliaceae	45	214
Acanthaceae	92	510	Araceae	29	126
Lamiaceae	72	435	Zingiberaceae	24	191
Rosaceae	40	432	Aricaceae	24	24
Apiaceae	72	288			
Brassicaceae	64	207			
Scrophulariaceae	62	368			
Asclepiadaceae	57	260			

In India the angiosperm wealth has key contribution to human welfare and economic upliftment. The great diversity of this group provides wide range of utilization options in being important sources of food, timber, medicine, vegetable oils, gums, resins, condiments and spices, fibres, dyes and tannin, narcotics, beverages, pesticides and many more to be accounted for.

The Indian angiospermic plants include 2,560 species with strong and durable wood that are often used as commercial and household timber. The majority of these species mostly belong to the families Euphorbiaceae, Lauraceae, Annonaceae, Rubiaceae, Moraceae and Fabaceae.

There are about 3000 angiosperm species which are regularly exploited for food as cereals, pulses, fruits, vegetables, beverages etc. Out of these species only 15-20 species are the major crops of India. 70% of the consumed Indian cereals are from *Triticum* spp., *Zea mays* and *Oryza* spp. only and the rest is supplied by other species like *Fagopyrum exculantum, Chenopodium* sp., *Amaranthus* sp., *Hordeum vulgare, Avena sativa, Secale cereal, Eleusine coracana, Setaria italica, Sorghum* sp., *Echinochloa* sp., *Panicum miliaceum, Pennisetum* sp. Second to the cereals as a source of human food in India are the pulses which have nearly 20 species under the family Fabaceae. The major Indian pulses are *Cicer arietium, Cajanus cajan, Dolichos* spp., *Phaseolus* spp., *Vigna* spp., *Glycine max, Lathyrus sativus, Pisum sativum, Vicia faba, Lens esculanta.*

The diversity of food crops in India is very remarkable both in wild and cultivated forms. The majority of food crops in India are obtained from only two families: Rosaceae, and Rutaceae. In Rutaceae, only *Citrus* has several species that supply fruits next to the family Rosaceae. Other importat families are Vitaceae, Anacardiaceae, Musaceae, Arecaceae, Bromeliaceae, Oleaceae, Sterculiaceae, Tiliaceae, Clusiaceae, Sapindaceae, Lauraceae, Myrtaceae, Rhamnaceae, Punicaceae, Moraceae, Annonaceae, Cucurbitaceae, Caricaceae etc. Another significant source of food which are rich in oils and fats are the nuts that are obtained mostly from the dicotyledonous families. The important nut yielding species in India are *Anacardium occidentale, Buchanania angustifolia, Juglans regia, Prunus communis, Terminalia catappa, Trapa natans, Aleurites moluccana* etc.

More than 1,200 species of angiosperms both cultivated and wild, supply vegetables to the people of India. Though the species of these vegetables belong to different families, still only four families contribute most of them. These are Cucurbitaceae, Solanaceae, Brassicaceae and Fabaceae. The other important families of vegetables are Dioscoreaceae, Euphorbiaceae,

Convolvulaceae, Araceae, Chenopodiaceae, Liliaceae, Apiaceae, Asteraceae, Amaranthaceae, Malvaceae, Polygonaceae etc.

Three major nonalcoholic beverages (tea, coffee and cocoa) are obtained from *Camellia chinensis, Coffea arabica and Theobroma cacao.* Other than these three, alcoholic beverages are obtained by using the fresh sap of *Borassus flabellifer, Phoenix sylvestris* and from the perianth of *Madhuca longifolia.* The sap of *Arundinaria* and *Bambusa* and the seeds of *Oryza* and *Elusine* are also used for making country liquor.

Sugar, an important gradient in food industry and alcoholic beverages is mainly obtained from *Saccharum* spp. and *Beta vulgaris.* Apart from these two genera a small proportion of sugar is also obtained from *Phoenix sylvestris, Borassus flabellifer, Cocos nucifera , Sorghum vulgare* var. *saccharatum.*

In the paper industry over 30 species, majority of which are the members of Poaceae & Bambusaceae, are used for paper pulp. The wood of many trees like *Boswellia serrata, Rhizophora apiculata* is also very important source of paper products.

Natural fibres obtained from different plant resources is very important for both large scale as well as cottage industries. Presently, near about 2,000 angiosperm species contribute their fibres to human kind. The major soft fibres of commercial importance are obtained from *Linum utitatissimum, Corchorus capsularis, C. olitorius, Crotalaria juncea, Cannabis sativa, Hibiscus cannabinus, H. sabdariffa, Helicteres isora, Boehmeria nivea* and others. Some important fibres are contributed by the members of the monocotyledonous families like Poaceae (*Eulaliopsis binata, Phragmites karka, Dinochloa andamanica,* species of *Saccharum, Dendroclamus, Arundinaria, Bambusa* etc.), Cyperaceae (*Cyperus corymbosus*), Liliaceae (*Agave sisalana, Yucca gloriosa*), Musaceae (*Musa textiles*), Arecaceae (*Borassus flabellifer, Calamus* spp., *Caryota urens*) and Pandaceae (*Pandanus odoratissimus*). Cotton is a very important commercial fibre obtained from *Gossypium* spp., *Ceiba pentandra, Bombax ceiba, B. malabaricum, B. insigne* etc. *Cocos nucifera* also contributes surface fibre to the industries.

Angiospermic species are important sources of vegetable oils. Though more than hundred species of India contribute essential oils, yet only 10-12 species provide commercially important oils. The major oil yielding plants are *Cocos nucifera, Brassica juncea, B. nigra, Helianthus annuus, Linum utitatissimum, Sesamum indicum, Ricinus communis, Olea europes, Arachis hypogea, Glycine max, Carthemus tinctorius* etc. Other comparatively less important oil yielding plants are *Pongamia pinnata, Calophyllum inophyllum, Madhuca longifolia, Schleichera oleosa, Garcinia indica, Jatropha karkas* etc.

About 370 species of angiosperms are exploited for providing essential oils often used in perfumery, flavouring agents and medicines.

India is a haven for spices, condiments and flavouring agents. More than 70 angiosperm species, both cultivated and wild, are used as spices in India. The major ones include *Piper nigrum, Zingiber officinale, Elettaria cardamomum, Cinnamomum zeylanicum, Eugenia caryophyllus, Myristica fragans, Capsicum annum, Cuminum cyminum, Coriandrum sativum, Foeniculam vulgare, Trigonella corniculata, Nigella sativa, Amomum* sp., *Vanilla planifolia, Allium* sp., *Mentha* sp., *Crocus sativus* etc.

Gums and resins are very important in different industries for their adhesive properties. In India 125 and 140 species of angiosperms contributing to yielding gums and resins respectively. *Acacia senegal, Astragalus fatmensis, A. gummifer, Cartonia siliqua* of the family Fabaceae, *Sterculia urens* of Sterculiaceae, *Cochlospermum religiosum* of Cochlospermaceae and *Anogeissus pendula* of Combretaceae are the significant ones for commercial gums. The resins are contributed by the members of Fabaceae, Dipterocarpaceae, Burseraceae and Apiaceae (for scented resins). Besides, Euphorbiaceae, Moraceae, Clusiaceae, Anacardiaceae, Apocynaceae, Solanaceae, Cannarbinaceae, Aricaceae etc. are also contributors of Indian gums.

More than 70 species of angiosperms under 80 families are used as source of rubber in India. But maximum amount of rubber is obtained from *Hevea brasiliensis, Manihot* spp. (Euphorbiaceae); *Castilla elastica, Ficus elastica* (Moraceae); species of *Taraxacum, Solidago* (Asteraceae); *Cryptostegia grandiflora, Asclepius* spp.(Asclepiadaceae); *Funtumia elastica, Couma macrocarpa* (Apocynaceae) and *Mimusops balata* (Sapotaceae). 95% of total production of natural rubber is contributed by *Hevea brasiliensis* only.

Wax is another important material for industrial and domestic use and is derived from the leaves of the palm, *Copernicia cerifera* (Arecaceae) and the stem of *Euphorbia antisyphilitica* (Euphorbiaceae).

About 550 species of angiosperms in India yield vegetable tannins and dyes which have been used since the early civilization. These dyes and tannin yielding plants are mostly dicotyledonous under the families of Fabaceae, Myrtaceae, Anacardiaceae, Polygonaceae, Rhizophoraceae, Sonneratiaceae and Combretaceae. Leaf galls and nut galls of *Quercus* spp. and *Pistacia vera* are very important rich sources of tannin. Monocots are very poor as sources of tannin except the family Arecaceae of which *Areca catechu* and *Phoenix sylvestris* are two important tannin sources.

The commercial dye yielding sources in India are mostly with three species: *Pterocarpus santalinus, Bixa orellana* and *Acacia catechu.* Other important dye yielding species are *Caesalpinia sappan, Artocarpus heterophyllus, Acacia* sp., *Mallotus phillippinensis, Morinda* spp., *Erythrina* spp., *Butea monosperma, Crocus sativus, Carthemus tinctorius, Lawsonia inermis, Indigofera* spp., *Uncaria gambier* and others.

There are several angiosperm species mostly belonging to Poaceae and Cyperaceae which are often used as fodder. Near about 1,250 species are the major sources of fodder. Some dicotyledonous plants like the species of *Ficus, Grewia, Terminalia, Zizyphus, Ipomoea, Quercus* are also important fodder resources of India.

India is a homeland of the wild relatives of cultivated plants and acknowledged as one of the world's 12 Vavilovian centres of origin. The rich diversity and distribution of the wild relatives are found in humid tropical and subtropical regions of the country. Table 6 shows distribution of wild relative of crop plants in different phytogeographical regions of India.

Table 6. Distribution of wild relatives of crop plants in different Phytogeographical regions of India

Phytogeographic region	*No. of Species represented*	*Genera represented*
Western Himalaya	125	*Pyrus, Prunus, Sorbus, Rubus, Ribes, Hordeum, Elymus, Eryemopyrum, Avena, Aegilops, Allium, Lepidium, Carum, Linum, Cicer, Cucumis.*
Eastern Himalaya	82	*Pyrus, Prunus, Sorbus, Rubus, Ribes, Hordeum.*
North-eastern region	132	*Citrus, Musa, Mangifera, Docynia, Elaeocarpus, Myrica, Morus, Vitis, Coix, Digitaria, Vigna, Canavalia, Mucuna,* Tuberous types, *Trichosanthes, Momordica, Cucumis, Solanum, Brassica, Corchorus, Piper, Amomum, Alpinia, Curcuma, Zingiber, Saccharum, Oryza.*
Gangetic plains	66	*Emblica, Syzygium, Artocarpus, Ziziphus*
Indus/N.W.Plains	45	*Ziziphus, Carissa, Capparis, Cordia, Grewia, Cucumis, Citrullus, Momordica, Sesamum.*
Malabar/ Western Peninsular region/ Western ghats	145	*Artocarpus, Garcinia, Syzygium, Diospyros, Euphoria, Mimusops, Mangifera, Spondias, Vitis, Chionachne, Trilobachne, Polytoca, Hygroryza, Oryza, Vigna, Atylosia, Dolichos, Mucuna, Abelmoschus, Solanum, Luffa, Cucumis, Momordica, Trichosanthes, Tuberous types, Sesamum, Curcuma, Piper, Myristica, Cinnamomum, Zingiber, Saccharum.*

Deccan/Eastern Peninsular region/ Eastern Ghats	91	*Artocarpus, Garcinia, Syzygium, Diospyros, Mimusops, Mangifera, Spondias, Vitis, Hygroryza, Oryza, Vigna, Atylosia, Dolichos, Mucuna, Sesamum, Saccharum.*

Source: Arora (1991)

Status

India has high endemicity of angiospermic flora contributing 83% of total flowering plant out of 11,058 floral endemic species in India, angiosperms attain 6,200 species. The major area of endemism are Eastern Himalaya and North-East (about 2500 species), Peninsular India including Western and Eastern Ghats (about 2,000 species), North Western Himalayas (about 8,000 species) and Andaman & Nicobar Island (250 species) (BSI, 2009). Acanthaceae and Poaceae are the families having highest endemic genera (17 each) whereas the genera *Pteracanthus* and *Niligirianthus* have highest number of endemic species (20 each). In the Himalayas, Barberideceae (98%), Saxifragaceae (92%), Ranunculaceae (72%) and Rosaceae (70%) are major dominant endemic families. In Peninsular India, the endemic families are Melastomaceae (56%), Balsaminaceae (44%), Acanthaceae (38%) and Asclepiadaceae (32%).

Ahmedullah (2000) made a list of prioritized endangered species of India on the basis of score value of each species. Out of 79 prioritized endangered species, 31 species are of high priority endangered species. Among the 10 top prioritized endangered species 9 are endemic to southern western ghat and only 1 that is *Kalanchoe roseus* from north eastern region. Out of 31 high priority endangered plant *Acer oblongum* var. *membranaceum* and *Saussurea costus* are from the western himalaya, *Kalanchoe roseus* , *Euonymus assamicus, Elaegnus conferta* ssp. *dendoidea* are from north eastern region and rest of all are endemic to Western Ghats.

Following the revised IUCN criteria and CAMP (Conservation and Management Planning) process, the FRLHT has so far assessed the conservation status of 139 medicinal plant species in south India. Out of these, the number of species encountered as extinct, critically endangered, endangered and vulnerable are respectively 4, 17, 33 and 49. In total 74.4% of the medicinal plants assessed in south India are threatened compared to 92% of plants assessed in central & north-east India and north-west Himalaya. The critically endangered medicinal plants are also greater percentage (45.3%) in north than the south (12.8%) [Kumar et al. 2000]. Among the mangrove and mangrove associated plants in India, 59 so for assesed, 57 species are of threatened category (Kumar et al. 2000).

Red Data books (Nayar & Sastry, 1987-1990) published by Botanical Survey of India enlisted around 600 species of threatened floral species of India. As per IUCN Red list (2008), India has 246 globally threatened floral species which constitute approximately 2.9% of the world's total number of threatened floral species (8457) [MoEF 2009]. Among these 246 floral species, the number of critically endangered, endangered and vulnerable species are respectively 45, 112 & 89, besides seven species are extinct and two species extinct in wild (MoEF 2009).

PLANT DIVERSITY, VIS-À-VIS, TRADITIONAL KNOWLEDGE

The plant diversity in India in terms of species richness is complemented by an enormous genetic diversity found within individual species. This rich biodiversity is in harmony with an equally rich cultural diversity and a remarkable wealth of indigenous and traditional knowledge of the local inhabitants who have utilized and sustained these resources since ages through various means, rituals and customs. The Convention on Biological Diversity (CBD) is the first legal mechanism at the global level which categorically highlights that plant genetic resources are not to be treated as heritage of humankind, but as properties of the countries in which they exist and are hence tradable commodities in the global market. The CBD not only provides a legal framework for the conservation of biodiversity, its sustainable utilization, and fair and equitable sharing of benefits arising from the use of these resources, but has advanced beyond the conventional intellectual property rights regime to recognize the role of traditional knowledge of local and indigenous communities in the conservation, development and utilization of biological and genetic resources and the need to share the benefits, thus arising with such knowledge holders. Notwithstanding the dearth of models worldwide related to benefit sharing arising out of the use of genetic resources with the holders of traditional knowledge or indigenous knowledge, India has set an example and has earned the distinction of being the first in the world in establishing a benefit-sharing model, with the 'Kani' tribals – the traditional knowledge holders and the scientists at Tropical Botanic Garden and Research Institute, Thiruvanthapuram as parties to the mechanism. Grant of patents on innovations based on/or which are part of the existing traditional knowledge is a cause of grave concern in countries like ours with rich bioresources as well as noteworthy traditional and indigenous knowledge. But India has taken a positive stance on the matter and has restricted the grant of patents based on TK/IK of the nation. To strengthen the mechanism, India has not only enacted The Biological Diversity Act, 2002 but has also undertaken the development of a Traditional Knowledge Digital Library (TKDL). The TKDL provides information on traditional knowledge existing in the country, in languages (English, German, French, Japanese and Spanish) and format understandable by patent examiners at International Patent Offices to safeguard the nation's TK against biopiracy. India is also the first nation to develop a *sui generis* system of protecting the rights of farmers, breeders and traditional communities by enacting the Protection of Plant Varieties and Farmers' Rights Act, 2001 (PPV&FR Act) in response to the TRIPS regulation, keeping in mind the nation's interests and the WTO-TRIPS agreement. The PPV& FR Act provides a legal mechanism of protection to newly bred, extant, farmers' and essentially derived varieties. In the absence of this mechanism, India would not have been able to protect its rich array of farmers' and extant varieties under the patenting system for plant varieties, since these are already in the public domain. The Indian Patent Law has also kept provisions for the disclosure of source of genetic resources and also the involved TK, if such resources or TK constitute part of the invention.

Reference

Adhikary, S.P. (2006). *Blue green algae: Survival strategies in diverse environment*. Pointer Publishers, Jaipur, pp. 1–201.

Amhedullah, M. (2000). Prioritization of Endangered Plants of India. In: *Setting Biodiversity Conservation Priorities for India*, Vol. 2. Singh et.al. (Ed.). pp. 445–456, WWF, India.

Arora, R.K. (1991). Plant diversity in Indian gene centre. In: *Plant Genetic Resources Conservation and Management: Concept and Approaches*. Paroda, R.S. and Arora, R.K. (Eds.). IBPGR Regional Office for South and South East Asia, New Delhi, pp. 25–54.

Bilgrammi, K.S., Jamaluddin, S. and Rizvi, M.A. (1991). *Fungi of India*. Today & Tomorrow Publishers, New Delhi.

BSI (2008). *Plant Discoveries 2008*: New Genera, Species and New Records. BSI, Kolkata.

BSI (2009). *Plant Discoveries 2009*: New Genera, Species and New Records. BSI, Kolkata.

Chandra, V., Kumar, D. and Srivastava, S.C. (1987). Calobryles: Distribution and phytogeographical discussion. *Geophytology*, 17: 227–232.

Dixit, R.D. (1984). *A census of the Indian pteridophytes.* Flora of India Series IV. Bot. Surv. India, Howrah.

Ghosh, S.R. and Ghosh, Basab (1997). Pteridophytes. In: *Floristic Diversity and Conservation Strategies in India, Vol. 1. Cryptogams and Gymnosperms.* Mudgal, V and Hazra, P.K. (Eds.). Botanical Survey of India, Calcutta, pp. 375–442.

Jayagopal (1985). Crustose coralline algae from Indian region. In: *Marine Plants.* Krishnamurthy and Untawale (Eds.). Seaweed Research and Utilisation Association, Madras.

Kumar, A., Walker, S. and Molur, S. (2000). Prioritization of Endangered Species. In: *Setting Biodiversity Conservation Priorities for India.* Singh et al. (Ed.). Vol. 2, pp. 342–425, WWF, India.

MoEF (2009). India's Fourth National Report to the Convention on Biological Diversity. Goyal, A.K. and Arora, Sujata. (Eds.). Ministry of Env. and Forests, Govt. of India, New Delhi.

Nayar, M.P. and Sastry, A.R.K. (1987–1990). *Red Data Book of Indian Plants,*Vol. 1–3. Botanical Survey of India, Kolkata.

Pande, S.K. (1958). Some aspects of Indian Hepaticology. *J. Indian Bot. Soc.,* 37: 1–26.

Rao, P.S.N. and Gupta, S.L. (1997). Fresh water algae. In: *Floristic Diversity and Conservation Strategies in India, Vol. 1 Cryptogams and Gymnosperms.* Mudgal, V. and Hazra, P.K. (Eds.). Botanical Survey of India, Calcutta, pp. 47–61.

Rao, P.S.N. and Mudgal, V. (1997). Marine algae. In: *Floristic Diversity and Conservation Strategies in India, Vol. 1 Cryptogams and Gymnosperms.* Mudgal, V. and Hazra, P.K. (Eds.). Botanical Survey of India, Calcutta, pp. 63–89.

Rath, J. (2010). Subaerial algal diversity in the biofilm of terracotta temple of Bishnupur, West Bengal, India and their biotechnological potential. Presented in the 12th International Conference of Culture Collection, Florianapolis, Brazil during 26th September, 2010–1st October, 2010.

Rodgers, W.A. and Panwar, H.S. (1988). *Planning Wildlife Protected Area Network in India*, Vol. 2. Project FO: IND/82/003, FAO, Dehradun.

Sen Sarkar, Neera (2011). Algae. In: *Biodiversity Conservation in the Sundarbans: State of Art Report.* Ghosh, A.K. (Ed.). WWF, India, New Delhi (in press).

Sharma, A.K. (2005). Biodiversity: Basic issues. In: *Biodiversity: Status and Prospects*. Tandon, P., Sharma, Manju and Swarup, Renu (Eds.). pp. 1–7. Narosa Publishing House, New Delhi.

Sharma, J.R. (1997). Fungi. In: *Floristic Diversity and Conservation Strategies in India, Vol. 1 Cryptogams and Gymnosperms*. Mudgal, V. and Hazra, P.K. (Eds.). Botanical Survey of India, Calcutta, pp. 91–149.

________. (1997). Woodrotting fungi of temperate timberline zone. In: *Floristic Diversity and Conservation Strategies in India, Vol. 1 Cryptogams and Gymnosperms*. Mudgal, V. and Hazra, P.K. (Eds.). Botanical Survey of India, Calcutta, pp. 151–193.

Singh, D.K. (1997). Liverworts. In: *Floristic Diversity and Conservation Strategies in India, Vol. 1 Cryptogams and Gymnosperms*. Mudgal, V. and Hajra, P.K. (Eds.). Botanical Survey of India, Calcutta, pp. 235-300.

Singh, D.K. and Semwal, R.C. (1995). Liverworts. In: *Bharat ki Vanaspati Vividhita*. Mudgal, V. and Hajra, P.K. (Eds.). pp. 225–246, Dehradun.

Singh, K.P. and Sinha, G.P. (1997). Lichens. In: *Floristic Diversity and Conservation Strategies in India, Vol. 1 Cryptogams and Gymnosperms*. Mudgal, V. and Hajra, P.K. (Eds.). Botanical Survey of India, Calcutta, pp. 195–234.

Singh, K.P. and Sinha, G.P. (2010). *Indian lichens: An annotated checklist*. Botanical Survey of India, Kolkata.

Singh, K.P. and Mudgal, V. (1997). Gymnosperms. In: *Floristic Diversity and Conservation Strategies in India, Vol. 1 Cryptogams and Gymnosperms*. Mudgal, V. and Hazra, P.K. (Eds.). Botanical Survey of India, Calcutta, pp. 443–472.

Simpson, M.G. (2005). *Plant systematics*. Academic Press.

Swarup, Renu (2008). Biodiversity characterization, conservation and sustainable utilization. In: *Biodiversity: Issues & Concerns*. SenSarkar et al. (Ed.). West Bengal Biodiversity Board, Kolkata.

Takhtajan, A. (1969). *Flowering plants, origin and dispersal*. Edinburgh: Tr. Jeffrey.

Venu, P. (1998). A review of floristic diversity inventory and monitoring methodology in India. *PINSA*, B64 No. 5 & 6, pp. 281–292.

Vohra, J.N. and Aziz, M.N. (1997). Mosses. In: *Floristic Diversity and Conservation Strategies in India, Vol. 1 Cryptogams and Gymnosperms*. Mudgal, V. and Hazra, P.K. (Eds.). Botanical Survey of India, Calcutta, pp. 301–374.

Catenella sp.

Enteromorpha intestinalis (L.) Nees

Russula lepida Pers.

Termitomyces microcarpus (Berk. & Broome) Heim

Lycopodium sp. Linn.

Helminthostachys zeylanica Hook.

Nymphaea pubescens Burm f.

Ipomoea aquatica Forsk.

Gloriosa superba Linn.

Drocera indica Linn

Mangrove vegetation- Sunderban

Chilika lagoon- A Ramsar site

Western Ghat- the biodiversity hotspot

Laterite Shal forest

Acrostichum aureum L.

Ganoderma sp.

11

A Note on Some Threatened Plants

M. Sanjappa, T.K. Paul and P. Lakshminarasimhan

INTRODUCTION

There are *c.* 422,000 species of flowering plants on earth, of which nearly 75,800 species are threatened (Bramwell, 2002). In India there are 17,712 species of flowering plants (Sanjappa & Singh, 2009), of which nearly 1700 are threatened. West Bengal lies between the Himalayas in the north and the Bay of Bengal in the south. It is the only state in India whose flora ranging right from the most impressive littoral forests of the Sundarbans to the luxuriant forests of the Himalayan foothills and the vegetation upwards culminating to the temperate and the alpine zone in the district of Darjeeling. The state has two natural divisions-the Himalayan in the northern part of the state comprising the districts of Darjeeling, Jalpaiguri and Cooch Behar, and the alluvial plain that lies in southern of it.

The perusal of the literature reveal that the flora of West Bengal comprises *c.* 3580 species belonging to 1333 genera and 200 families of angiosperms (Chakraverty et al. 1999), 21 species of gymnosperms, 416 species of pteridophytes, 771 species of bryophytes, 873 species of algae, 539 species of fungi and 329 species of lichens.

Now-a-days the flora of this state has been under great pressure due to various factors leading to the destruction of the biodiversity. Urbanization and industrialization in the Bengal plains has increased the pressure on dry deciduous 'sal' forests in the districts of Burdwan, Birbhum, Bankura, Purulia and West Midnapur district. The increase of sea level in the Bay of Bengal led to the increase of salinity of South 24 Parganas causing the destruction of vegetation living on sweet water. Landslides, urbanization, construction of dams for hydro-electric power projects, extension of commercial plantations like Tea, Cinchona are causing the destruction of important floristic elements of Darjeeling hills. Tourist activities and the large scale collection of materials by the teachers and students as well as by the material suppliers on commercial basis have almost finished the rare and interesting plant species of Darjeeling and suburbs.

Very little work on the Rare, Endemic and Threatened plants of West Bengal has been done. Giri & Nayar (1983) listed 7 species of Orchids from Sundarbans, 4 other species from the plains and 6 taxa described from West Bengal and represented by only type collections. Chakraverty *et al.* (1999) listed 37 taxa which are rare and threatened in West Bengal state. Rao *et al.* (2003) listed 9 threatened taxa from West Bengal. In the Indian Plant Red Data Book (Jain & Sastry, 1984) 2 species and in the Red Data Book of Indian Plants (Nayar & Sastry, 1987, 88, 90) 17 species have been included from West Bengal.

In spite of the poor representation of the endemics in the Indo-gangetic region (Chatterjee,1940), a number of new taxa have been described in the recent past. This paper intends to focus on some taxa described from West Bengal and which have not been collected after type. Further, it aims in creating awareness among interested plant collectors to relocate these taxa or arrive at a definite decision on the threatened status of them conserve them *in-situ* or *ex-situ*.

Species not collected after type collection in West Bengal

S. No.	*Name of Taxa*	*Last Collected*
1.	*Acer hookeri* Miq. var. *majus* Pax (Aceraceae)	T. Anderson from Darjeeling & Sikkim in 1862
2.	*Aspidopterys balakrishnanii* R.C. Srivast. (Malpighiaceae)	H.H. Haines from Lower Tendu forests, West Duars, Jalpaiguri district in last century; also in Uttar Pradesh
3.	*Caldesia janaki-ammaliae* Guha & M.S. Mondal ["janaki-ammalii" (Alismataceae)]	S. Kurz from Lower Bengal and Kanchrapara, Nadia district in 1867
4.	*Casearia sikkimensis* N. Mukh. (Flacourtiaceae)	J.S. Gamble from Dhobijha, Kurseong, Darjeeling in 1882
5.	*Cuscuta sharmanum* Mukerjee & P.K. Bhattach. (Convolvulaceae)	P.K. Bhattacharyya from Burdwan in 1964; in the protologue it is given that it occurs also in Midnapur

6.	*Dalbergia duarensis* Thoth. (Leguminosae - Papilionoideae)	E.A. Heawood from Mahakalguri, Alipurduar, Jalpaiguri district in 1891
7.	*Dicliptera paniculata* (Forssk.) I. Darbysh. var. *subaequibracteata* (Bennet) Karthik. & Moorthy [=*Peristrophe bicalyculata* (Retz.) Nees var. *subaequibracteata* Bennet] (Acanthaceae)	S.S.R. Bennet from Shyampur, Howrah in 1965
8.	*Gastrochilus corymbosa* A.P. Das & Chanda (Orchidaceae)	A.P. Das from Jalapahar, Darjeeling in 1982
9.	*Hygrophila heinei* Sreem. var. *birbhumensis* (B.P. Guha) H.B. Naithani & S. Biswas [=*Cardanthera uliginosa* (L.f.) Buch.-Ham. ex Benth. var. *birbhumensis* B.P. Guha] (Acanthaceae)	B.P. Guha from Nanur, Birbhum in 1965
10.	*Hypericum benghalense* S.N. Biswas (Hypericaceae)	S.K. Mukherjee from Kalimpong, Darjeeling in 1959
11.	*Hypericum humifusum* L. subsp. *suborbiculatum* S.N. Biswas (Hypericaceae)	S.K. Mukherjee & party from Darjeeling [Paratype; date of collection is not given], Holotype from Nepal
12.	*Ixora athroantha* Bremek. (Rubiaceae)	M.V. Laurie from Sukna, Darjeeling in 1940
13.	*Liparis breviscapa* A.P. Das & D. Lama (Orchidaceae)	A.P. Das from Darjeeling in 1981
14.	*Liparis tigerhillensis* A.P. Das & Chanda (Orchidaceae)	A.P. Das from Tiger hill, Darjeeling in 1982
15.	*Osbeckia darjeelingensis* G.S. Giri & M.P. Nayar (Melastomataceae)	J.S. Gamble from Jhenaikuri, Darjeeling in 1879
16.	*Premna jalpaiguriana* T.K. Paul (Verbenaceae)	J.K. Sikdar from Indang, Jalpaiguri district in 1975
17.	*Pycreus sanguinolentus* (Vahl) Nees subsp. *sacer* (Korlahalli) Karthik. [=*Cyperus sanguinolentus* Vahl subsp. *sacer* Korlahalli] (Cyperaceae)	B.C. Korlahalli from Sodhpur in 1964
18.	*Spiraea darjeelingensis* Panigrahi & K.M. Purohit (Rosaceae)	J.S. Gamble from Darjeeling in 1877
19.	*Uvaria lurida* Hook.f. & Thomson var. *sikkimensis* King (Annonaceae)	G. King from Mongpoo, Darjeeling in 1881

References

Bramwell, D. (2002). How many plant species are there? *Plant Talk.* 28: 32–34.

Chakraverty, R.K., Srivastava, R.C., Mitra, S., Bandyopadhyay, S. and Bandyopadhyay, S. (1999). In: *Floristic Diversity and Conservation Strategies in India.* Mudgal V. and Hajra P.K. (Eds.). BSI, Calcutta, 3: 1575–1621.

Chatterjee, D. (1940). Studies on the endemic flora of India and Burma.J. *Asiat. Soc. Bengal,* 5: 19–67.

Giri, G.S. and Nayar, M.P. (1983). Some threatened plants of Bengal plains. In: *An Assessment of Threatened Plants of India.* (Proceedings of the seminar hall at Dehradun, 14–17 September, 1981). Jain, S.K. and Rao, R.R. (Eds.). 91–93. BSI, Kolkata.

Jain, S.K. and Sastry, A.R.K. (1984). *The Indian Plant Red Data Book*–I, POSSCEF, BSI, Howrah.

Nayar, M.P. and Sastry, A.R.K. (2000). *Red Data Book of Indian Plants.* Vols. 1–3, BSI, Calcutta (1987, 1988, 1990); Reprinted.

Rao, C.K., Geetha, B.L. and Suresh, G. (2003). *Red List of Threatened Vascular Plant Species in India,* BSI ENVIS Centre, Ministry of Environment & Forests, Howrah.

Sanjappa, M. and Singh, P. (2010). Plant discoveries 2009: New genera, species and new records, BSI, Kolkata.

Fig. 1. *Acer hookeri* Miq. var. *majus* Pax (Aceraceae)

Fig. 2. *Aspidopterys balakrishnanii* R.C. Srivast. (Malpighiaceae)

Fig. 3. *Caldesia janaki-ammaliae* Guha & M.S. Mondal ["janaki-ammalü" ( )]

Fig. 4. *Dalbergia duarensis* Thoth. (Leguminosae-Papilionoideae)

Fig. 5. *Hygrophila heinei* Sreem. var. *birbhumensis* (B.P. Guha) H.B. Naithani & S. Biswas [*Cardanthera uliginosa* (L.f.) Buch. -Ham.ex Benth. var. *birbhumensis* B.P. Guha] (Acanthaceae)

Fig. 6. *Hypericum benghalense* S.N. Biswas (Hypericaceae)

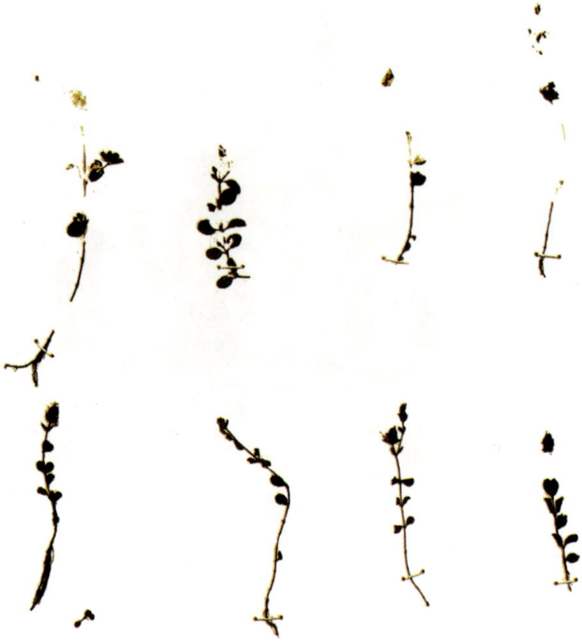

Fig. 7. *Hypericum humifusum* L. subsp. *suborbiculatum* S.N. Biswas (Hypericaceae)

Fig. 8. *Osbeckia darjeelingensis* G.S. Giri & M.P. Nayar (Melastomataceae)

Fig. 9. *Premna jalpaiguriana* T.K. Paul (Verbenaceae)

Fig. 10. *Uvaria lurida* Hook.f. & Thomson var. *sikkimensis* King (Annonaceae)

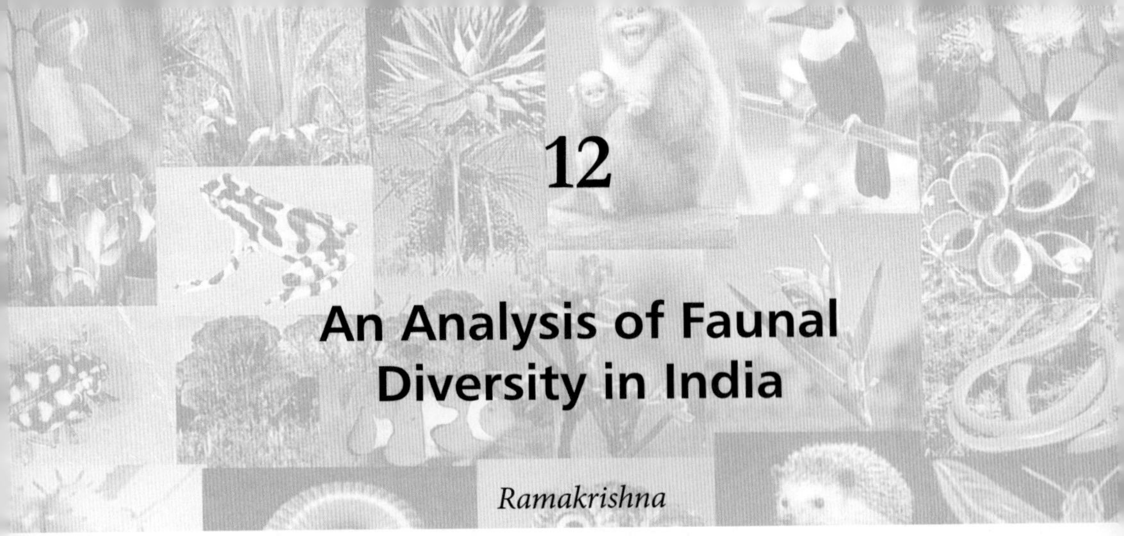

12

An Analysis of Faunal Diversity in India

Ramakrishna

INTRODUCTION

Biological Diversity or Biodiversity, defined as variety and variability within species, between species and of ecosystems, is crucial to poverty alleviation and development, as envisaged in Millennium Development Goals (MDGs) established by the United Nations to achieve a significant reduction in the rate of biodiversity loss by 2010 and to achieve the international commitment by the year 2015. Genetic diversity contributes to resilience and help farmers adapt to climate change by providing crop and livestock varieties that can cope with new conditions such as rising temperatures, drought and flooding. Species diversity provides goods that can be used for subsistence, barter and trade. Some examples of these goods include:

- foods such as fruits, vegetables, grains, nuts, fish and meat, milk and eggs,
- fibres, leather and pelts,
- medicinal plants,
- firewood and other fuels,
- building materials such as wood for furniture, grasses and reeds for baskets and thatch, etc.

Species diversity often contributes to development by supplying material used for small-income generating activities, such as the sale of craft items, local foods or traditional medicines. "Ecosystem" means a dynamic complex of plant, animal and micro-organism communities and their nonliving environment interacting as a functional unit. Ecosystem services improve local well-being and provide maintenance of soil fertility and pasture regeneration, clean water and productive agricultural systems, watershed protection, erosion control, maintains diversity of local varieties, pest control, food web, predator – prey interaction, crops and agricultural systems contributes to risk distribution. Increased levels of organic matter in soil contributes to increased harvests and improved ecosystem diversity that can contribute to development by encouraging nature-based tourism. Thus the ecosystem approach is a strategy for the integrated management of Land, Water and Living resources *viz.*, variety of genes, species and habitats.

Biological diversity must be treated more seriously as a global resource, to be indexed, used, and above all, preserved, the rationale for such idea is based on three important criteria. First, exploding human population is degrading the environment at an accelerating rate. Second, science is discovering new uses of biological diversity in ways that can relieve both human suffering and environmental destruction. Third, much of the diversity is being irreversibly lost through extinction caused by the destruction of natural habitats. Each species is the repository of an immense amount of genetic information. The number of genes range from about 1,000 in bacteria and 10,000 in some fern to 400,000 or more in many flowering plants and a few animals. A typical mammal such as the house mouse (*Mus musculus*) has about 100,000 genes (Hinegardner 1976). Species richness (i.e., the number of species) is the simplest and the most intuitive concept for characterizing community diversity. This is also referred to as alpha diversity in ecological science. Documentation of biotic resource of a country therefore focuses on developing 'indicators' to be used in the assessment which are reflected in a combination of ecological fundamentals, such as biodiversity, critical habitat and key ecological relationships; site-specific considerations, environmental stress and potential impacts. Assessments and documentations also provide biodiversity values that are recognized and taken into consideration in the planning and decision making process. It is for this reason the documentation of faunal resources, which will enhance the effective performance of planning and management and also assist in competencies or compensable factors found in evaluating the diversity of the country, has been taken up in the present instance.

India is endowed with an immense variety of natural resources in its rich animal and plant heritage. According to world bio-geographic classification, India represents two of the major realms (the Palaearctic and Indo-Malayan) and three biomes (Tropical Humid Forests, Tropical Dry/Deciduous Forests and Warm Deserts/Semi-Deserts). These include 12 bio-geographic regions. However, the Wildlife Institute of India has proposed a modified classification – which divides the country into 10 bio-geographic zones and 20 biotic provinces. The vegetation wealth and diversity in the country is enormous due to the variety of climatic and altitudinal variations coupled with diverse ecological habitats. Champion and Seth (1968) recognized 16 major forest types comprising 221 minor types; of these tropical moist deciduous forms the major percentage (37%), followed by tropical dry deciduous (28.6%), tropical wet evergreen (8%), tropical thorn forest (2.6%) and others of minor values. Recent Forest Cover Assessment by Forest Survey of India (2009) indicates forest and tree cover of the country is 78.37 million hectares amounting to a dense forest cover of 8.35 million hectares (2.54% of geographic area), medium dense forest cover of 31.90 m ha (9.71 %) and open forest of 28.84 m ha (8.77%), thus totalling 21.02%; while the non forest cover constitute a major portion (77.72%) with scrub forest depicting only 1.26% of the total area.

India's major biogeographic habitats

No.	*Biogeographic Zone*	*Biotic Provinces*
1.	Trans-Himalaya	Upper reaches of Himalaya
2.	Himalaya	North-west Himalaya, West Himalaya, Central Himalaya, Eastern Himalaya
3.	Desert	Kutchh, Thar, Ladak
4.	Semi-arid	Central India, Gujarat Rajwara
5.	Western Ghats	Malabar Coast, Western Ghats Mountains
6.	Deccan Peninsula	Deccan Plateau, South Central Plateau, Eastern Plateau, Chhota Nagpur, Central High lands
7.	Gangetic Plain	Upper Gangetic Plain, Lower Gangetic Plain
8.	North-East India	Brahmaputra Valley, North-Eastern Hills
9.	Islands	Andaman Islands, Nicobar Islands, Lakshadweep Islands
10	Coasts	West Coast, East Coast

Vegetative coverage in India

No.	*Vegetation Type*	*Location*
1.	Tropical wet evergreen	West of the Western Ghats, Orissa coast, east of Indo-Gangetic plain, Assam and the Andamans
2.	Tropical semi-evergreen	Assam, West Bengal, Orissa, Maharashtra, Karnataka and Kerala
3.	Tropical dry evergreen	Coast of Tamil Nadu
4.	Tropical moist deciduous	Assam, West Bengal, Bihar, Orissa, eastern Uttar Pradesh, Madhya Pradesh, Maharashtra, Tamil Nadu, Karnataka and Kerala.

5.	Tropical dry deciduous	Indo-Gangetic plain from Bihar to Punjab, Maharashtra, Madhya Pradesh, Orissa, Andhra Pradesh, Tamil Nadu and Karnataka.
6.	Tropical Thorn	Western Ghats,west of Uttar Pradesh & Rajasthan
7.	Littoral & swamp	Ganga-Brahmaputra delta, the Mahanadi the Godavari, the Krishna & the Cauveri.
8.	Sub-tropical wet hill	Eastern Himalaya and Assam hills
9.	Sub-tropical moist hill	Western and central Himalaya, Meghalaya, Manipur & Nagaland
10.	Sub-tropical dry Evergreen	Western Himalaya and Kashmir hills.
11.	Montane wet temperate	Eastern Himalaya, Assam & Assam hills.
12.	Himalayan moist Temperate	Himalaya of Punjab, Kashmir, Uttar Pradesh, Sikkim & West Bengal
13.	Himalayan dry temperate	Kashmir, Garhwal, Uttar Pradesh, Sikkim & inner Himalaya.
14.	Alpine	Himalaya in higher reach
15.	Savanna	Strip of grassland from Karnataka to Rajasthan through Maharashtra and Madhya Pradesh.

RESOURCE AREAS

Forests are an important reservoir of biodiversity. Ancient and frontier forests, because of their long standing and relatively lower levels of human disturbance, are typically richer in biodiversity than other natural or semi-natural forests. An illustration of the conservation importance of forests related to biodiversity is found in the recent analysis of biodiversity hotspots (Myers et al. 2000). Loss of forest will inevitably result in reduction of biodiversity as a direct result of loss of habitat. There are many anthropocentric reasons to preserve biodiversity (direct use value, option value, etc.), but the principal such reason considered here is the indirect use value in the form of ecosystem services. A loss in biodiversity affects the stability of an ecosystem resulting in a reduction of its resistance to disruption of the food web (by loss of the weak interaction effect), resistance to species invasion and resilience to global environmental change (McCann, 2000; Chapin et al. 2000). Therefore, the argument for conserving biodiversity is two-fold:

1. Biodiversity is essential for maintaining ecosystem services and
2. Biodiversity increases the stability and resilience of an ecosystem to a disturbance

PROTECTED AREA NETWORK

The conventional approach to maintain biological diversity generally has been to proceed species by species and threat by threat (Scott et al. 1993). This reactive approach to conservation has proven to be difficult, expensive, biased, and inefficient (Franklin 1993, Scott et al. 1993) and, therefore, contributes to an unpredictable ecological and economic environment. In this regard Parties to the Convention on Biological Diversity, in adopting a Strategic Plan, have committed themselves to achieving, by 2010, a significant reduction in the rate of biodiversity loss at the global, national and regional levels, as a contribution to poverty alleviation and to the benefit of all life on earth. National plans have been developed to provide interim measures to protect highly threatened or highly valued areas wherever this is necessary based on the relevant criteria such as irreplaceability of target biodiversity components, minimum effective size and viability, species migration requirements, integrity, ecological processes and ecosystem services. This is achieved based on the "*Gap Analysis*" which envisages a series of scientific, social and political principles such as representation to choose focal biodiversity across biological scales (species and ecosystems) and realms (terrestrial, freshwater, and marine), resilience to high quality examples to withstand change, redundancy to have high quality examples to withstand change, multiple functional gaps and management, and participation of stakeholders to capture the full array of biodiversity in the protected area system. The selection criteria are also based on threatened, endemic, habitat builders, keystone and migratory species.

Protected areas are recognized as the most important core 'units' for its resource and *in-situ* conservation. It is also a geographically defined area which is designated or regulated and managed to achieve specific resource conservation objectives. Protected area coverage was endorsed by the seventh Conference of the Parties (COP-7) of the Convention on Biological Diversity (CBD) as an indicator for immediate testing in relation to the adopted target of significantly reducing the rate of biodiversity loss by 2010. However, the concept of protection is not new to this country. The concept of sacred groves, operation of closed period for fishing and hunting and the establishment of wildlife reserves were known during the Mauryan kings in India in the second and third centuries (Grove, 1995). It is therefore, "*Protected areas are a cultural response to the perceived threats to nature*" (McNeely, 1998). But, the establishment of Yellowstone National Park in US in1872 is often cited as the starting point of the modern era of protected areas. The term 'protected area' as used refers to "An area of land and/or sea especially dedicated to the protection and maintenance of biological diversity, and of natural and associated cultural resources, and managed through legal or other effective means" (IUCN 1994). Gap analysis is therefore a nationwide program with the following two objectives: (1) identify areas rich in species diversity and measure the extent to which they are protected in conservation areas, and (2) map and identify vegetation types not adequately represented in the current reserve systems and operates on a relatively large geographic area, and at a coarse spatial scale (1:100,000) (Scott et al. 1993).

In India, the State Government may, by notification, declare its intention to constitute any area other than an area comprising within any reserve forest or the territorial waters as a sanctuary if it considers that such an area is of adequate ecological, faunal, floral, geomorphologic, natural or zoological significance, for the purpose of protecting, propagating or developing wildlife or its environment. The declaration of a National Park by the state government also falls with the same principle of protection, propagation and developing wildlife in an area, whether within a sanctuary or not. The other conservation strategies adopted by the government are the initiation **Project Tiger** a programme that was launched in 1973 in order to ensure the maintenance of available population of tigers in India and to increase the species numbers. **Project Crocodile** is the second Breeding Programme initially started in Orissa was then extended to several other States in 1975. The major objective of the project is to protect the three endangered species of crocodiles, *Gavialis gangeticus, Crocodylus palustris and Crocodylus porosus.* The third programme, **Project Rhino** was launched in 1987 in Kaziranga Wildlife Sanctuary in Assam to save the Lesser One-horned Rhinoceros from extinction. Lastly in 1991-92 the **Project Elephant** was initiated to ensure long-term survival of identified population of elephants in their natural habitats.

Securing the integrity of nearly a total of 89 National Parks with an area of 37530.76 km^2 or 1.14% of the country's geographic area and 489 Wildlife Sanctuaries with an area of 117042.04 km^2 or 3.56% have been established. This gives a total of 578 Protected Areas, with an area of 154,572.80 km^2 or 4.70% of the country's geographic area, within the forest ecosystem to preserve, mainly through the preparation of management plans and also through provision of support for surveillance and monitoring activities (Table).

Existing Protected Area Status

State & Union Territory	*Area (km²)*	*No. of NPs*	*Area (km²)*	*% of state*	*No. of WLS*	*Area (km²)*	*% of state*
Andhra Pradesh	275045	4	373.23	0.14	21	12530.09	4.56
Arunachal Pradesh	83743	2	2290.82	2.74	11	7606.73	9.08
Assam	78438	5	1968.60	2.51	15	883.16	1.13
Bihar	94163	1	335.65	0.36	11	2949.17	3.13
Chhattisgarh	135194	3	2929.50	2.17	10	3419.46	2.53
Goa	3702	1	107.00	2.89	6	647.96	17.50
Gujarat	196024	4	479.67	0.24	21	16422.71	8.38
Haryana	44212	1	1.43	0.00	7	162.73	0.37
Himachal Pradesh	55673	2	1429.40	2.57	32	5770.85	10.37
Jammu & Kashmir	222236	4	4650.07	2.09	15	10219.65	4.60
Jharkhand	79714	1	231.67	0.29	10	1863.23	2.34
Karnataka	191791	5	2472.18	1.29	21	3888.22	2.03
Kerala	38863	3	536.52	1.38	12	2143.36	5.52
Madhya Pradesh	308252	9	3656.36	1.19	25	7198.19	2.34
Maharashtra	307713	5	955.93	0.31	35	14376.56	4.67
Manipur	22327	1	40.00	0.18	3	393.30	1.76
Meghalaya	22429	2	267.48	1.19	3	34.20	0.15
Mizoram	21081	2	250.00	1.19	4	771.00	3.66
Nagaland	16579	1	202.02	1.22	3	20.34	0.12
Orissa	155707	2	990.70	0.64	18	6971.15	4.48
Punjab	50362	0	0.00	0.00	10	316.73	0.63
Rajasthan	342239	4	3856.53	1.13	24	5712.83	1.67
Sikkim	7096	1	1784.00	25.14	5	265.10	3.74
Tamil Nadu	130058	5	307.84	0.24	19	2539.82	1.95
Tripura	10486	0	0.00	0.00	4	603.62	5.76
Uttaranchal	53484	6	4077.00	7.62	6	2395.52	4.48
Uttar Pradesh	241335	1	490.00	0.20	23	5333.47	2.21
West Bengal	88752	5	1693.25	1.91	15	1203.34	1.36
Andaman & Nicobar	8249	8	901.00	10.92	94	372.13	4.51
Chandigarh	114	0	0.00	0.00	1	25.42	22.30
Dadra & Nagar Haveli	491	0	0.00	0.00	1	92.16	18.77
Lakshadweep	32	0	0.00	0.00	0	0.00	0.00
Pondicherry	492	0	0.00	0.00	0	0.00	0.00
Daman & Diu	112	0	0.00	0.00	1	2.18	1.95
Delhi	1483	0	0.00	0.00	1	13.20	0.89
India	**3287263**	**89**	**37530.76**	**1.14**	**489**	**117042.04**	**3.56**

(Courtesy: http://wiienvis.nic.in)

BIOSPHERE RESERVES

Biosphere reserve is a specific type of conservation area, which accommodates and benefits both the natural environment and the communities living in and around it. This is possible because a biosphere reserve consists of three different but associated zones *viz.*, Core, Buffer and Transition. Core zone, a most ecologically sensitive and pristine area where nature conservation is a priority with allowed low-impact activities. Buffer zone a less ecologically sensitive but mostly natural area where recreation and sustainable utilisation of natural products can be accommodated and finally, the Transition zone, a less ecologically sensitive area where a great variety of land uses occur. All zones are interdependent and are managed and protected according to the definitions above. By linking conservation, development and the sustainable use of natural resources, a harmony and balance between nature and people can be achieved. List of Indian Biosphere Reserves:

1. Nilgiri (Tamil Nadu, Kerala and Karnataka)
2. Nanda Devi (Uttrakhand),
3. Nokrek (Meghalaya)
4. Manas (Assma)
5. Dibru-Saikhowa (Assam),
6. Sundarban (West Bengal),
7. Gulf of Mannar (Tamil Nadu),
8. Great Nicobar (Andaman & Nicobar Islands),
9. Similipal (Orissa),
10. Dehang-Debang (Arunachal Pradesh),
11. Khangchendzonga (Sikkim),
12. Pachmarhi (Madhya Pradesh),
13. Achanakmar-Amarkantak (Chattisgarh and Madhya Pradesh),
14. Agasthyamalai (Tamil Nadu and Kerala),
15. Katchchh (Gujarat) and
16. Cold Desert (Himachal Pradesh).

INDIAN HOTSPOTS

Conservation priorities are assigned based on the analysis of areas of 'hotspots', or areas featuring exceptional concentrations of endemic species and experiencing exceptional loss of habitat, the other analysis being hotspots, featuring richness of rare or taxonomically unusual species

(Myers et al. 2000). Elements of Conservation Value include

- Diversity
- Endemism
- Rarity
- Endangerment
- Representivity
- Uniqueness

The boundaries of such area have been determined by `biological commonalities'. Each of the areas features a separate biota or community of species that fits together as a bio-geographic unit. The criteria for such analysis are also based on the distribution of endemic plant species comprising at least 0.5% of all plant species and the number of endemic vertebrate groups (mammals, birds, reptiles and amphibians). The second criteria for such analysis is the degree of threat through habitat loss, to qualify, a hotspot should have lost 70% or more of its primary vegetation, this being the form of habitat that usually contains the most species, especially endemics. In several hotspots there is species congruence insofar as high counts for endemic plants are matched by high counts for endemic vertebrates, and in the Indian context can be seen in Andaman and Nicobar group of Islands (Indo- Burmese and Sunderland Hotspots). Islands are considered to be ecologically vulnerable for fact that it includes a large number of endemic species or subspecies, small populations, ecological implications of island size, evolution in absence of predators, ecologically "looser" – easier for invasive species to establish population if immigrate to island. Based on the numbers of endemics, endemic species/area ratio for both plants and vertebrates, and habitat loss certain areas are further designated as 'The hottest hotspots'. The hotspots findings accord well with several other priority setting analyses, *viz.*, there is a 68% overlap with Birdlife International's Endemic Bird Areas, 82% with IUCN/WWF International's Centres of Plant Diversity and Endemism and 92% with the most critical and endangered eco-regions of WWF/US's Global 200 List.

The Western Ghats range starts near the border of Gujarat and Maharashtra, south of the River Tapti (22° N Lat), and runs approximately 1600 km , interrupted only by the 30 kilometers Palghat Gap runs through the states of Maharashtra, Goa, Karnataka, Tamil Nadu and Kerala ending at Kanyakumari (8°N lat), at the southern tip of India. About sixty percent of the Western Ghats are located in the state of Karnataka. These hills cover 1,60,000 km^2 are strikingly similar in their geology, climate and evolutionary history. The area exhibits complex topography, wide range of microclimatic and soil conditions, mosaic of plant communities and animal associations and has richest biological resources with about 6000 vascular plants are in Western Ghats and Sri Lanka hotspot, of which more than 3,000 (52 percent) are endemic. Western Ghats is the only area in India having an admixture of the following biotic elements, resultant of the biogeographic evolution of the country such as Gondwana derivatives, Peninsular endemics, Madagascaran elements, Indo-Chinese derivatives, Malayan elements and Palaearctic elements.

The Western Ghats mediates the rainfall regime of peninsular India by intercepting the southwestern monsoon winds. The western slopes of the mountains experience heavy annual rainfall (with 80 percent of it falling during the southwest monsoon from June to September), while the eastern slopes are drier; rainfall also decreases from south to north. Dozens of rivers originate in these mountains, including the peninsula's three major eastward-flowing rivers. Thus, they are important sources of drinking water, irrigation, and power. The wide variation of rainfall patterns in the Western Ghats, coupled with the region's complex geography, produces a great variety of vegetation types. These include scrub forests in the low-lying rainshadow areas and the plains, deciduous and tropical rainforests up to about 1,500 meters, and a unique mosaic of montane forests and rolling grasslands above 1,500 meters.

The Himalaya Hotspot is home to the world's highest mountains, including Mt. Everest. The mountains rise abruptly, resulting in a diversity of ecosystems that range from alluvial grasslands and subtropical broadleaf forests to alpine meadows above the tree line. Vascular plants have even been recorded at more than 6,000 meters. The hotspot is home to important populations of numerous large birds and mammals, including vultures, tigers, elephants, rhinos and wild water buffalo.

Stretching in an area over 3,000 kilometers of northern Pakistan, Nepal, Bhutan and the northwestern and northeastern states of India, the Himalaya hotspot includes all of the world's mountain peaks higher than 8,000 meters, (this includes the world's highest mountain, Mt. Everest) as well as several of the world's deepest river gorges.

This immense mountain range, which covers nearly 750,000 km^2, has been divided into two regions: the Eastern Himalaya, which covers parts of Nepal, Bhutan, the northeast Indian states of West Bengal, Sikkim, Assam, and Arunachal Pradesh, southeast Tibet (China), and northern Myanmar; and the Western Himalaya, covering the Kumaon-Garhwal, northwest

Kashmir, and northern Pakistan. While these divisions are largely artificial, the deep defile carved by the antecedent Kali Gandaki River between the Annapurna and Dhaulagiri mountains has been an effective dispersal barrier to many species.

The abrupt rise of the Himalayan Mountains from less than 500 meters to more than 8,000 meters results in a diversity of ecosystems that range, in only a couple of hundred kilometers, from alluvial grasslands (among the tallest in the world) and subtropical broadleaf forests along the foothills to temperate broadleaf forests in the mid hills, mixed conifer and conifer forests in the higher hills, and alpine meadows above the treeline.

The Indo-Burma hotspot encompasses 2,373,000 km^2 of tropical Asia east of the Ganges-Brahmaputra lowlands. Formerly including the Himalaya chain and the associated foothills in Nepal, Bhutan and India, the Indo-Burma hotspot has now been more narrowly redefined as the Indo-Chinese subregion. The hotspot contains the Lower Mekong catchment. It begins in eastern Bangladesh and then extends across north-eastern India, south of the Bramaputra River, to encompass nearly all of Myanmar, part of southern and western Yunnan Province in China, all of the Lao People's Democratic Republic, Cambodia and Vietnam, the vast majority of Thailand and a small part of Peninsular Malaysia. In addition, the hotspot covers the coastal lowlands of southern China (in southern Guangxi and Guangdong), as well as several offshore islands, such as Hainan Island (of China) in the South China Sea and the Andaman Islands (of India) in the Andaman Sea.

The Sundaland hotspot covers the western half of the Indo-Malayan archipelago, an area of some 17,000 equatorial islands, and is dominated by two of the largest islands in the world: Borneo (725,000 km^2) and Sumatra (427,300 km^2). More than a million years ago, the islands of Sundaland were connected to mainland Asia. As sea levels changed during the Pleistocene, this connection periodically disappeared, eventually leading to the current isolation of the islands. The topography of the hotspot ranges from the hilly and mountainous regions of Sumatra and Borneo, where Mt. Kinabalu rises to 4,101 meters, to the fertile volcanic soils of Java and Bali, the former dominated by 23 active volcanoes. Granite and limestone mountains rising to 2,189 meters are the backbone of the Malay Peninsula.

Politically, Sundaland covers a small portion of southern Thailand (provinces of Pattani, Yala, and Narathiwat); nearly all of Malaysia (nearly all of Peninsular Malaysia and the East Malaysian states of Sarawak and Sabah in northern Borneo); Singapore at the tip of the Malay Peninsula; all of Brunei Darussalam; and all of the western half of the megadiversity country of Indonesia, including Kalimantan (the Indonesian portion of Borneo, Sumatra, Java, and Bali). The Nicobar Islands, which are under Indian jurisdiction, are also included.

METHODS OF MEASURING BIODIVERSITY

Wilson (1988) estimates a total number of globally described species of plants and animals at 13,92,485. Minelli (1993) estimated 18,00,000 as the total number of living species, almost corresponding to the earlier estimate of 18,20,000 species by Stork (1981). To this, an estimated nearly 3,00,000 (Stern, 1981) or 2, 50,000 (Raupp, 1988) fossil species are to be added. Nearly 1.5 – 1.7 million species of plants, animals and microbes have been named and described so far (Wilson, 1985; Groombridge, 1992). Assuming the global biodiversity to be 30 million species, the present knowledge of 1.75 million species amounts to a mere 5.67% of the total, IUCN (2004). Currently documented rates of species introduction to different regions of the world are greater than documented rates of extinction. Thus, while the total number of species on the planet is decreasing due to extinctions, the total number of species on each individual continent is actually increasing and the distribution of species on earth is becoming more homogenous, *i.e.* suites of species are becoming more similar across the planet. (Duraiappah et al. 2005). Consequence of this, the documentation of the biological resource of the country and measuring biodiversity extend beyond the species level to encompass entire habitats and ecosystems, and can also focus more narrowly on the details of populations and genes becoming more important in recent years. Effective methods of measuring biodiversity are needed to monitor changes in the state of nature and to measure progress towards the target, set by the World Summit on Sustainable Development, of achieving 'a significant reduction in the current rate of biodiversity loss by 2010'.

The assessment of biological diversity depends on the distribution and diversity, of species, in different areas, habitats or ecosystems. Conservation efforts can be usefully assigned only on the basis of Species Richness, (the total number of given species in a quantified area); Species Evenness, the degree to which the number of individual organisms are evenly divided between different species of the community; Disparity, measures the phenotypic differences among species resulting from the differences among genes within a population; Species Rarity, the rarity of individual organisms within a quantified area and Genetic Variability, each population of a species contributes to additional biodiversity due to variations between genes. Further the criteria for assigning the priorities for conservation of biological diversity depends on the number of occurrences of a genotype, species or ecosystem (Rarity), degree of separation of population, species or ecosystem from its comparable analog (Distinctiveness), degree of occurrence (Representation), imminent danger or harm (Threat) and the presence or absence of key stone resources (Function).

An understanding of the biological diversity and ecosystem function rests on the identification of the biota in their habitats. The biodiversity assessment on the estimates of number of vertebrate and invertebrate species is the major task owing to the complexity related to their abundance, diversity and interactions in the varied habitats/ecosystems. Often only reasonable predictions on broad estimates of number of living species can be made depending on the available data pertaining to reference collection and literature and also on evaluations made on their ecological preferences. The process is also complex and challenging, invariably testing the competence and knowledge of taxonomists involved. Magurran (1988) gave a comprehensive account of methods for measuring biodiversity based on variety and abundance. Abundance is easily defined as total numbers of individuals or the density of individuals, though biomass or percentage ground cover (for terrestrial plants) may also be appropriate measures. If we choose to measure abundance as biomass rather than number of individuals, we may observe very different trends over time if we combine our measure across species. Variety is less easily defined. Hence, it is unrealistic to expect to encapsulate biodiversity in a single measure or indeed in just two (Buckland et al.).

Biological diversity is identified as the number of taxa within an area, as the number of different life-history stages, the different species interactions within an area, or simply the number of individuals of a selected species or group of species and the geographic distribution of those individuals. The different ecological roles that species may play at different stages in their lifecycles can be considered or the level of endemism and/or biodisparity (range of morphologies and reproductive styles in a community, determined by heterogeneity and unpredictability of habitats) (RAMSAR, 1999, in Derous, 2004). The simplest method for target areas for conservation action is to identify countries with the highest number of species (greatest species richness, for e.g., the concept of Mega diversity countries based on the species of vertebrates, swallowtails butterflies and higher plants). Various methodologies are adopted to estimate the faunal resources of the area, among them the following are important.

Species/Area Relationships

The global distribution of biological diversity is that the tropics at lower latitudes harbour relatively more species per unit area than the temperate zones at higher latitudes and even within tropics the biological diversity does not exhibit any distinctly recognizable patterns. The number of species in an area increases with the size of the area and this increase follows a predictable pattern known as Arrhenius Relationship, Log S = C + Z Log A, Where S = No. of species, A = Area and C & Z are constants. According to this *Species/Area* relationship more species occur in tropical communities than in the temperate and arctic communities, e.g., a small region at 60° latitude may have 10 species of ants, at 40° there may be 50 – 100 species and within 20° of the equator 100 – 200 species. Within the latitudinal limits (belts) the number of species varies widely among the habitats according to the productivity of the area and degree of structural heterogeneity. Therefore, tropics show higher diversity and rate of speciation, as they are older and more stable (Non equilibrium Hypothesis), and less stringent competition due to greater productivity, more resources, increased spatial heterogeneity and increased control over competing populations exercised by predators (Equilibrium hypothesis). The differences between habitats are referred as beta (β) diversity while diversity within a site or habitat is known as alpha (ά). Differences in the site diversity over large areas, such as continents, are sometimes referred to as gamma (ý) diversity.

The simplest measure of species diversity is the number of species or the species richness, which is calculated after Margalef (1968) as modified by Brower & Zar (1977): D_a = (s-1)/log N, Where, D_a = Margalef's index; s = no. of species and N = total number of individuals. Usually, Shannon – Weiner Diversity Index and Simpson diversity indices are used to describe the species diversity. This measure of species diversity, based on the information theory or related to the concept of "uncertainty" was calculated after Shanon & Weiner (1949): H' = Σs P_i log P_i = 1. Where, H' = measure of Shannon – Weiner diversity, S = number of species and P_i = proportion of the total number of individuals occurring in species i. However, in the present context, H' was calculated, using its algebraic manipulation as the equivalent equation after Brower & Zar (1977): H' = (N log N - Σn_i log n_i)/N; Where, n_i = total number of individuals occurring in species, $_i$ = abundance of species. The maximum possible diversity H' or H_{max} was calculated using the following formula: H_{max} = log s where s is the number of species or category. The most common indicator of trends in species abundance is species richness within a prescribed area. But monitoring the number of species in an area can be misleading, because it is generally the assemblage of organisms in an area that is important, rather than simply the number of species present (Derous, 2004). The species richness is valued as the common currency of the diversity of life, the "face" of biodiversity –the problem with this emphasis on using species for quantifying biodiversity is that such a focus on species may provide patterns that mask many important patterns and trends for other properties of biodiversity beyond taxonomy (Norse, 1994). For example, when the sizes of the areas chosen are substantially smaller than biogeographic regions, this will result in a high degree of species replication and will fail to protect those species most at risk (rare species), which mostly have unique habitat preferences (Derous, 2004). It is for this reason, a combined index of species richness, rarity and vulnerability was proposed by Rey Benayas & de la Montaña (2003) as a measure of the diversity of an area with species. Applying this index had a much better performance to reveal the areas of high diversity than when the constituting criteria were used. It should be noted that this index was developed for terrestrial ecosystems and that it has not been tested on its applicability in the marine environment (Derous, 2004). In marine ecosystems, intensification of fishing beyond sustainable levels has lead to ecological extinction of some species in some regions, and major imbalances in the entire marine food web (Watson et al. 2003) and it is for this reason Marine Trophic Index is adopted to measures the change in mean trophic level of fisheries landings.

FAUNAL RESOURCE

India is very rich in terms of biological diversity due to its unique biogeographical location, diversified climatic conditions and enormous ecodiversity and geodiversity. In fact, within only about 2% of world's total land surface, India is known to have over 7.45% of the species of animals that the world holds and this percentage accounts nearly for 91,559 species so far known, of which insects alone include 61,151 species. Indeed, those groups that are known to occur in India have not received adequate attention, while some other invertebrates (like Nemertinea, Nematomorpha, Priapulida, Pogonophora and Pentastomida) are not known at all in India. Further studies will not only reveal many more species but also certainly enhance the information on endemic species.

Estimated number of described species

Taxonomic group	*No. of species*		*% in India*
	World	*India*	
PROTISTA (Protozoa)	**31250**	**2577**	**8.24**
ANIMALIA			
Mesozoa	70	10	14.08
Porifera	4562	500	10.70
Cnidaria	9916	842	8.49
Ctenophora	100	12	12.00
Platyhelminthes	17500	1622	9.22
Nemertinea	600	-	-
Rotifera	2500	330	13.20
Gastrotricha	3000	100	3.33
Kinorhyncha	100	10	10.00
Nematoda	30000	2850	9.50
Nematomorpha	250	-	-
Acanthocephala	800	229	28.62
Sipuncula	145	35	24.14
Mollusca	66535	5072	7.62
Echiura	127	43	33.86
Annelida	12700	840	6.61
Onychophora	100	1	1.00
Arthropoda	**987949**	**68389**	**6.90**
Crustacea	35534	2934	8.26
Insecta	867351	61151	6.90
Arachnida	73440	5818	7.90
Pycnogonida	600	16	2.67
Pauropoda	360	–	–
Chilopoda	3000	100	3.33
Diplopoda	7500	162	2.16
Symphyla	120	4	3.33
Merostomata	4	2	50.00
Phoronida	11	3	27.27
Bryozoa (Ectoprocta)	4000	200	5.00
Entoprocta	60	10	16.66
Brachiopoda	300	3	1.00
Pogonophora	80	–	–
Priapulida	8	–	–
Pentastomida	70	–	–
Chaetognatha	111	30	27.02
Tardigrada	514	30	5.83
Echinodermata	6223	765	12.29
Hemichordata	120	12	10.00
Chordata	**48451**	**5199**	**10.40**
Protochordata	2106	119	5.65
Pisces	21723	2589	11.72
Amphibia	5150	297	4.66
Reptilia	5817	460	7.91
Aves	9026	1232	13.66
Mammalia	4629	402	8.58
Total (Animalia)	1228103	91559	7.45

Estimated number of Endemic species

Group	No. of species		Percentage
	Total	Endemic	
Protozoa			
Free living	1247	90	7.21
Parasitic	1330	550	41.33
Mesozoa	10	10	100.00
Porifera Freshwater	31	13	41.93
Cnidaria	842	*10	–
Platyhelminthes	1622	1160	71.88
Rotifera	330	23	7.00
Gastrotricha	100	64	64.00
Kinorhyncha	10	7	70.00
Nematoda	2850	*400	–
Acanthocephala	229	203	88.64
Mollusca			
Terrestrial	1487	498	33.50
Freshwater	201	77	40.00
Echiura	43	12	28.00
Annelida			
Oligochaeta	473	368	77.80
Hirudinea	59	25	42.37
Arthropoda			
Crustacea	2934	501	17.07
Insecta	61151	20765	34.86
Arachnida	5818	2623	45.08
Phoronida	11	1	1.00
Bryozoa	4000	**12	–
Entoprocta	10	1	1
Chaetognatha	111	3	2.70
Chordata			
Pisces	2546	223	8.75
Amphibia*	298	156	52.34
Reptilia*	460	214	46.52
Aves	1232	176	14.28
Mammalia*	402	44	10.94

** Incomplete data and hence % not given,

* Data updated 2009

ANIMAL BREEDS IN INDIA

A "breed" is mostly the product of human intervention; it can only develop when a sub-population of a domestic animal species is separated and sexually isolated from the rest of the gene-pool. This can happen as a result of conscious selection for certain phenotypic or performance characteristics, as is typically the case in herd book registered livestock breeds. Local breeds play a multi-functional role in rural livelihoods, contributing not only cash products but also manure and traction. They are of social benefit as "insurance" against natural disasters or economic bottlenecks. Furthermore, indigenous breeds may have valuable genes with commercial potential. Especially the breeds kept by pastoralists, are regularly exposed to stressful condition.

In India, over 250 breeds of farm animals including cattle, buffalo, sheep, goat, camel, horse, pig and poultry are 'recognised', though there are also a number of other breeds not yet described or recognised. Each of these breeds are indigenous to a specific region, have certain specialisations, are hardy and adapted for conditions in their area of origin, require very little management when compared to exotic breeds and cross-breeds, especially veterinary costs; and can subsist on a far inferior diet during the leaner months.

Table 1. Farm Animal /Poultry Breeds in India

Sl. No.	*Species*	*Local*	*Regional Trans boundary*	*International Trans boundary*	*Total*
1.	Buffalo	17	4	2	23
2.	Cattle	52	6	6	64
3.	Goat	28	3	3	34
4.	Sheep	58	4	0	62
5.	Yak	5	0	0	5
6.	Pig	7	1	0	8
7.	Horse	6	1	0	7
8.	Ass	3	0	0	3
9.	Dromedary Camel	9	0	0	9
10.	Bactrian Camel	1	0	0	1
11.	Rabbit	3	0	0	3
12.	Chicken	23	1	4	28
13.	Duck	4	0	2	6
14.	Quail	2	0	0	2

(*Source:* SOW – AnGR, FAO-2007 with addition of some more breeds and deletion of duplication/synonyms)

Details of Breeds

Buffalo-23 Local: Assamese, Banni, Bhadawari, Chilka, Diyara, Godavari, Jerangi, Kalahandi, Manda, Manipuri, Marathwada, Nagpuri, Pandharpuri, Parlekhemundi, Sambalpur, South Kanara, Toda.

Regional Transboundary: Mehsana, Nili -Ravi, Surti, Tarai.

International Transboundary: Jaffarabadi, Murrah

Cattle- 64 Local: Alambadi, Amritmahal, Bachaur, Bargur, Binjharpuri, Cutchi, Dangi, Deoni, Frieswal, Gangatiri, Gaolao, Ghumsuri, Gujamavu,

Hallikar, Jersind, Kankrej, Kappiliyan, Karan Fries, Karan Swiss, Kenkatha, Khamala, Khasi, Kherigarh, Krishna Valley, Kumauni, Ladakhi, Malnad Gidda, Malvi, Mampati, Manapari, Mewati, Mhaswad, Motu, Nagami, Nagori, Nimari, Ponwar, Pullikulam, Punganur, Red Purnea, Ramgarhi, Rathi, Red Kandhari, Sanchori, Shahabadi, Son Valley, Tarai, Taylor, Tho Tho, Umblachery, Vechur, Zosial.

Regional Transboundary: Hariana, Hissar, Kangayam, Khillari, Siri,Sunandini.

International Transboundary: Sahiwal, Red Singhi, Gir, Ongole, Nelore,Tharparkar

Goat – 34 : Local: Andaman Feral Goat, Assamese Hill, Attapady Black, Baigani , Barren goat, Bidari, Changthangi, Chegu, Dalua, Ganjam, Gohilwadi, Himalayan Ibex, Himalayan Tahr, Indian Mohair, Jhakrana, Kannai Adu, Kutchi, Malabari, Marwari, Mehsana, Nilgiri Tahr, Osmanabadi, Ramdhan, Sangamneri, Sirohi, Surti, Teressa, Zalawadi.

Regional Transboundary: Bengal, Gaddi, Tibetan.

International Transboundary: Barbari, Beetal, Jamunapari

Sheep – 62 Local : Avikalin, Avimanns, Avivastra, Bagri, Balangir, Bellary, Berari, Bhakarwal, Bharat Merino, Biangi, Bonpala, Changthangi, Chokla, Chotanagpuri, Coimbatore, Dakshini, Deccani, Desi, Dhamda, Dumba, Gaddi, Ganjam, Garole, Godavari, Gurez, Hassan, Indian Karakul, Jaisalmeri, Jalauni, Karanah, Kashmir Merino, Kashmir Valley, Kenguri, Kheri, Kilakarsal, Magra, Malpura, Mandya, Marathwada, Marwari, Mecheri, Munjal, Muzaffarnagri, Nali, Nellore, Nilgiri, Patanwadi, Pugal, RamnadWhite, Rampur Bushair, Sangamneri, Sardarsamand, Shahabadi, Shapo, Sonadi, Telengana, Tiruchy Black, Vembur.

Regional Transboundary: Hissardale, Madras Red, Poonchi, Tibetan

Yak – 5 Local: Arunachal, Chour-gau, Himachal, Ladakhi, Sikkim

Pig – 8 Local: Andaman wild pigs, Ankamali, Boer, Deshi, Doom, Ghungroo, Nicobari

Regional Transboundary: Ghori

Horse – 7 Local: Deccani, Kathiawari, Manipuri Pony, Marwari, Spiti Pony, Zaniskari Pony.

Regional Transboundary: Bhutia

Ass – 3 Local: Indian, Indian wild, Kiang

Dromedary (Camel)- 9 Local: Bikaneri, Jaisalmeri, Kutchi, Malvi, Marwari, Mewari, Mewati, Shekhawati, Sindhi

Bactrian Camel – 1 Local: Indian Bactrian

Rabbit – 3 Local: British Angora, Grey Giant, Russian Angora

Chicken – 28 Local: Cari Gold, Debendra, Dhanraja, Gramalakshmi, Gramapriya, Kadaknath, Kalinga Brown, Kashmir Faverolla, Krishna-J, Miri, Mrityunjay, Nicobari, Vanaraja, Yamuna, Ankleshwar, Bursa, Chittagong, Danki, Daothigir, Ghagus, Haringhatta black, Kalasthi, Punjab brown

Regional Transboundary: Girraja

International Transboundary: Aseel, Brahma, Cochin, Indian game

Duck – 6 Local: Chemballi, Kuttanadu Chara, Nageshwari, Sythetemete

International Transboundary: Indian Runner, Khaki Campbell

INDIAN INITIATIVE ON ANIMAL CONSERVATION

Community based Programmes has been incorporated into the conservation strategy by giving importance to the ***sacred groves***.

Sacred groves are defined as the species rich habitats that were consecrated to a deity or spirit, where extraction of any living material is forbidden. These sacred groves serve as prey refugia, where prey populations could maintain a minimal population. Such sacred groves investigated recently in Andhra Pradesh, Kerala, Karnataka, Madhya Pradesh, Bihar, and some Northeastern states are identified as the species rich area of the primary forests as in Western Ghats, or area with rare species (*Kunstleria keralensis*) endemic frog (*Philaetus sanctisilvaticus*) as in Amarkantak, M.P. sacred ponds having a *refuge* of the black turtle (*Chelonys nigricans*) in Chittagong, Bangladesh. The conservation idiom of the tribals is also seen in their hunting practices. The Jenu Kurubas in Southern India never extract honey completely from the honeycomb in order to ensure that the bees do not suffer and the production does not come to an abrupt end. Similarly, the tubers and other roots on which tribes sustain are never uprooted fully and are used ensuring that the mother plants survive and reproduce.

Forest based programmes by conservation of forests through the Forest Act assures economically, ecologically, socially and culturally sustainable utilisation of the forests. It aims at promoting a rational use of forest resources. A key element of the Forest Act, with regards to safeguarding biodiversity, is defining certain habitats of special importance and giving guidelines as to how these habitats may be managed. Joint Forest (Protected Area) Management, an Act under Participatory Management Under Wildlife Act, Silvicultural practices, Co-operation between forest dwellers and forest authorities are added to the forest protection & conservation.

Use of remote sensing for data gathering, allied to the introduction of Geographic Information System (GIS) is a powerful tool to process that data in conjunction with information collected using traditional field techniques. This has been further substantiated by Peter Raven (2004) that locally driven indigenous efforts incorporating participatory approach, conflict resolution, and mutual learning are more likely to garner the relatively broad support and participation of local communities and strengthening of the local institutions for conservation.

Taxonomy is the science of identification, classification and naming of organisms. Taxonomic work involves study of morphological characteristics and phylogenetic relationship of organisms, which is essential for applied biological sciences including medicine, agriculture, forestry and fisheries. A sound taxonomic knowledge based is a pre-requisite for environmental assessment; ecological research, effective conservation, management and sustainable use of biological resources. At this crucial juncture, when the need for a taxonomic stock taking of the earth's biodiversity is becoming increasingly important and urgent, taxonomic experts are ageing and declining in number, both nationally and globally. This decline of taxonomic expertise is a critical problem that needs to be addressed urgently. Current requirements of taxonomic work and available expertise and studies indicate a dire need to encourage excellence and motivate experts to do work on hitherto neglected groups of organisms e.g., microbes, lower groups of plants, animals etc. The challenges are quite serious because while on the one hand the existing experts are ageing and retiring, on the other not many young scholars are opting for studies in taxonomy. Hence, the Ministry of Environment and Forests set up a Technical Group to develop the All India Project and after inter-ministerial consultations, the project was approved. The project envisaged establishment of centres for research in identified priority gap areas in the field of taxonomy, education and training (fellowships, scholarships, chairs, career awards etc.) and strengthening of National Taxonomic Institutions The modalities of implementing the All India Coordinating Project and prioritising activities under the project was decided after detailed consultations with experts. The Ministry launched the All India Coordinated Project on Capacity Building in Taxonomy (AICOPTAX) since 1999-2000.

Regulations through Laws: Conservation of biological diversity is taken up through the introduction of various legal procedures especially the Wildlife Protection Act, 1972, Environmental Protection Act, 1991 and declaration of *Ecologically Sensitive Areas* in the country. The Indian Wildlife

(Protection) Act, 1972 intended to provide a comprehensive National legal framework for wildlife protection, with conservation of species as the main criteria. The strategy includes total environmental protection and conservation with the assumption that all such areas should be free from human activities. The act prohibits hunting of wildlife, protects their habitats, and restrains trade in wild animals, trophies etc. The scope of this act was slightly ambiguous in the initial stages, as the definition of wildlife included only selected wild animals and birds. However, the scope was broadened in Wildlife Protection Amendment Act 1991 to include flora as well as fauna. The two-pronged approaches of this act are:

- Specified endangered species are protected regardless of its location and
- All species are protected in specified areas

Another important criterion for conservation is the endemic status of the fauna. India has many endemic plants and vertebrate and invertebrate species. Among plants, species endemism is estimated at 33% with c. 140 endemic genera but no endemic families (Botanical Survey of India, 1983). Areas rich in endemism are northeast India, the Western Ghats and the north-western and eastern Himalaya. A small pocket of local endemism also occurs in the Eastern Ghats. The Gangetic plains are generally poor in endemics, while the Andaman and Nicobar Islands contribute at least 220 species to the endemic flora of India (Botanical Survey of India, 1983). Threatened Plants Unit (TPU) is in the preliminary stages of cataloguing the world's centres of plant diversity; approximately 150 botanical sites worldwide are so far recognised important for conservation action, but others are constantly being identified (IUCN, 1987). Five locations have so far been issued for India: the Agastyamalai Hills, Silent Valley and New Amarambalam Reserve and Periyar National Park (all in the Western Ghats), and the Eastern and Western Himalaya.

Among various faunal groups, 396 higher endemic vertebrate species have been identified by WCMC from source references. Endemism among mammals and birds is relatively low. Only 44 species of Indian mammals have a range that is confined entirely to within Indian territorial limits. Four endemic species of conservation significance occur in the Western Ghats. They are the Lion-tailed macaque *Macaca silenus*, Nilgiri leaf monkey *Trachypithecus johni* (locally better known as Nilgiri langur *Presbytis johnii*), Brown palm civet *Paradoxurus jerdoni* and Nilgiri tahr *Hemitragus hylocrius*.

Only 55 bird species are endemic to India, with distributions concentrated in areas of high rainfall. The areas have been mapped by Bird Life International (formerly the International Council for Bird Preservation). They are located mainly in eastern India along the mountain chains where the monsoon shadow occurs, southwest India (the Western Ghats), and Andaman and Nicobar Islands (ICBP, 1992). In contrast, endemism in the Indian reptilian and amphibian fauna is high. There are around 187 endemic reptiles, and 156 endemic amphibian species. Eight amphibian genera are not found outside India. They include, among the caecilians, *Indotyphlus, Gegeneophis* and *Uraeotyphlus*; and among the anurans, the toad Bufoides, and the microhylid *Melanobatrachus*, and the frogs *Ranixalus, Nannobatrachus* and *Nyctibatrachus*. Perhaps most notable among the endemic amphibian genera is the monotypic *Melanobatrachus* which has a single species known only from a few specimens collected from the Anaimalai Hills in the 1870s. It is possibly most closely related to two relict genera found in the mountains of eastern Tanzania (Groombridge, 1983).

Further, the 'Washington' Convention on International Trade in Endangered Species of Wild Fauna and Flora, more commonly known as CITES, aims to protect certain plants and animals by regulating and monitoring their international trade to prevent it reaching unsustainable levels. The Convention entered into force in 1975. There are more than 180 Parties to the Convention. The CITES Secretariat is administered by the United Nations Environment Programme (UNEP). *(Refer:* checklist of CITES species and annotated cites appendices and reservations, Published by: CITES Secretariat/UNEP World Conservation Monitoring Centre)

India's Environmental policy, including International Agreements and implementation is principally made by the Ministry of Environment & Forests (MoEF), Government of India. The Convention of Biological Diversity (1992) to which the Government of India is a signatory, dwell a great deal on the need to document the biological diversity existing with in the Indian territory. Being conscious of the intrinsic value of the biological diversity and of the ecological, genetic, social, economic, scientific, educational, cultural, recreational and aesthetic values, especially Article – 7 of the Convention proposes the member country for the "***Identification and Monitoring***". It also calls for the signatories to monitor through sampling and other techniques, besides identifying the process and categories of activities, which are likely to have significant adverse impact on the conservation and sustainable use of resources. This involves a combination of collection and collating the existing data, generation of new data accessible easily for the conservation of resources.

Inventorisations provide and develop data on natural resources and help to develop biological bases, reference collections, gene banks, technologies and services required by Government in the execution of its regulatory, developmental and protection functions. These inventories are the benchmarks of databases of reference collections of fauna from different geographic areas and provide density and volumes of different classes of natural resources of the country. The inventorisation will aid in timely warning of impending threats to natural resource base and the impact of human activities thereon in the maintenance of biological diversity in all ecosystems and areas

ACKNOWLEDGEMENT

The author wish to express sincere thanks to the Secretary, Ministry of Environment and Forests, Government of India for all support.

REFERENCES

Anonymous, (1992). Animal Genetic Information Resource, UNEP, FAO.

Anonymous, (2003). Measuring biodiversity for conservation, The Royal Society, 6–9 Carlton House Terrace, London SW1Y 5AG.

Alfred, J.R.B, Das, A.K., and Sanyal, A.K. (1998). *Faunal diversity in India.* ENVIS Centre, Zoological Survey of India, p. 497.

Biodiversity Hotspots http://www.biodiversityhotspots.org/.

Buckland S.T., Magurran, A.E., Green, R.E. and Fewster, R.M. (2005). Monitoring change in biodiversity through composite indices. *Phil. Trans. R. Soc. B.*, 360: 243–254.

CITES: Convention on International Trade on Endangered Species of Wild Fauna and Flora. www.cites.org/.

Convention on Biological Diversity. www.biodiv.org.

Derous, S. (2004). Background paper for "BWZee – A biological valuation map for the Belgian continental shelf: Expert workshop on marine biological valuation", held December 4–6 2004, University of Ghent, Ghent.

Dinerstein, E. et al. (1996). *The global 200: Key ecoregions for saving life on earth.* World Wildlife Fund, US, Washington, D.C.

Duraiappah, A.K., Naeem, S., Agardy, T. et al. (2005). *The Future of Biodiversity*: Synthesis of the findings of the millennium assessment for the convention on biological diversity. (UNEP Publ.)

Franklin, J.F. (1993). Preserving biodiversity: Species, ecosystems, or landscapes. *Ecological Applications*, 3: 202–205.

Frissell, C.A. (1997). Ecological principles. In: *Watershed Restoration: Principles and Practices*. Williams, J.E., Wood, C.A. and Dombeck, M.P. (Eds.). pp. 96–115. American Fisheries Society, Bethesda, Maryland, USA.

Groombridge, B. *Global biodiversity*. Chapman and Hall, London, 1992.

Grove, R.H. (1995). *Green imperialism*: Colonial expansion, tropical island Edens and the origins of environmentalism, pp. 1600–1860. Cambridge, UK: Cambridge University Press.

Hinegardner, R. (1976). Evolution of genome size. In: *Molecular Evolution*. Ayala, F.J. (Ed.). pp. 179–199. Sinauer Associates, Sunderland, Mass.

IUCN (1980). *World conservation strategy*: Living resource conservation for sustainable development. International Union for Conservation of Nature.

Magurran, A.E, Phillip, D.A.T. (2001). Implications of species loss in freshwater fish assemblages. *Ecography*, 24, 645–650.

Magurran, A.E. (1988). *Ecological diversity and its measurement*. Princeton University Press, Princeton, New Jersey.

McNeely, J.A. (1998). How protected areas can respond to the changing nature of society. In: *Protected Areas in the 21st Century: From Islands to Networks*, pp. 189–202, Gland, Switzerland: IUCN.

________. (1998). *Economics and biological diversity*, pp. 1–235, Gland, Switzerland: IUCN.

Millennium Ecosystem Assessment (2005). Ecosystems and human wellbeing: Biodiversity synthesis. World Resources Institute, Washington, D.C.

Minelli, A. (1993). *Biological systematics*: The state of art xvi, p. 387. Chapman and Hall, London.

Moyle, P.B. and Yoshiyama, R.M. (1994). Protection of aquatic biodiversity in California: A five-tiered approach, *Fisheries*, 19(2): 6–18.

Myers, N. (1988). Threatened biotas: 'Hotspots' in tropical forests. *Environmentalist*, 8, 187–208.

________ (1990). The biodiversity challenge: Expanded hotspots analysis. *Environmentalist,* 10, 243–256.

Myers. N., Russell, A. Mittermeier, Cristina, G. Mittermeier, Gustavo, A.B., da Fonseca and Jennifer Kent, (2000). Biodiversity hotspots for conservation priorities. *Nature,* Vol. 403: www.nature.com.

Norse, E.A. (Ed.). (1994). *Global marine biological diversity*: A strategy for building conservation into decision-making. Island Press: Washington, D.C., p. 383.

Penny, F. Langhammer, Mohamed I. Bakarr, Leon A. Bennun, Thomas M. Brooks, Rob P. Clay, Will Darwall, Naamal De Silva, Graham J. Edgar, Güven Eken, Lincoln D.C. Fishpool, Gustavo A.B. da Fonseca, Matthew N. Foster, David H. Knox, Paul Matiku, Elizabeth A. Radford, Ana S.L. Rodrigues, Paul Salaman, Wes Sechrest and Andrew W. Tordoff Peter Valentine, Series Editor, (2007). *Identification and gap analysis of key biodiversity areas,* IUCN, Gland, Switzerland.

Ramakrishna and Alfred, J.R.B. (2007). *Handbook on Faunal Resources of India.* Publ.: Zoological Survey of India, 1–377.

Ramakrishna, Radhakrishnan, C. and Gopi, K.C. (2001). Western ghats in perspective of its zoogeography and biodiversity richness. *ENVIS Neswsletter,* No. 7: 2–4.

Raven, P.H. (2002). Science, sustainability, and the human prospect. *Science,* 297, 954–958.

Rey Benayas, J.M., De la Montana, E. (2003). Identifying areas of high-value vertebrate diversity for strengthening conservation. *Biological Conservation,* 114(3): 357–370.

Scott, M.J., Davis, F.W., Cusuti, B., Noss, R., Butterfield, B., Groves, C., Anderson, H., Caicco, S., Erchia, F.D., Edwards, T.C., Ulliman, J. and Wright, R.G. (1993). GAP analysis: A geographic approach to protection of biological diversity. *Wildlife Monographs,* 123: 1–41.

Stork, N.E. (1988). Insect diversity: Facts, fiction and speculation. *Biol. J. Linn. Soc.,* 35: 321–337.

The Royal Society (2003). *Measuring Biodiversity for Conservation.* Policy document 11/03. The Royal Society, London.

United Nations World Summit on Sustainable Development: Plan of Implementation. www.un.org/esa/.

Watson, Tyedmars, R.P., Kitchingman, A. and Pauly, D. (2003). Whats left? The emerging shape of the global fisheries crisis. *Conservation Biology in Practice,* 4(3): 30–31.

Wilson, E.O. (1984). *Biophilia.* Harvard University Press, Cambridge, Mass, p. 176.

__________. (1988). *Biodiversity.* National Academy Press, Washington, D.C.

__________. (1999). *The Diversity of Life.* Norton: New York.

13

The Natural History of the Hoolock Gibbon *Hoolock hoolock*: India's only Ape

Anwaruddin Choudhury

INTRODUCTION

The Hoolock gibbon *Hoolock hoolock* is the only ape found in the Indian subcontinent as well as South Asia. Its close relatives, at least 14 species of gibbons including the Siamang, are found in South-east Asia. Other apes, all larger in size, hence, also known as the great apes are the Orang-utan (*Pongo pygmaeus*) of Sumatra and Borneo, and the Gorillas (*Gorilla gorilla* and *G. beringei*) and Chimpanzees (*Pan troglodytes* and *P. paniscus*) of Africa. In captivity, Hoolock gibbons have lived up to an estimated 40 years (Weigl 2005).

A fair amount of published information is now available on hoolock gibbons in Assam and other areas of India (McCann 1933, Tilson 1979, Choudhury 1987, 1990, 1991, 1996, 1998, 2000, 2003, 2006, 2009ab; Mukherjee et al. 1988, Alfred 1992, Sati and Alfred 2001, Chhetri et al. 2007, Kakati et al. 2009) and there are number of synoptic works on primates or wildlife in general which also mention the species (Pocock 1939, 1941; Prater 1948; Choudhury 1983, 1988, 1997, 2001, Chivers 2001, Menon 2003). Some unpublished theses and reports also provide important data (Choudhury 1989, Misra et al. 1994, Kakati 1997, Das 2002). On Bangladesh gibbons, there are a number of publications such as Gittins and Akonda (1982), Gittins (1984), Feeroz (1991), Islam and Feeroz (1992), Ahsan (1994, 2001), and Muzaffar et al. (2007). Information on Myanmar are found in Geissmann et al. (2008) while that of China in Yang et al. (1987), Ma et al. (1988), Lan (1994) and Zhang (1998).

In this article, the natural history of the Hoolock gibbon covering many of its aspects have been described. A comprehensive bibliography is given at the end.

My first encounter with the hoolocks was in early 1960s when I was a child. It was in the Inner Line Reserved Forest, southern Cachar in Assam. However, scientific observation started since 1986 and continued sporadically till date. I carried out field surveys in areas where hoolock gibbons occur in North-east India as part of primate research as well as broader survey of wildlife in general. During this period visits were also made to some of the gibbon habitats of Myanmar (Burma) and Bangladesh. Although visits were made to China as well but the gibbon areas could not be explored. The presence of the gibbon was ascertained by direct sighting or by hearing their calls, as well as through finding preserved skulls in the tribal villages and interviews of local forest staff, villagers, and hunters, using visual aids such as photos and drawings. Direct observations and censuses were carried out along trails (mostly during foot-transect, and at a few sites, from elephant-back), roads (by car), and rivers (by boat).

The data were obtained during numerous field surveys. Intensive field works were done between February 1986 and August 1994. Subsequently, sporadic visits were made every year. Due to dense vegetation, the visibility was poor at places.

LOCAL NAMES. *Holou bandor* (Assamese), *Shinu* (Ao Naga), *Ulluck, Uluman* (Bengali, especially in Barak Valley region), *Amepa* (Idu Mishmi), *Kepe/Hegei* (Jemi Naga), *Hoidu* (Karbi/Mikir), *Saio* (Khiamniungan Naga), *Hauhuk* (Mizo), *Ten gun* (Rengma Naga), *Agu-hu* (Sema Naga), *Saka-huli, Chekkah* (Wancho), *Huri* (Yimchunger Naga), *Kepeng* (Zeliang Naga; Liangmai dialect).

TAXONOMY

The type locality of the species is the Garo Hills in Meghalaya (originally recorded as Assam), India (Harlan 1831). Formerly in the genus *Hylobates*, Prouty et al. (1983a, 1983b) argued for the placement of the Hoolock gibbon in a separate subgenus, *Bunopithecus* Matthew and Granger, 1923, based on its distinct karyotype. Groves (2001) accepted its distinctiveness and placed it in the subgenus *Bunopithecus* accordingly, but doubted the validity of the name. Brandon-Jones et al. (2004) and Groves (2005) placed it in the genus *Bunopithecus* based on the findings of Takacs et al. 2005, while still doubting the validity of the name. Eventually, Mootnick and Groves (2005) showed that *Bunopithecus* was not applicable to the species (or to gibbons at all), and named instead a new monotypic genus, *Hoolock* Mootnick and Groves 2005.

SUBSPECIES. Two subspecies are generally recognised (Groves 1967, 2001), of which one is found. The other subspecies is *leuconedys,* which occurs east of the Chindwin River. There are intergradations between the two subspecies near the upper areas of Chindwin and some of these may also occur within Indian limits. Das et al. (2006) surmised them to be *leuconedys.* The completely isolated population of the north and west of the Lohit River could well be a new subspecies, may be named as *mishmiensis* (after Mishmi Hills where their entire range is located). The variations among three forms are not significant enough to warrant full specific rank. In fact, these are truly geographic variations generally separated by zoogeographic barriers in the form of larger rivers.

SYNONYMS. *choromandus, fuscus, golok, hulok, leuconedys, scyritus.*

MORPHOLOGY

The Hoolock gibbon is a small primate with long limbs and without any tail (a basic characteristic of any ape). Its fur is dense and shaggy. Adult males and juveniles of both sexes of hoolocks are black. The young gibbons are greyish to brownish with face, palms and soles black. When subadult, at around four years of age, the black pelage of the females starts fading and slowly changes to greyish and then to a tan color (buff or blonde to yellowish grey-brown), which they retain as adults. The male and young females have a silvery-white band above their eyebrows, for which it is also called the "White-browed" gibbon. There is also a beard like tuft in the groin of the males, which is very nicely seen in silhouetted gibbons. There is a genital tuft of hair which may be black, whitish or brown (up to 3 inches long).

Size and weight:

Head and body length (cm) : 45.0 – 63.0

Weight (kg) : male 6.0 – 8.0; female 6.0 – 6.5

There is no visible sexual dimorphism; however, in maximum weight, males are slightly heavier. According to Schultz (1969), the male : female ratio of weight is 100:94.

DISTRIBUTION

The range of Hoolock gibbon is between the Brahmaputra and the Salween Rivers covering parts of North-east India, eastern Bangladesh, northern Myanmar (Burma) and a small area of southern China (McCann 1933, Groves 1972, Choudhury 1987). In India it is confined to the northeast, where it is restricted to the south of the Brahmaputra River and east of the Dibang River (Parsons 1941, Choudhury 1987). Its range in North-east India was not shown correctly in Corbet and Hill (1992).

Arunachal Pradesh

It is found from plains to above 2000m in the Mishmi Hills. Recorded in Lower Dibang Valley, Lohit, Tirap and Changlang districts. Protected areas where the gibbon occurs are listed in Table 1. Outside the protected areas, there are large habitats in Digaru, Kamlang, Manabhum, Tengapani and Turung Reserved Forests (all in Lohit district). Some other notable areas (all reserved forests) are Diyun, Honkap, Miao, Namchik, Namdang, Namphuk, Pangsau, Rima (all in Changlang district), Borduria and Namsang (Tirap district).

Assam

Not uncommon in the forests south of the Brahmaputra River where the species has been recorded in all the districts except one. Large population occurs in Karbi Anglong (Choudhury 2009a), Cachar and Tinsukia (Choudhury 2009b) districts. Other districts where the gibbon occurs are, Dibrugarh, Goalpara, Golaghat, Hailakandi, Jorhat, Kamrup, Kamrup (Metro), Karimganj, Morigaon, Nagaon, North Cachar Hills (Dima Hasao) and Sivasagar. Protected areas where the gibbon occurs are listed in Table 1. Outside the protected areas, there are large habitats in Dhansiri (Karbi Anglong district), Inner Line (Cachar and Hailakandi districts), Joypur (Dibrugarh district), Katakhal (Hailakandi district), Langting-Mupa (North Cachar Hills district), Lumding (Nagaon district), and Upper Dihing (west block) (Tinsukia district) Reserved Forests. Other noteworthy areas are Barak (Cachar district), Badshahitilla (Karimganj district), Krungming (North Cachar Hills district), Suang (Nagaon district), Kholahat (Morigaon district), Ajgar Hills (Goalpara district), Garbhanga, Rani (Kamrup Metropolitan district), Bogaikhas (Kamrup district), Upper Dihing (east block), Dum Duma, Kakojan, Torani, and Tinkopani (all five in Tinsukia district) Reserved Forests. In Tinsukia district, there are gibbons in some village woodlands also.

Manipur

It is rare all over. Mainly confined to five hill districts (Chandel, Churachandpur, Senapati, Tamenglong and Ukhrul) and Jiribam subdivision of Imphal East district. Recorded up to 2500m in Shiroi. In the valley proper it is extirpated long back. Protected areas where the gibbon occurs are listed in Table 1. Outside, there are sizeable habitat in Tolbung Reserved Forests (Churachandpur district), Moreh proposed Reserved Forests (Chandel district), and Ango/Anko Range and Shiroi (both in Ukhrul district). There is a specimen (male) at ZSI from Kalanaga, Barail Range in Senapati district, which was obtained in 1936.

Meghalaya

The type of species was obtained from Garo Hills. The hoolock gibbon occurs in all the districts, i.e., East Garo Hills, West Garo Hills, South Garo Hills, East Khasi Hills, West Khasi Hills, Jaintia Hills and Ri-Bhoi. Protected areas where the gibbon occurs are listed in Table 1. Outside the protected areas, there are sizeable habitats in Narpuh Reserved Forests (Jaintia Hills district), and in W Khasi Hills district. In smaller pockets such as Darugiri, Songsak and Dambu Reserved Forests (East Garo Hills district), also gibbons are there. Many Garos do not molest the ape and hence, there are groups in some village woodlands as well. There are a number of sizeable, privately owned forests in the catchment of the Jadukata River in West Khasi Hills district. A few gibbons still survive in sacred groves such as those in Chiehruphi in the Jaintia Hills district. There are specimens in the collection of Zoological Survey of India from Ri-Bhoi district [Manihar Basti, 13 km north of Nongpoh at 760 m a.s.l., from 8 km east and 11 km west of Nongpoh] and Garo Hills (obtained in 1870).

Mizoram

Overall it is rare and scattered. Recorded in all the districts, i.e., Aizawl, Champhai, Kolasib, Lawngtlai, Lunglei, Mamit, Saiha and Serchhip. Protected areas where the gibbon occurs are listed in Table 1. Outside, the gibbon occurs in Inner Line Reserved Forest (Kolasib and Aizawl district), in Tuirial project site, and in the forests in the extreme south covering southern parts of Lunglei (Ngengpui Reserved Forest) Lawngtlai and Saiha districts. Elsewhere, stray groups are found.

Nagaland

It is rare all over. Recorded in all the districts, *viz.*, Dimapur, Kiphire, Kohima, Longleng, Mokokchung, Mon, Peren, Phek, Tuensang, Wokha and Zunheboto. Protected areas where the gibbon occurs are listed in Table 1. Any survivor in Pulie Badge Wildlife Sanctuary is doubtful. It has also vanished from Rangapahar Wildlife Sanctuary. In Saramati, it occurs above 2600m in summer (Choudhury 2006). There is a specimen in the collection of Zoological Survey of India from Chumukedima (originally recorded as Samaguting), Dimapur district, which was obtained in 1872.

Tripura

It is recorded in all the four districts: North, Dhalai, West and South. Protected areas where the gibbon occurs are listed in Table 1. Elsewhere, scattered groups are found.

Bangladesh

It is restricted to the good quality evergreen and semi-evergreen forests of undivided Sylhet, Chittagong and Chitttagong Hill Tracts. It was also reported from Balichari Range of Mymensingh (Khan 1981). The current reorganized districts where gibbons occur are Moulvi Bazar, Habiganj, Khagrachari, Rangamati, Bandarban and Cox's Bazar. Protected areas where the gibbon occurs are listed in Table 1.

Myanmar

Hoolock Gibbons occur in northern and western parts of the country, west of the Salween River (Groves 2001). Both the subspecies and their intergrades (in north only) occur in the country with the river Chindwin being the zoogeographic barrier. The subspecies *luconedys* ranges up to 2,700 m elevation (IUCN 2010).

China

Hoolock Gibbons are restricted to a small area in western Yunnan bordering Myanmar, west of the Salween River (Haimoff et al. 1987). The subspecies *leuconedys* occurs in the country.

Table 1. Protected Areas with known hoolock gibbon population

Name of the Country/State/Protected Area	*Area (km²)*
INDIA	
Assam	
Amchang WS	78.6
Barail WS	326.2
Bherjan-Borajan-Podumoni WS	7.2
Dibru-Saikhowa NP	340
Dihing-Patkai WS	111.2
East Karbi Anglong WS	222
Garampani WS	6
Hollongapar Gibbon Sanctuary	21
Kaziranga NP	850
Marat Longri WS	451
Nambor WS	37
Nambor-Doigrung WS	97
North Karbi Anglong WS	96
Arunachal Pradesh	
Kamlang WS	783
Mehao WS	281.5
Namdapha NP	1985
Manipur	
Bunning WS	115.8
Jiri-Makru WS	198
Namdapha NP	1985
Kailam WS	157.8
Yangoupokpi-Lokchao WS	184
Zeilad WS	21
Meghalaya	
Balpakram NP	312 (actual area is less than 200 km² due to anomaly in computing)
Nokrek NP	68 C
Nongkhyllem WS	35 C
Siju WS	5.2
Mizoram	
Dampa WS	500
Khawnglung WS	41
Lengteng WS	80

Murlen NP	150
Ngengpui WS	110
Phawngpui NP	50
Tawi WS	35.7
Thorangtlang WS	50
Nagaland	
Intanki NP	202
Fakim WS	6.4
Tripura	
Gumti WS	389.5
Sepahijala WS	18.5
Trishna WS	194.7
BANGLADESH	
Himchari NP	17.3
Lawachara NP	12.5
Kaptai NP	54.6
Satchari NP	2.4
Pablakhali WS	420.9
Chunati WS	77.6
Rema-Kalenga WS	17.9
Teknaf GR	116.1

HABITAT

A strictly forest-dwelling primate, the Hoolock gibbon is found in three major types of habitats, tropical evergreen or rain forest, tropical moist deciduous and subtropical broadleaf. Marginally they also occur in temperate broadleaf forest in the mountains. However, the deciduous forests of most parts of northeastern India are not pure stands of deciduous tress but contain large number of evergreen trees. There are also patches of semi-evergreen forests within the deciduous biotope. The Hoolock has also been observed in bamboo thickets amidst semi-evergreen or deciduous forests. The 'hollong' *Dipterocarpus macrocarpus*, 'gurjan' *Dipterocarpus turbinatus*, *Terminalia myriocarpa*, *Duabanga sonneratoides*, *Artocarpus chaplasa* and *Mesua ferrea* are some of the notable trees of rain forest. Tropical semi-evergreen forest also occurs elsewhere in patches. Large part of the gibbon's range in Central Assam and northern Meghalaya are covered with tropical moist deciduous forest. The 'sal' (*Shorea robusta*) and the emergent *Tetrameles nudiflora*, *Gmelina arborea*, *Dillenia scabrela*, and *Bombax ceiba* are some of the species. The main bamboo species are *Dendrocalamus hamiltonii* and *Oxytenanthera nigrociliata/parvifolia*. In the deciduous forests, the hoolocks evidently prefer the mixed patches and pockets of evergreen areas. Bamboo and scrub has invaded large areas in the abandoned *jhums*. Gibbons can also be found in small forest patches in plantations of deciduous species such as 'teak' (*Tectona grandis*) but not in the pure teak areas. The subtropical and temperate broadleaf forests occur mostly in Mishmi Hills, Saramati, Barail, and other high elevation areas, normally above 1500m.

Most of the present habitat of the Hoolock gibbon is in the mountains, hills or in rugged and low undulating terrain. This is mainly because of destruction of forest and intensive paddy cultivation in the low lying plain areas.

The Hoolock has been recorded at altitudes of less than 30 m to more than 2600 m. The known "area of occupancy" (according to IUCN criterion) of hoolock gibbons in India is about 27,000 km^2 (down from more than 30,000 km^2 in 1990s) of which about 12,000 km^2 supports a fairly dense gibbon population. In the remaining area, only scattered individuals or isolated groups are found. Elsewhere, in Bangladesh, known "area of occupancy" is about 600 km^2. However, data from Myanmar and China are still scanty but is fairly large in the former country.

POPULATION STATUS

Although quite widespread, the Hoolock gibbon has become rare all over its range except in some protected areas. It is nowhere abundant because of hunting or other disturbance. In places, lone animals or groups can be isolated by 5 or 10 km from other groups. A general assessment on the basis of densities worked at different sites (Choudhury 2006, 2009ab) indicates that there are some 9,000 – 11,500 gibbons in India, about 250 in Bangladesh (Islam et al. 2006) and an unknown but apparently several thousands (owing to very large habitat) in Myanmar and China. Geissmann et al. (2008) stated that "at present, Myanmar potentially supports the largest remaining populations of either hoolock species". The total population of the eastern subspecies, *H. h. leuconedys* in Myanmar may be over 10,000 individuals, and perhaps up to 50,000 individuals (Brockelman, pers. comm., cited in Geissmann, 2007). The population of western subspecies is not known but could be few thousand as the available habitat indicates. Haimoff et al. (1987) had estimated about 100 – 200 while Zhang (1998) fewer than 150.

SOCIO-BEHAVIOURAL ACTIVITIES

Hoolock gibbons are arboreal animals and come down to the ground to cross clearings only. Their locomotion is mostly through brachiation in the forest canopy. On large tree trunks, they also walk bipedally. Leaping and climbing are other modes of locomotion. During travel, the group usually moves in a single-file and the adult of either sex can lead. While on the ground to cross clearings, they walk bipedally, often with raised forearms. Their home range is normally 22 to 30 hectares in ideal condition; however, due to fragmentation and degradation, it could have extreme variation, from 3 to 90 hectares. They defend their territory by their characteristic loud call *whoku, huku, huku* . . . normally uttered in the morning and forenoon, and also by threat displays but rarely actual physical contact during confrontation. Calls can usually be heard up to a kilometer away (Choudhury 1989b) although often audible from a distance of 3km if from a higher elevation. The overlapping of territory varies from 5 to 25%. The daily range usually varies from 500 to 1.2 km; however, there could be extremes as well, from less than 200 m to more than 3 km.

Gibbons differ from other sympatric diurnal primates in taking final roosting positions. While all the macaques and langurs take it late (evening or around dusk), hoolocks take it fairly early, in the mid-afternoon (between 1300 and 1500hrs). During the hot hours of the day they retire to shady trees and rest. They may sleep in sitting posture, with head buried between the knees, or by lying down on horizontal tree branches (somewhat like humans). During the morning in the winter, the gibbons 'sunbask' in high branches exposing their backs to the sun for several minutes at a time (Tilson 1979; Choudhury 1989a,1996).

Gibbons live in small monogamous family groups of a mated pair and their offspring. The largest group ever recorded was of seven animals (McCann 1933). The average group size is 3.1 (Choudhury 1990). The largest group observed by this author was six gibbons about 1 km west of Raja-ali Beat office in Upper Dihing (east block) Reserved Forest in Assam's Tinsukia district on 19 August 1993. It contained an adult pair, another adult male which seemed to be younger than the alpha male, two sub-adults of different age groups and a juvenile.

Although hoolock gibbons take an omnivorous diet that includes insects, spiders, bird eggs, leaves, and flowers; it may not be inappropriate to term them frugivorous as their diet includes more than 60% fruits including figs.

Some trees species play a particularly important role in Hoolock gibbon ecology, by providing a major component of food supply in certain seasons, e.g., *Dillenia scabrella* (fruit; in March, April, May), *Anthocephalus cadamba* (fruit; in September, October), *Sapium baccatum* (fruit; in August, September, October), *Artocarpus chaplasha* (fruit; in June and July), *Castanopsis indica* (fruit; in September, October), and various figs (*Ficus* spp.).

Mating in hoolock gibbons normally takes place during summer-monsoon rainy season while the young ones are born in autumn and winter, usually between September and February. However, there are instances of birth in summer as well. The gestation period is 183–225 days in captivity (Geissmann 1991) and 195–210 days in the wild (Sati and Alfred 2001). The inter-birth interval is 2–3 years. An infant gibbon remains with mother for the first six months during which it does not take any outside food. Then slowly it starts taking other food. The young leaves its dependence on mother by around two and half years and remain in the family group as an 'independent' member till it forms a separate family group. The age of sexual maturity in both sexes is approximately seven years. Emigration from natal groups is of both sexes when the gibbons become mature adults; however, for quite some time, individuals may have to remain alone.

CONSERVATION PROBLEMS

Habitat Destruction and Fragmentation

Forest destruction through tree felling, encroachment, *jhum*, and monoculture tree plantations is a major threat to the survival of the hoolock gibbon in its range. The forest cover all over its range is disappearing at an alarming rate. More than 1,000 km^2 of forest was destroyed annually in the northeastern region of India (including Assam, Arunachal Pradesh, and other states) during the 1970s and 1980s (data from the National Remote Sensing Agency). In Assam, dense forest cover has declined from 19.9% of the geographical area in 1980–82 to 16.4% in 2004–05 (India, NRSA 1983; India, FSI 2005). In Manipur, dense forest cover has declined from 50.5% of the geographical area in 1980–82 to 25.6% in 2001. During the same period, the decrease in Meghalaya was from 33.1% to 25.3%, in Mizoram from 62.6% to 42.4%, and in Nagaland from 42.8% to 32.5% (India, NRSA 1983; India, FSI 2003).

Encroachment is a major problem in the reserved forests, and *jhum* cultivation is an important cause of forest loss and fragmentation in the hilly areas throughout these states. For instance, *jhum* currently covers more than 1,800 km^2, or 8.2% of the 22,327 km^2 of the small state of Manipur. Opencast mining for coal has affected the sacred grove of Chiehruphi in Jaintia Hills. Coal and limestone mining in the Garo Hills in Meghalaya and in Tinsukia district in Assam has also destroyed important gibbon habitat. Even protected areas and reserved forests have fragmented into parts. Yangoupokpi-Lokchao Wildlife Sanctuary in Manipur and Nambor-Garamnapi complex of sanctuaries are cut into two by national highways. The small reserved forest of Darugiri in Meghalaya is divided into three parts by major roads, which the gibbons are unable to cross. The number of fragmented units state by state is as follows: >130 in Assam, >25 in Arunachal Pradesh, >16 in Manipur, >20 in Meghalaya, >22 in Mizoram, >10 in Nagaland, and >15 in Tripura. In Bangladesh, there are 22 fragments (Muzaffar et al. 2007). This is excluding the scattered groups and individuals spread all over in the abandoned *jhums* and heavily degraded tracts, for which estimate of fragmentation was virtually impossible.

Poaching

The hoolock gibbon is hunted for food by many of the tribes in its range. However, there are tribes such as the Karbis (formerly called the Mikirs) and Mishmis (Idu, Digaru and Miju) do not normally kill gibbons as the local folk lores say that it was a human once upon a time. But a few new generation youth occasionally kills an odd animal. Sometimes poaching is also done for supposed medicinal value of its body parts. In the past, traditional weapons such as snares and self-made muzzle-loaders were used, but the last two decades has seen the increased use of automatic firearms.

Trade

Commerce in primates is not significant, but occurs. Smoked gibbon meat on sale are occasionally seen the local markets of Nagaland, Manipur and northern and western Myanmar. Young gibbons are occasionally captured for pets, and small numbers are also trapped (illegally) to supply zoos.

Other Problems

Other conservation issues include the destructive harvesting of bamboo for paper mills, open-cast coal mining, and various developmental activities.

CONSERVATION MEASURES TAKEN

The hoolock gibbon is protected under Schedule-I of the Wildlife (Protection) Act of India, which prohibits its killing or capture, dead or alive. Enforcement, however, is virtually nonexistent, even in the protected areas. Most locals are unaware of its legal status. IUCN (2010) has listed it as "Endangered." The Hoolock occurs in at least 39 protected areas in India and eight in Bangladesh (Table 1), and several others in its range in Myanmar and China. In addition, it also occurs in more than 120 reserved forests in India.

DISCUSSION

Being strictly a forest-dweller, the Hoolock gibbon needs continued existence of good quality forest cover in its range. The dense forest is vanishing fast owing to felling of trees for log-based industries, and encroachment. In much of the dense forest where the habitat is still ideal, gibbons have long since vanished due to hunting. On the other hand, a few groups do still survive in degraded areas. Hence, unlike Assam, and except in parts of Meghalaya, the extent of dense forest may not have much bearing on gibbon abundance and distribution. In the Garo Hills, Karbi Anglong and parts of Mishmi Hills they are generally not molested and hence still occur even in small patches near villages, but due to *jhum*, there are regular micro-level changes in gibbon home ranges there. Elsewhere in the hilly tracts all over the range, individuals isolated by or exposed due to *jhum* are hunted down within a short time.

Except for parts of the Garo Hills, Karbi Anglong, parts of Mishmi Hills and in the plains of assam and Bangladesh where there is some degree of community protection, the gibbons have survived in a number of areas in mountainous tracts in their range merely because of inaccessible and difficult terrain.

Throughout its range, the gibbon is sympatric with other primates including the Assamese macaque (*Macaca assamensis*), Stump-tailed macaque (*M. arctoides*), Pig-tailed macaque (*M. nemestrina*), Rhesus macaque (*Macaca mulatta*), Capped langur (*Trachypithecus pileatus*), and Slow loris (*Nycticebus coucang*). In Mizoram, southern Assam, Bangladesh and in southwest Myanmar, it is also sympatric with Phayre's leaf monkey (*Trachypithecus phayrei*).

Large contiguous habitats for long-term conservation are fewer. However, all such tracts should be protected as these are beneficial for other species as well and for conservation of water sources. The ultimate cause of habitat destruction is, however, the very rapid growth of the human population across its range. Since the bulk of the rural population practice agriculture and *jhum* as their main occupation, and new villages and hamlets appear constantly, the large-scale destruction of natural habitat seems inevitable. Despite this depressing prognosis, we believe, however, that if hunting can be reduced through community awareness, and if protective measures are successfully put in place in the sanctuaries and parks, then the forests remaining could support viable populations of gibbons in the long term.

RECOMMENDATIONS

A number of important known habitats for gibbons, which are outside the protected area network, should be declared as wildlife sanctuaries. Small pockets should be declared sanctuaries for the development of eco-tourism with community involvement. Awareness campaigns should involve the churches and the village headmen to promote conservation measures and programs should be set up for the regular monitoring of the gibbon populations in select sites in the four states.

ACKNOWLEDGMENTS

During the field study, I was given considerable support and assistance from many civil and forest officials all over; NGO, and a large number of villagers, relatives, and friends, and I thank them all collectively. I thank Assam Science technologyb & Environment Council, and American Society of Primatologists for supporting some of the field trips in 1990s. Thanks are also to Prof. Colin Groves and David Chivers, renowned primatologists with experience on hoolocks for the valuable discussions with them. They include My thanks too to my late parents who provided valuable inputs, my wife and relatives, friends, and other well-wishers who gave me the benefit of their advice.

For declaration of the sanctuaries and accepting the recommendations made in Choudhury (1988, 1989ac, 1993 and 1996), I thank late Nagen Sharma, Pradyut Bordoloi, both former Forest Ministers of Assam; Jotson Bey, then Chief Executive Member of the Karbi Anglong Autonomous Council. The sanctuaries declared at my recommendation (I had the opportunity to process them and also issue notification of some of these as Joint Secretary in Environment & Forest department) include: Bherjan-Borajan-Podumoni, East and North Karbi Anglong, Marat Longri, Nambor, Nambor-Doigrung, Dihing-Patkai, Amchang and Barail.

References

Ahsan, M.F. (1994). Behavioral ecology of the hoolock gibbon in Bangladesh. Ph.D. Dissertation. University of Cambridge, Cambridge, UK.

Ahsan, M.F. (2001). Socio-ecology of the Hoolock Gibbon (*Hylobates hoolock*) in two forests in Bangladesh. In: *The Apes: Challenges for the 21st Century*. Conference proceedings, pp. 286–299. Chicago Zoological Society, Brookfield, Illinois.

Alfred, J.R.B. (1992). The hoolock gibbon: *Hylobates hoolock*. *Primate Report*. 34: 65–9.

Brandon-Jones, D., Eudey, A.A., Geissmann, T., Groves, C.P., Melnick, D.J., Morales, J.C., Shekelle, M. and Stewart, C.B. (2004). Asian primate classification. *Int. J. Primatol.*, 25: 97–164.

Chhetri, R., Chhetri, D and Bhattacherjee, P. (2007). *Amar holou*. Gibbon Conservation Centre, Mariani, India.

Chivers, D.J. (2001). The swinging singing apes: Fighting for food and family in far-east forests. In: *The Apes: Challenges for the 21st Century*. Conference proceedings, pp. 1–28. Chicago Zoological Society, Brookfield, Illinois.

Choudhury, A.U. (1983). Plea for a new wildlife-refuge in eastern India. *Tigerpaper*, 10(4): 12–15. Bangkok.

__________. (1987). Notes on the distribution and conservation of Phayre's leaf monkey and hoolock gibbon in India. *Tigerpaper*, 14(2): 2–6.

__________. (1988). Priority ratings for conservation of Indian primates. *Oryx.*, 22: 89–94.

__________. (1989a). Primates of Assam: Their distribution, habitat and status. Ph.D. Thesis Gauhati University. Guwahati.

__________. (1989b). Vocalization in hoolock gibbon (*Hylobates hoolock*). *Prim. Rep.*, 25: 53–4.

__________. (1989c). Campaign for Wildlife Protection: National Park in the Barails. *WWF-Quarterly*. No. 69, 10(2): 4–5.

__________. (1990). Population dynamics of hoolock gibbons in Assam, India. *Am. J. Primatol.*, 20: 37–41.

_________. (1991). Ecology of the hoolock gibbon, a lesser ape in the tropical forests of NE India. *J. Trop. Ecol.*, 7: 147–153.

_________. (1992). Dhansiri Tiger Reserve. A proposal submitted to the Karbi Anglong District Council and the Government of India. p. 8 + map.

_________. (1993). *A Naturalist in Karbi Anglong*. Gibbon Books, Guwahati.

_________. (1996). A survey of hoolock gibbon (*Hylobates hoolock*) in southern Assam, India. *Prim. Rep.*, 44: 77–85.

_________. (1997). *Checklist of the Mammals of Assam*. 2nd Edition. Gibbon Books and ASTEC. Guwahati.

_________. (1998). A survey of primates in the Jaintia Hills. *ASP Bull.* 22(3): 8–9.

_________. (2000). A survey of hoolock gibbon (*Hylobates hoolock*) in Dibru-Saikhowa National Park, Assam, India. *Prim. Rep.*, 56: 61–66.

_________. (2001). Primates in NE India: An overview of their distribution and conservation status. *ENVIS Bull: Wildl. Protected Areas*, 1(1): 92–101.

_________. (2003). *The Mammals of Arunachal Pradesh*. Regency Publications. New Delhi.

_________. (2006). The distribution and status of hoolock gibbon, *Hoolock hoolock*, in Manipur, Meghalaya, Mizoram, and Nagaland in Northeast India. *Primate Conservation,* 20: 79–87.

_________. (2009a). The distribution, status and conservation of hoolock gibbon, *Hoolock hoolock* in Karbi Anglong District, Assam, Northeast India. *Primate Conservation,* 24: 117–126.

_________. (2009b). The hoolock gibbon (*Hoolock hoolock*) in Tinsukia and Dibrugarh districts of Assam, India. *Asian Primates Journal,* 1(2): 24–30.

Corbet, G.B., Hill J.E. (1992). *The Mammals of the Indomalayan Region: A Systematic Review*. Oxford University Press. Oxford.

Das, J. (2002). Socioecology of hoolock gibbon *Hylobates hoolock hoolock* in response to habitat change. Ph.D. Thesis Gauhati University. Guwahati.

Das, J., Feeroz, M.M., Islam, M.A., Biswas, J., Bujarboura, P., Chetry, D., Medhi, R. and Bose. J. (2003). Distribution of hoolock gibbon (*Bunopithecus hoolock hoolock*) in India and Bangladesh. *Zoos' Print J.*, 18: 969–976.

Das, J., Biswas, J., Bhattacharjee, P.C., Mohnot, S.M. (2006). First distribution records of the eastern hoolock gibbon *Hoolock hoolock leuconedys* from India. *Zoos' Print Journal,* 21(7): 2316–20.

Feeroz, M.M. (1991). The ecology and behavior of hoolock gibbons (*Hylobates hoolock*) of Bangladesh. M.Sc. Dissertation. Jahangirnagar University, Bangladesh.

Feeroz, M.M. (2001). Species diversity and population density of non-human primates in north-east and south-east of Bangladesh. *Ecoprint,* 8: 53–57.

FSI (1999). *State of Forest Report 1999.* Forest Survey of India (FSI). Dehradun.

FSI (2003). *State of Forest Report 2003.* Forest Survey of India (FSI). Dehradun.

FSI (2005). *State of Forest Report 2005.* Forest Survey of India (FSI). Dehradun.

Geissmann, T. (1991). Reassessment of age of sexual maturity in gibbons (*Hylobates* spp.). *American Journal of Primatology,* 23: 11–22.

Geissmann, T., Grindley, M., Momberg, F., Lwin, Ngwe, and Moses, Saw. (2008). Hoolock gibbon and biodiversity survey and training in southern Rakhine Yoma, Myanmar: Preliminary report. Myanmar Primate Conservation Program, BANCA, FFI, PRCF and Yangon University, Yangon, Myanmar, p. 31.

Gittins, S.P. (1984). The distribution and status of hoolock gibbons in Bangladesh. In: *The Lesser Apes: Evolutionary and Behavioural Biology.* Preuschoft, H., Chivers, D.J., Brockelman, W.Y. and Creel, N. (Eds.). pp. 13–15. Edinburgh University Press, Edinburgh, UK.

Gittins, S.P. and Akonda, A.W. (1982). What survives in Bangladesh? *Oryx.,* 16: 275–281.

Groves, C.P. (1967). Geographic variation in the hoolock or white-browed gibbon (*Hylobates hoolock* Harlan 1834). *Folia Primatologica,* 7: 276–83.

_________. (1972). Systematics and phylogeny of gibbons. In: *Gibbon and Siamang.* Rumbaugh, D.M. (Ed.). pp. 1–89. Basel:Karger.

_________. (2001). *Primate Taxonomy.* Smithsonian Institution Press, Washington, D.C.

_________. (2005). Order primates. In: *Mammal Species of the World: A Taxonomic and Geographic Reference.* 3rd Edition. Vol 1. Wilson, D.E., Reeder, D.M. (Eds.). pp. 111–184 Johns Hopkins University Press. Baltimore.

Haimoff, E.H., Xiao-Jun, Y., Swing-Jing, H. and Nan, C. (1987). Conservation of gibbons in Yunnan Province, China. *Oryx.,* 21: 168–173.

Harlan, R. (1831). Description of a new species of orang (*Simia hoolock*) from the north-eastern province of British east India, lately the kingdom of Assam. *Trans. Am. Phil. Soc.*, 4(1): 52–59.

Islam, M.A., and Feeroz, M.M. (1992). Ecology of hoolock gibbon of Bangladesh. *Primates,* 33: 451–464.

Islam, M.A., Feeroz, M.M., Muzaffar, S.B., Kabir, M.M. and Begum, S. (2006). Conservation of the hoolock gibbons (*Hoolock hoolock*) of Bangladesh: Population estimates, habitat suitability and management options, *Technical Report, U.S. Fish* and *Wildlife Service.*

IUCN (2010). *IUCN Red List of Threatened Species.* Version 2010.1. <http://www.iucnredlist.org/>. Downloaded on 16 March 2010.

Kakati, K. (1997). Food selection and ranging in the hoolock gibbon in Borajan reserve forest, Assam. M.Sc. dissertation. Wildlife Institute of India, Dehradun.

Kakati, K., Raghavan, R., Chellam, R., Qureshi, Q. and Chivers, D. (2009). Status of western hoolock gibbons (*Hoolock hoolock*) populations in fragmented forests of eastern Assam. *Primate Conservation,* 24: 127–137.

Khan, M.A.R. (1981). The non-human primates of Bangladesh. *Tigerpaper,* 8(1): 12–15.

Lan, D. (1994). Progress of surveys of hoolock gibbon in Yunnan: Distribution, population size, habitat and conservation. *Chin. Prim. Res. Conserv. News,* 3(1): 8–10.

Ma, S., Wang, Y. and Poirier, F.E. (1988). Taxonomy, distribution, and status of gibbons (*Hylobates*) in southern China and adjacent areas. *Primates,* 29(2): 277–86.

McCann, C. (1933). Notes on the colouration and habits of the white-browed gibbon or hoolock (*Hylobates hoolock* Harl.). *J. Bombay Nat. Hist. Soc.,* 36: 395–405.

Menon, V. (2003). *A field guide to Indian mammals.* Dorling Kindersley (India), Delhi.

Misra C., Raman, T., Johnsingh, A. (1994). Survey of primates, serow, and goral in Mizoram. Report, Wildlife Institute of India, Dehradun.

Mootnick A., Groves C.P. (2005). A new generic name for the hoolock gibbon (Hylobatidae). *Int. J. Primatol.,* 26(4): 972–976.

Mukherjee, R.P., Chaudhuri, S. and Murmu, A. (1988). Hoolock gibbons in Arunachal Pradesh, northeast India. *Primate Conservation,* 9: 121–3.

Muzaffar, S.B., Islam, M.A., Feeroz, M.M., Kabir, M., Begum, S., Mahmud, M.S., Chakma, S. and Hasan, M.K. (2007). Habitat characteristics of the endangered hoolock gibbons of Bangladesh: The role of plant species richness. *Biotropica.,* 39(4): 539–545.

NRSA (1983). *Mapping of Forest Cover in India from Satellite Imagery 1972–75 and 1980–82. Summary Report, North Eastern States/Union Territories.* National Remote Sensing Agency (NRSA), Government of India. Hyderabad.

Parsons, R.E. (1941). Rivers as barriers to the distribution of gibbons. *J. Bombay Nat. Hist. Soc.,* 42: 434–926.

Pocock, R.I. (1939, 1941). *The Fauna of British India: Mammalia. Primates and Carnivora.,* Taylor and Francis. London.

Prater, S.H. (1948). *The Book of Indian Animals.* Bombay Natural History Society. Bombay (Mumbai).

Prouty, L.A., Buchanan, P.D., Pollitzer, W.S., Mootnick, A.R. (1983a). A presumptive new hylobatid subgenus with 38 chromosomes. *Cytogenet. Cell Genet.,* 35: 141–142.

__________. (1983). *Bunopithecus*: A genus-level taxon for the hoolock gibbon (*Hylobates hoolock*). *Am. J. Primatol.,* 5: 83–87.

Sati, J.P. and Alfred, J.R.B. (2001). Hoolock gibbon (*Bunopithecus hoolock*). *Envis. Bull. Wildl. Protect Area.,* 1(1): 60–7.

Schultz, A.H. (1969). *The Life of Primates.* Weidenfeld and Nicolson, London.

Takacs, Z., Morales, J.C., Geissmann, T., Melnick, D.J. (2005). A complete species-level phylogeny of the Hylobatidae based on mitochondrial ND3-4 gene sequences. *Molec. Phylogenet. Evol.,* 36: 456–467.

Tilson, R.L. (1979). Behaviour of hoolock gibbons (*Hylobates hoolock*) during different seasons in Assam. *J. Bombay Nat. Hist. Soc.,* 76: 1–16.

Weigl, R. (2005). *Longevity of Mammals in Captivity*; from the living collections of the world. Stuttgart (DE): E. Schweizerbartsche, p. 214.

Yang, D.H., Zhang, J. and Li, C. 1987. Preliminary survey on the population and distribution of gibbons in Yunnan Province. *Primates,* 28(4): 547–9.

Zhang, S.Y. (1998). Current status and conservation strategies of primates in China. *Primate Conservation,* 18: 81–4.

14

Genomic Diversity of Ethnic Populations of India

Analabha Basu and Partha P. Majumder

IMPORTANCE OF STUDYING GENETIC DIVERSITY OF INDIAN POPULATIONS

South Asia, including India, constitutes about one-sixth of the world population. The contemporary population of this region is structured as an array of sub-populations that are largely isolated in terms of mating, with minimal admixture between sub-populations. Understanding the genomic diversity of ethnic populations of India holds a major key to the understanding of global genetic diversity and human population structure. This understanding is also of critical importance to understanding the nature and extent of human migration from out-of-Africa to Oceania. Recent genetic data also show that India may have served as the primary access route to the peopling of East Asia (Abdulla et al. 2009). However, India has been grossly under-represented in recent, large-scale genomic studies of global human diversity and evolution (Reich et al. 2009).

SOCIAL STRUCTURE OF INDIAN ETHNIC POPULATIONS AND ITS IMPLICATION ON GENETIC STRUCTURE

India occupies the center-stage of human evolution (Cann, 2001). Contemporary India is a rich tapestry of largely intra-marrying ethnic populations. These populations belong to a diverse set of culture and language groups. There are four distinct language families in India, namely, Austro-Asiatic, Dravidian, Tibeto-Burman and Indo-European. The geographical distribution of the language groups within India is largely non-overlapping. The Dravidian speaking groups inhabit southern India,

Indo-European speakers inhabit northern India and Tibeto-Burman speakers are confined to northeastern India. By contrast, the numerically small group of Austro-Asiatic speakers, who are exclusively tribal, inhabit fragmented geographical areas of eastern and central India. Culturally, the vast majority of the people of India belong to either tribal or caste societies. The tribal populations are characterized by their traditional modes of subsistence: hunting and gathering, unorganized agriculture, slash and burn agriculture and nomadism (practiced by a limited number of groups). They also have no written form of language and speak a variety of dialects. On the other hand, the Hindu society in India comprises castes, who carry out a wide range of occupations and have written forms of language. There is a long-standing debate about the genesis of the caste and tribal populations of India. One model suggests that the tribes and castes share considerable Pleistocene heritage, with limited recent gene flow between them (Kivisild et al. 2003), whereas an exact opposite view concludes that caste and tribes have independent origins (Cordaux et al. 2004). Another under-appreciated factor in some studies is the innate complexity of caste origins, including the recognition that some castes have tribal origins (Basu et al. 2003, Majumder, 2010).

There are over 400 tribal groups and over 4,000 caste groups in India. Though caste and tribe have been administrative social categories in British India; their existence as a social construct probably predates any other similar social categories found elsewhere in the world. Populations belonging to the caste fold have a ranked social order. There are four broad caste groups; however, commonly the caste populations are now ranked as high, middle and low. While the usage of the ranks as 'high', 'middle' and 'low' can appear to have a value judgment attached to it, but it is scientifically imperative to take cognizance of this hierarchy, since, as D.D. Kosambi has pointed out "stratification of Indian society reflects and explains a great deal of Indian history, if studied in the field without prejudice". What is important from a scientific point of view is that there is virtually no exchange of genes among tribal populations or between tribal and caste populations. There is also little exchange of genes between castes, primarily due to strict social rules of marriage within the caste system. Social stratification and norms governing mate-exchange between social strata impact on genetic structures of populations. Therefore, human geneticists have studied the genetic structures, similarities and dissimilarities of ranked caste groups in India aiming to shed light on human history, without prejudice. Historical and anthropological studies suggest that in the establishment of the caste system in India there have been varying levels of admixture between the tribal people of India and the later immigrants bringing along with them knowledge of agriculture, artisanship and metallurgy from central and west Asia. The migrants from central and west Asia who likely entered India through the north-western corridor, spread to most areas of northern India, but not to southern India. There is a distinct gradient of decreasing genetic similarity (representing a cline) of Indian populations with the West- and Central-Asian gene pools as we move eastward or southward from the north-western corridor (Reich et al. 2009, Basu et al. 2003, IGVC 2008, Sengupta et al. 2006). In other words, southern and northern India have had differential inputs of genes from central and west Asia. This differential admixture is expected to have differential impacts on the genetic structures of castes of different ranks.

MOLECULAR GENETICS IN THE STUDY OF HUMAN EVOLUTION AND DISPERSAL

Paleontology and archeology have traditionally provided inferences on early human origin and dispersal, particularly about periods prior to written and/or oral history. However, such disciplines often rely heavily on scanty, ill-preserved evidence, such as extremely small fragments of skeletal remains buried under the soil for centuries, or even exposed to the ravages of nature. Recently, molecular genetics has come to make increasing contributions to our understanding of human evolution, migration and population structure. Since we pass on our genes from one generation to the other, and since random mutations occur and accumulate in our genomes over time, the extant variation we carry in our genomes comprises footprints of the history of our species. The pattern in which these mutations and polymorphisms (mutations that increase to high frequencies in populations) accumulate in our genomes is indicative of the dynamics of the population history (Kingman 2000, Rosenberg and Nordborg 2002, Nordborg 1997). Besides the random accumulation of mutations, the gene pool of a population is also systematically shaped by demographic changes and differential selective effects (Kingman 2000, Nordborg 1997 and Barreiro et al. 2008).

Traditionally the maternally-inherited (transmitted by a mother to all her children) mitochondrial DNA (mtDNA) and the paternally-inherited Y-chromosomal DNA (transmitted by a father to all his sons), particularly the non-recombining part of the Y chromosome (NRY), were popular genetic materials to draw inferences about our past (Cann et al. 1987, Cavalli-Sforza

1966, Ke et al. 2001 and Paabo 1995). Both mtDNA and NRY have their own merits of usage. The uniparental lineages provide a tool to separately investigate the history of males and females; these provide easy and direct ways of studying gender-specific migrations, which are often different, and contrast them. It is inherently simple to analyze because it is non-recombining and the only variation that we see is because of mutations that have accumulated over time. Both mtDNA and NRY harbor regions which are 'hypervariable' or have high mutation rates, and also regions which rarely vary or allow mutations to accumulate. The more stable regions are used to draw inferences about distant history, while the fast-mutating regions are used to draw more recent historical inferences.

However, mitochondria and Y-chromosomes are strongly impacted upon by a non-genetic phenomenon, known as genetic drift, that leads to chance fluctuations in frequencies of mutants over generations. By contrast, the impact of genetic drift on autosomal chromosomes is less pronounced, indicating that frequencies of mutations over generations are dictated to a smaller extent by chance. Thus, dates of population divergence and migration events that are usually estimated on the basis of frequencies of genetic markers can potentially show wide differences depending on whether mtDNA/Y-chromosomal or autosomal markers are used.

Molecular genetic diversity studies on Indian ethnic populations have primarily used mtDNA and NRY markers (Basu et al. 2003, Sengupta et al. 2006, Bamshad et al. 2001, Kivisild et al. 1999, Kumar et al. 2008 and Underhill et al. 2010); the number of studies that have used autosomal markers have been comparatively fewer (IGVC 2008 and Rosenberg et al. 2006). Most past studies that have used autosomal markers, have either been based on a restricted set of loci, or have been based on a large set of markers but on a small number of individuals or populations, although there are some exceptions (Abdulla et al. 2009, Reich et al. 2009 and IGVC 2008).

WHICH EXIT ROUTE?

The hypothesis that humans have a single origin (monogenesis) was proposed by Charles Darwin in *Descent of Man* (1871). The concept was speculative until the 1980s, when it was corroborated by a study of present-day mitochondrial DNA (mtDNA), combined with evidence based on physical anthropology of archaic human remains. According to genetic and fossil evidence, archaic *Homo sapiens* evolved to anatomically modern humans solely in Africa, between 200,000 and 100,000 years ago, It is now established beyond reasonable doubt that anatomically modern humans evolved in Africa and a small population of about 1,000 individuals, most probably from East Africa, expanded and spread throughout much of Africa about 100,000 years ago (Cann, 1983 Ramachandran et al. 2005 and Li et al. 2008). Genetic data indicate further migration into Asia, Europe and later to other parts of the world (Ramachandran et al. 2005 and Li et al. 2005).

Which route did the first humans take when they moved out of Africa? The easiest would have been a walking trail along the river Nile, across the Sinai Peninsula ('northern exit route'). Then, modern humans could have gone towards Europe or towards the fertile- crescent in West Asia. This seems logical because this feat could be achieved with the most rudimentary advancement of technology, as well as, it was walking over flat river valley. However, the modern theory of human dispersal 'out of Africa' postulates that there were at least two waves (Foley 2007, Lahr and Foley 1998, 1999): one via a northern exit route through the Nile valley and another, via a southern exit route from the Horn of Africa across the mouth of the Red Sea along the coastline of India to southeastern Asia and Australia (Macaulay et al. 2005 and Quintana-Murci et al. 1999). The Southern-exit route calls for knowledge of riding the sea and intuitively would be more recent. Paradoxically, archeological and other records provide evidence in favour of the exact opposite. Fossil records and evidence from genetic data shows that islands in South-East Asia and Southern Pacific including Australia were populated at least as early as the oldest record suggestive of the presence of modern humans outside Africa via the Northern exit route. It is also noted that, the archeological evidence of the northern exit route is associated with the Upper Paleolithic blade-dominated technology, whereas the remnants of the Southern-exit route are associated with much simpler Middle Paleolithic technology.

The reason for the above 'paradox' possibly is because, modern humans taking the northern exit route, probably encountered archaic humans (particularly Neanderthals). On the other hand, the southern exit route was probably free of such problems. Excavation sites in Israel and Lebanon, dated between 45 and 50 kya (i.e., 45,000 to 50,000 years ago), which till date are the earliest evidence of migration of modern humans out of the African continents via the northern exit route, shows the proximity and the territorial overlap of these modern humans with archaic humans, primarily the Neanderthals. Perhaps, the technological superiority of the modern humans, i.e., their ability to use stone tools more efficiently, over Neanderthals possibly played an important role in their successful dispersal via the northern exit route.

The probable date of dispersal through the southern exit route was substantially earlier (about 70–80 kya) (Lahr and Foley 1998). Archaeological evidence to resolve this controversy has been scanty, primarily because the coastlines of that period have become deeply submerged because of the rapid rise of sea levels (Stringer 2000).

Some recent archaeological finds from India have indicated that the major route of dispersal to India from out-of-Africa was through the southern route (Mellars 2006). Molecular genetic studies, primarily those based on mtDNA and NRY polymorphisms, have favored a southern dispersal route (Forster 2009 and Underhill et al. 2001). The strongest genetic evidence in favour of an early southern exit into India comes from the mtDNA signatures of Indian and other Asian populations. A vast majority of the populations of this region harbor derivatives of the mitochondrial DNA lineages (M and N), which are closely related to the Africa specific L3 lineage (Quintana-Murci et al. 1999). The southern exit hypothesis is also supported by analyses of mtDNA data from Andaman Islands (Kivisild et al. 2003, Endicott et al. 2003 and Thangaraj et al. 2005) and New Guinea (Forster et al. 2001). Further, the Y-chromosomal lineages (C and D) are found only in Asian continent and Oceania (Kivisild et al. 2003 and Underhill et al. 2001), but not in Eurasia or north Africa. The absence of the Y and mtDNA haplogroups (groups of chromosomes bearing similar sets of mutations) in Eurasia and northern Africa pose a strong argument in favour of an early southern exit route of migration and settlement into India, but there are strong caveats to the argument. The virtual absence of the mtDNA M haplogroup in the Eurasian as well as north African populations and its extreme high frequency in India and Oceania supports the southern exit route hypothesis. Although available genomic data strongly indicate that India was peopled on one of the earliest was of human migration from Africa and the dispersal took place via the southern exit route, a major limitation of more direct genetic evidence is the absence of reliable and comparable data from populations which lie on the southern exit route (Stringer 2000).

THE EARLY SETTLERS

Who are the earliest inhabitants of India. The Austro-Asiatic speaking populations of India, who are exclusively tribal, show the highest frequencies of the ancient mtDNA lineage M. They also show the highest frequency (about 20%) of sub-lineage M2, which has the highest nucleotide diversity within a fast evolving segment (HVS1) of the mtDNA compared to other sub-lineages. Based on these patterns, it has been suggested that the Austro-Asiatic speakers are possibly the earliest settlers of India. Recent results on Y-chromosomal markers provide further support for this inference. The Y lineage OM95, found in high frequency in India, had originated in the Indian Austro-Asiatic populations around 65 kya. These findings are consistent with linguist Colin Renfrew's observation that the present distribution of the Austric language group is due to the initial dispersal process out of Africa, whereas later agricultural dispersal can account for distribution of the Elamo-Dravidian or Sino-Tibetan languages (to which family Tibeto-Burman languages belong). However, recent studies have found that many Dravidian tribal populations also have M2 frequencies comparable to those of the Austro-Asiatic tribal people (Kumar et al. 2008 and Chandrasekar et al. 2009). The phenomenon of "elite dominance" through which newer settlers can "overpower" pre-existing groups can obliterate pre-existing genomic signatures, thereby presenting difficulties in genomic inferences on antiquities of populations. The antiquity of Dravidian speakers in India, who are not all tribal but also belong to the organized caste system, has been extremely enigmatic. There is evidence that they were widespread over nearly the entire landmass of India, which may have overlapped with regions inhabited by the modern Austro-Asiatic speakers, and were possibly pushed to their predominant current habitat in the southern region of India probably after the advent of the Indo-European speakers (Basu et al. 2003).

SOCIAL, GEOGRAPHICAL AND LINGUISTIC STRUCTURES AS EVIDENT FROM mtDNA AND NRY MARKERS

Analysis of genetic structure has shown that Indian ethnic populations when grouped as tribal versus non-tribal, or by geographical region of habitat, or by linguistic affiliation, have resulted from admixture of four or five ancestral populations (Abdulla et al. 2009 and IGVC, 2008). The Tibeto-Burman speaking tribals separate from the non Tibeto-Burman speakers as the most distinct cluster (Abdulla et al. 2009, Basu et al. 2003 and IGVC, 2008). As the ancestral source of the Indian gene pool, in addition to the original African source population, west Asia (by demic diffusion of agriculture) and central Asia have been the major contributors, although migrations from Asia may have taken place only in historical times, perhaps not earlier than 8 kya. The extent of variation of female lineages (mtDNA) in India is rather restricted (Basu et al. 2009 and Roychoudhury et al. 2001) indicating a small founding group of females. By contrast, the variation of male lineages

(Y-chromosomal) is very high (Basu et al. 2009 and Sengupta et al. 2006). This pattern may be indicative of sex-biased ancient gene flow into India with more male immigrants than female (Bamshad et al. 1998), possibly occurring within the last 5000 years through invasions and wars. This phenomenon obscures ancient genetic signatures and results in quick introduction of high genetic variability, often mimicking extreme natural selection (Zerjal et al. 2003). The success of some of the Y-chromosomal haplotypes that arose in central Asia to spread over vast regions of Eurasia (Zerjal et al. 2003), as well as south and south-east Asia is indicative of the 'success' of the cultural and technological dominance of west Eurasia and central Asia (Underhill et al. 2010 and Zerjal et al. 2003).

The mtDNA lineage U, which is likely to have arisen in central Asia, has a high frequency in India implying that large-scale migration brought a large number of copies of this lineage into India. However, this lineage was shown to comprise two deep sub-lineages, U2i and U2e, with an estimated split around 50 kya. The sub-lineage U2i is found in high frequencies in India (particularly among tribes~77%) but not in Europe ~0%, whereas U2e is found in high frequencies in Europe (>10%) but not in India (except at very low frequencies among castes, but not among tribals). Thus, a substantial fraction of the U lineage – specifically, the U2i sublineage – may be indigenous to India (Basu et al. 2003, Sengupta et al. 2006 and Kivisild et al. 1999).

Analysis of the complete mtDNA genome sequence has revealed a large number of sequence variants within major haplogroups in Indian populations, many of which, however, occur sporadically. This indicated a common spread of the root haplotypes of haplogroups M, N and R around 70–60 kya along the southern exit route. The analysis has further revealed that entry of the haplogroup U2 post-dates the earliest settlement along the southern route.

Central Asian populations are supposed to have been the major contributors to the Indian gene pool, particularly to the northern Indian gene pool, and the migrants had supposedly moved into India through what is now Afghanistan and Pakistan. Using mitochondrial DNA variation data collated from various studies, we have shown that populations of Central Asia and Pakistan show the lowest genetic distance with the north Indian populations (F_{ST} = 0.017), higher (F_{ST} = 0.042) with the south Indian populations, and the highest (F_{ST} = 0.047) with the northeast Indian populations. Northern Indian populations are genetically closer to Central Asians than populations of other geographical regions of India (Basu et al. 2003 and Bamshad et al. 2001).

Although considerable cultural impact on social hierarchy and language in south Asia is attributable to the arrival of nomadic central Asian pastoralists, studies using NRY polymorphisms reveal that the influence of central Asia on the pre-existing gene pool was minor. The ages of accumulated microsatellite variation in the majority of Indian haplogroups exceed 10-15 kya, which attests to the antiquity of regional differentiation. Therefore, NRY data do not support models that invoke a pronounced recent genetic input from central Asia to explain the observed genetic variation in south Asia. Haplogroups R1a1 and R2 indicate demographic complexity that is inconsistent with a recent single history. Associated microsatellite analyses of the high-frequency R1a1 haplogroup chromosomes indicate independent recent histories of the Indus Valley and the peninsular Indian region. These data are also more consistent with a peninsular origin of Dravidian speakers than a source with proximity to the Indus and with significant genetic input resulting from demic diffusion associated with agriculture; rather it indicates that pre-Holocene and Holocene-era – not Indo-European – expansions have shaped the distinctive South Asian Y-chromosome landscape (Sengupta et al. 2006).

ANALYSES USING LARGE SETS OF GENOMIC MARKERS PROVIDES BETTER ESTIMATES AND NEW PARADIGMS

Analyses of data on 405 SNPs from 75 genes and a 5.2 Mb region on chromosome 22 in 1871 individuals from diverse 55 endogamous Indian populations (IGVC, 2008), have revealed that Indian populations form a continuum of genetic spectrum bridging Caucasoid and Mongoloid populations. The HUGO Pan Asian SNP Consortium's study (Abdulla et al. 2009) also showed that most of the Indian populations shared ancestry with European populations, which is consistent with the recent observations and our understanding of the expansion of Indo-European speaking populations. The study also provided a strong evidence that the peopling of India (and also of southEast Asia) was via a single primary wave of migration out-of-Africa (Abudlla et al. 2009). In the Indian Genome Variation Consortium study (IGVC, 2008), *FST* values between pairs of populations were found to vary from 0.000 to 0.111, with a mean of 0.03 ± 0.0005, suggesting that the extent of differentiation overall was low. Maximum *FST* values were observed among the tribal populations of different linguistic lineages. On a pan-India level, when populations were grouped by language or by geographical region of habitat, the extent of genetic differentiation among linguistic or geographical groups was not statistically significant However, grouping by ethnicity (caste and tribe) indicated significant differentiation, possibly due to antiquity and isolation of the tribal compared to the caste populations. Although no clear geographical groupings of populations was found, ethnicity (tribal/nontribal) and language were the major determinants of genetic affinities between the populations of India.

Using over 500,000 biallelic autosomal markers, Reich et al. 2009) have also found a north to south gradient of genetic proximity of Indian populations to western Eurasians/central Asians. In general, the central Asian populations are genetically closer to the higher-ranking caste populations than to the middle- or lower-ranking caste populations. Among the higher-ranking caste populations, those of northern India are, however, genetically much closer (F_{ST} = 0.016) than those of southern India (F_{ST} = 0.031). Phylogenetic analysis of Y-chromosomal data collated from various sources yielded a similar picture.

By analyzing DNA samples of 132 individuals drawn from 25 population groups of India for about 500,000 autosomal biallelic markers, (Reich et al. 2009) have recently proposed that extant populations of India were 'founded' by two hypothetical ancestral populations, one ancestral north Indian (ANI) and another ancestral south Indian (ASI). Presumably, these ancestral populations were derived from ancient humans who entered India via the southern and the northern exit routes from out of Africa. All extant Indian populations are derived from admixture between the two ancestral populations, with the ANI contribution being higher among extant north Indian populations and that of ASI being higher among extant south Indian populations. Since a large number of unbiasedly selected autosomal markers were used, this study does not suffer from the shortcomings of those studies that have used data of only one locus (mitochondrial or Y-chromosomal). This simple model is attractive, but requires validation especially because of small numbers of individuals drawn from each population and inclusion of a restricted number of populations.

References

Abdulla, M.A., Ahmed, I., Assawamakin, A., Bhak, J., Brahmachari, S.K., Calacal, G.C., Chaurasia, A., Chen, C.H., Chen, J. and Chen, Y.T., et al. (2009). Mapping human genetic diversity in Asia. *Science,* 326: 1541–1545.

Bamshad, M., Kivisild, T., Watkins, W.S., Dixon, M.E., Ricker, C.E., Rao, B.B., Naidu, J.M., Prasad, B.V., Reddy, P.G. and Rasanayagam, A. et al. (2001). Genetic evidence on the origins of Indian caste populations. *Genome Research,* 11: 994–1004.

Bamshad, M.J., Watkins, W.S., Dixon, M.E., Jorde, L.B., Rao, B.B., Naidu, J.M., Prasad, B.V., Rasanayagam, A., and Hammer, M.F. (1998). Female gene flow stratifies Hindu castes. *Nature,* 395: 651–652.

Barreiro, L.B., Laval, G., Quach, H., Patin, E. and Quintana-Murci, L. (2008). Natural selection has driven population differentiation in modern humans. *Nature Genetics,* 40: 340–345.

Basu, A., Mukherjee, N., Roy, S., Sengupta, S., Banerjee, S., Chakraborty, M., Dey, B., Roy, M., Roy, B. and Bhattacharyya, N.P. et al. (2003). Ethnic India: A genomic view, with special reference to peopling and structure. *Genome Research,* 13: 2277–2290.

Cann, R.L. (2001). Genetic clues to dispersal in human populations: Retracing the past from the present. *Science,* 291: 1742–1748.

Cann, R.L., Stoneking, M., and Wilson, A.C. (1987). Mitochondrial DNA and human evolution. *Nature,* 325: 31–36.

Cavalli-Sforza, L.L. (1966). Population structure and human evolution. *Proceedings of the Royal Society of London. Series B,* Containing papers of a biological character, 164: 362–379.

Chandrasekar, A., Kumar, S., Sreenath, J., Sarkar, B.N., Urade, B.P., Mallick, S., Bandopadhyay, S.S., Barua, P., Barik, S.S., Basu, D. et al. (2009). Updating phylogeny of mitochondrial DNA macrohaplogroup in India: Dispersal of modern human in South Asian corridor. *PloS one,* 4: e7447.

Cordaux, R., Aunger, R., Bentley, G., Nasidze, I., Sirajuddin, S.M., and Stoneking, M. (2004). Independent origins of Indian caste and tribal paternal lineages. *Curr. Biol.,* 14: 231–235.

Endicott, P., Gilbert, M.T., Stringer, C., Lalueza-Fox, C., Willerslev, E., Hansen, A.J., and Cooper, A. (2003). The genetic origins of the Andaman Islanders. *American Journal of Human Genetics,* 72: 178–184.

Foley, R.A., and Lahr, M.M. (2007). Ancient DNA closes on human uniqueness: The base nature of Neanderthals. *Heredity,* 98: 187–188.

Forster, P. (2004). Ice ages and the mitochondrial DNA chronology of human dispersals: A review. *Philosophical Transactions of the Royal Society of London,* 359: 255–264; discussion 264.

Forster, P., Torroni, A., Renfrew, C. and Rohl, A. (2001). Phylogenetic star contraction applied to Asian and Papuan mtDNA evolution. *Molecular Biology and Evolution,* 18: 1864–1881.

Indian Genome Variation Consortium (2008). Genetic landscape of the people of India: A canvas for disease gene exploration. *Journal of Genetics,* 87: 3–20.

Ke, Y., Su, B., Song, X., Lu, D., Chen, L., Li, H., Qi, C., Marzuki, S., Deka, R. and Underhill, P. et al. (2001). African origin of modern humans in East Asia: A tale of 12,000 Y-chromosomes. *Science,* 292: 1151–1153.

Kingman, J.F. (2000). Origins of the coalescent. 1974–1982. *Genetics,* 156: 1461–1463.

Kivisild, T., Bamshad, M.J., Kaldma, K., Metspalu, M., Metspalu, E., Reidla, M., Laos, S., Parik, J., Watkins, W.S. and Dixon, M.E., et al. (1999). Deep common ancestry of indian and western-Eurasian mitochondrial DNA lineages. *Curr. Biol.,* 9: 1331–1334.

Kivisild, T., Rootsi, S., Metspalu, M., Mastana, S., Kaldma, K., Parik, J., Metspalu, E., Adojaan, M., Tolk, H.V. and Stepanov, V. et al. (2003). The genetic heritage of the earliest settlers persists both in Indian tribal and caste populations. *American Journal of Human Genetics,* 72: 313–332.

Kumar, S., Padmanabham, P.B., Ravuri, R.R., Uttaravalli, K., Koneru, P., Mukherjee, P.A., Das, B., Kotal, M., Xaviour, D. and Saheb, S.Y. et al. (2008). The earliest settlers' antiquity and evolutionary history of Indian populations: Evidence from M2 mtDNA lineage. *BMC Evolutionary Biology,* 8: 230.

Lahr, M.M., and Foley, R. (1994). Multiple dispersals and modern human origins. *Evolutionary Anthropology: Issues, News, and Reviews,* 3: 48–60.

Lahr, M.M., and Foley, R.A. (1998). Towards a theory of modern human origins: Geography, demography, and diversity in recent human evolution. *American Journal of Physical Anthropology Suppl.,* 27: 137–176.

Li, J.Z., Absher, D.M., Tang, H., Southwick, A.M., Casto, A.M., Ramachandran, S., Cann, H.M., Barsh, G.S., Feldman, M. and Cavalli-Sforza, L.L. et al. (2008). Worldwide human relationships inferred from genome-wide patterns of variation. *Science,* 319: 1100–1104.

Macaulay, V., Hill, C., Achilli, A., Rengo, C., Clarke, D., Meehan, W., Blackburn, J., Semino, O., Scozzari, R. and Cruciani, F. et al. (2005). Single, rapid coastal settlement of Asia revealed by analysis of complete mitochondrial genomes. *Science,* 308: 1034–1036.

Majumder, P.P. (2010). The human genetic history of South Asia. *Curr. Biol.,* 20: R184–187.

Mellars, P. (2006). Going east: New genetic and archaeological perspectives on the modern human colonization of Eurasia. *Science,* 313: 796–800.

Nordborg, M. (1997). Structured coalescent processes on different time scales. *Genetics,* 146: 1501–1514.

Paabo, S. (1995). The Y-chromosome and the origin of all of us (men). *Science,* 268: 1141–1142.

Quintana-Murci, L., Semino, O., Bandelt, H.J., Passarino, G., McElreavey, K., and Santachiara-Benerecetti, A.S. (1999). Genetic evidence of an early exit of Homo sapiens from Africa through eastern Africa. *Nature Genetics,* 23: 437–441.

Ramachandran, S., Deshpande, O., Roseman, C.C., Rosenberg, N.A., Feldman, M.W., and Cavalli-Sforza, L.L. (2005). Support from the relationship of genetic and geographic distance in human populations for a serial founder effect originating in Africa. *Proceedings of the National Academy of Sciences of the United States of America,* 102: 15942–15947.

Reich, D., Thangaraj, K., Patterson, N., Price, A.L., and Singh, L. (2009). Reconstructing Indian population history. *Nature,* 461: 489–494.

Rosenberg, N.A., and Nordborg, M. (2002). Genealogical trees, coalescent theory and the analysis of genetic polymorphisms. *Nature Reviews,* 3: 380–390.

Rosenberg, N.A., Mahajan, S., Gonzalez-Quevedo, C., Blum, M.G., Nino-Rosales, L., Ninis, V., Das, P., Hegde, M., Molinari, L. and Zapata, G., et al. (2006). Low levels of genetic divergence across geographically and linguistically diverse populations from India. *PLoS genetics,* 2: e215.

Roychoudhury, S., Roy, S., Basu, A., Banerjee, R., Vishwanathan, H., Usha Rani, M.V., Sil, S.K., Mitra, M., and Majumder, P.P. (2001). Genomic structures and population histories of linguistically distinct tribal groups of India. *Human Genetics,* 109: 339–350.

Sengupta, S., Zhivotovsky, L.A., King, R., Mehdi, S.Q., Edmonds, C.A., Chow, C.E., Lin, A.A., Mitra, M., Sil, S.K. and Ramesh, A. et al. (2006). Polarity and temporality of high-resolution Y-chromosome distributions in India identify both indigenous and exogenous expansions and reveal minor genetic influence of Central Asian pastoralists. *American Journal of Human Genetics,* 78: 202–221.

Stringer, C. (2000). Palaeoanthropology. Coasting out of Africa. *Nature,* 405: 24–25, 27.

Thangaraj, K., Chaubey, G., Kivisild, T., Reddy, A.G., Singh, V.K., Rasalkar, A.A., and Singh, L. (2005). Reconstructing the origin of Andaman Islanders. *Science,* 308: 996.

Underhill, P.A., Passarino, G., Lin, A.A., Marzuki, S., Oefner, P.J., Cavalli-Sforza, L.L., and Chambers, G.K. (2001). Maori origins, Y-chromosome haplotypes and implications for human history in the Pacific. *Human Mutation,* 17: 271–280.

Underhill, P.A., Passarino, G., Lin, A.A., Shen, P., Mirazon Lahr, M., Foley, R.A., Oefner, P.J., and Cavalli-Sforza, L.L. (2001). The phylogeography of Y-chromosome binary haplotypes and the origins of modern human populations. *Annals of Human Genetics,* 65: 43–62.

Underhill, P.A., Myres, N.M., Rootsi, S., Metspalu, M., Zhivotovsky, L.A., King, R.J., Lin, A.A., Chow, C.E., Semino, O. and Battaglia, V. et al. (2010). Separating the post-Glacial coancestry of European and Asian Y-chromosomes within haplogroup R1a. *Eur. J. Hum. Genet.,* 18: 479–484.

Zerjal, T., Xue, Y., Bertorelle, G., Wells, R.S., Bao, W., Zhu, S., Qamar, R., Ayub, Q., Mohyuddin, A. and Fu, S., et al. (2003). The genetic legacy of the Mongols. *American Journal of Human Genetics,* 72: 717–721.

15

Biodiversity and Food Security in India

P.L. Gautam, C. Thomson Jacob, Y. Singh, C.A. Reddy and K.P. Raghuram

INTRODUCTION

Human beings always depend on biomass for their food and have been exploiting biological diversity to meet their requirements of food, health and livelihood security in various ways depending upon the need and means at their disposal to exploit it. Over a period of time, demographic pressure, human needs and threats to biodiversity have increased which has to be conserved, protected and regenerated and also used in a sustainable manner so that future generations continue deriving benefits from this precious and fundamental resource provided by the nature (www.cbd.int).

BIODIVERSITY

Biological diversity, encompasses variety and variability of all life on earth, forms the web of life of which we are an integral part and upon which we so fully depend (www.cbd.int). It provides us all with critical goods and services on which our lives depend. Provision of food, fibers, energy and medicines, purification of air and water, moderation of floods and droughts, stabilization of climate—these are just some of the vital services provided by biodiversity. The goods and services provided by biodiversity constitute the basis upon which the economy, including trade, is built. As such, biodiversity has acted as a unique ingredient of sustainable development and is essential for achieving the U N Millennium Development Goals (UNESCO, 2010). Biodiversity and development are closely linked: biodiversity sustains development, and development has an impact on biodiversity, either positive or negative. Although biodiversity may not contribute directly to all sectors of development, sustainable development cannot be achieved if biodiversity is compromised by development efforts.

Agricultural biodiversity offers crucial raw material for improving in perpetuity, productivity and quality of crops, livestock and fish. It also offers immense opportunities, notably for the landless poor for entrepreneurial initiatives that generate employment and income for a range of value added foods, medicines and other products. It provides common opportunities for developing decentralized and locale-specific community food security system involving field gene banks, seed banks and grain banks developed and managed by local people. This approach will help to enlarge the food security basket by including nutrition-rich under-utilized crops. This may be the most sustainable and affordable pathway to achieving the U N Millennium Development Goals in relation to elimination of hunger and poverty.

BIODIVERSITY IN INDIA

India is known for its rich heritage of biodiversity, an outcome of over 3.5 billion years of evolutionary development, shaped by natural processes and more by human interference. Globally, so far 1.75 million species have been identified against the estimates ranging from 3 to 100 million(www.cbd.int). The Global Biodiversity Assessment (1995) estimated that about 13 to 14 million species may exist on our planet. India a mega diversity country with only 2.4% of land area, but accounts for 7-8% recorded species of the world. India is also one of the eight primary centres of origin of cultivated plants and is acknowledged centre of crop diversity including about 375 closely related wild species including rice, pulses, millets, vegetables, fruits and fiber plants.

India possesses a distinct identity, because of its varied geography and great diversity in its natural ecosystems that have immense richness of agricultural biodiversity including diversity in crop-plants, wild-plants, livestock, aquatic species, below ground biota, including microbes. This Indian gene centre is one of the 12 mega diversity centres of the world and has Himalayas, Western Ghats-Sri Lanka and Indo-Burma as three "biodiversity hot spots" among the 34 identified across the world. The Indian sub-continent has been divided into 8 regions, including biodiversity-rich zones; Western Himalayas, North-eastern region and Western Ghats and which largely superimpose over the phyto-geographical regions (India's 4th National Report to CBD, 2009). The PPV & FR Authority identified 22 "agro-biodiversity hot spots" across the country (Protection of Plant Varieties & Farmers' Right Authority, 2010).

India hosts about 49,000 species of flowering and non-flowering plants (18.8%), out of 260,000 described across the globe. The country has 91,364 animal species, representing 7.28% of the total global faunal species. Of these, the vast majority are insects, with over 68,000 species (Arora et al. 2006)· Domesticated animals comprise accredited breeds of cattle, buffalo, sheep, goats, camel, horse and poultry, beside other species like pig, donkey, mithun, yak, and ducks. This constitutes about 8 per cent of the total world cattle breeds, 26.4% of buffalo breeds and 8.3% of goat breeds. Out of 27,800 fish species reported globally about 2,200 species occur in India comprising 73 (3.3%) from cold freshwater, 544 (24.7%) from warm fresh waters, 143(6.5%) from brackish waters and 1,440 (65.4%) from marine ecosystem(www.ces.iisc.ernet.in/energywater/paper/fishdiversity/diversity1.html)· Among amphibians, 62% of the known species are endemic to India, of which majority occur in Western Ghats(NBFGR, 2007). The country is endowed with enormous variability in fungi, bacteria, actinomycetes, viruses and cyanobacteria. These form an invaluable gene pool. There may be 1.5 million species of fungi, yet only 5% are described, while others are still undiscovered and their roles are yet to be known. Nonetheless, the rich biodiversity resource includes diversity of about 850 moneran, 2,577 protistan, 23,000 (18.2%of global list) fungal and 2,500 algal species (Khoshoo,1995).

Crop Genetic Diversity

Agriculture remains one of the dominant drivers and mainstay of economic growth in India. India is one of the eight Vavilov's centres of origin of cultivated plants in the world. India stands seventh in the world in terms of contribution of species to agriculture and animal husbandry. In qualitative terms too, the contribution has been significant. India has over 800 crop species and 320 wild relatives: millets (51); legumes (31); fruits (109); spices and condiments (27); vegetables (54); fiber crops (24); oil seeds, tea, coffee, tobacco and sugarcane (12); and, medicinal plants (3,000) (India's 4th National Report to CBD, 2009).

Animal Genetic Resources

India, endowed with varied forms of animal genetic resources, is traditionally considered as an important rearing centre for domesticated animals. India has vast resources of livestock (485 million) and poultry (489 million), which play a vital role in rural livelihood security. In terms of population, India ranks first in buffaloes, second in cattle and goats, third in sheep, fourth in ducks, fifth in chicken and sixth in camels in the world. The genetic resources of farm animals in India are represented by a broad spectrum of native breeds of cattle, buffaloes, goats, sheep, swine, equines, camel and poultry. There are around 140 listed breeds of livestock and poultry in India, with 30 breeds of cattle, 10 of buffalo, 42 of sheep, 20 of goat, 3 of pig, 6 of horse and pony, 8 of camel and 18 of poultry. Besides, there are breeds of yak, mithun, ducks, quails and several nondescript populations. The major achievements of NBAGR, among others, include, (i) development of conservation models and performance and economics of AGR; (ii) molecular characterization of genes responsible for immune response and milk performance of indigenous AGR; (iii) evaluation of genetic diversity and characterization of candidate genes involved in heat tolerance; and iv) building DNA fingerprinting profiles (India's 4th National Report to CBD, 2009).

Fish Genetic Resources

India is endowed with vast inland and marine bio-resources. It is the third largest producer of fish in the world and the second largest producer of inland fish. As such, fisheries and aquaculture play an important role in social development, economic upliftment of farmers and fisher folks, apart from contributing to the nutritional security of the country. Besides establishing live gene banks of fish genetic resources, NBFGR has developed sperm cryopreservation protocols of prioritized fish species, prepared national fish seed certification document, established IPR and Patents Cell and is set to further strengthen standardization of fish database, characterize and sequence prioritized species and develop diagnostic kits for exotic pathogens.(India's 4th National Report to CBD,2009).

Agriculturally Important Microorganisms

NBAIM has a repository of 2,517 cultures which includes filamentous fungi (2,077), bacteria (394), Actinomycetes (36) and yeasts (10). The sources of fungi collections include plants (1,212), soil insects (641), air flora (39) and others (185). The special collections of micro-organisms having importance in agriculture includes bio-control agent, bio-pesticides, bio-fertilizers, bio-remediation and industrial important microbes (India's 4th National Report to CBD, 2009).

Insect Genetic Resources

The National Bureau of Agriculturally Important Insects (NBAII) was established in 1993 and has 16 centres spread across the country for carrying out field studies on biosuppression of pests of crops. The Directorate at Bangalore has been carrying out basic research on the biosystematics of important groups of insect bio-agents such as trichogrammatids, chrysopids, coccinellids, and tachinids. The reference collection maintained at NBAII has been catalogued in the form of a technical bulletion and also available in a retrievable, electronic format. The holdings contain 3441 species under 216 families and 16 orders and a small collection of insect type material is also being maintained. The huge collection of unsorted and unidentified insects maintained at Project Directorate of Biological Control (PDBC) is a veritable store house for taxonomic study and unfortunately, for want of expertise in various groups, these collections have not been touched for many years. The National Biodiversity Authority (NBA) has recognized PDBC as one of the Designated National Repositories for various kinds of biological specimens, both dead and live (www.pdbc.res.in).

THREATS TO BIODIVERSITY AND ITS DECLINE

In 20th century the loss of biodiversity occurred at a rate thousands times higher than the average rate during the preceding 65 million years. The extinction of species and erosion of genes continue to occur at an alarming rate. The important threats to biodiversity and particularly agro biodiversity are described below:

Habitat Destruction

Reduced or degraded habitats threaten biodiversity at gene, species and ecosystem level, hampering the provision of key products and services. Most of the threatened species are land-based, with more than half occurring in forests. Freshwater and marine habitats, especially coral reefs are also very vulnerable.

Invasive Alien Species

Alien invasive species are posing a serious threat to biodiversity conservation. Such species include weeds, invertebrate and vertebrate pests, and plant and animal pathogens. The impact and costs of invasive species include crop losses, rangeland value decline, water resource depletion, livestock disease, genetic contamination and management and eradication cost. The water hyacinth, a native of the Amazonian region, has now spread worldwide and a good example of an aggressive alien species that displaces native aquatic flora and fauna. The alien species prevalent in north-east region like *Eupatorium*, *Mikenia*, and *Lantana* have spread in the post- shifting cultivation areas. Convention on Biological Diversity (CBD) 1992 recognized biological invasion as the second worst threat to biodiversity after habitat destruction (Jenkins, 1999);

Genetic Homogeneity

Genetic homogeneity associated with mono-cropping and modern agriculture leads to the replacement of large number of local varieties with a few high yielding strains. Covering large areas with one or two strains has dangerous consequences. We now largely depend on a few crops such as rice, wheat, maize, soybeans and potatoes for sustaining global food systems. However their genetic homogeneity increases their vulnerability to biotic and abiotic stresses. Depending on just few crops/varieties is a dangerous situation. If a new pest or new disease arises, it can become a Pan-India/Pan-world disease with catastrophic effects on food availability.

Genetic Erosion

Genetic erosion in agricultural and livestock biodiversity is the loss of genetic diversity, including the loss of individual genes, and the loss of particular combinants of genes (or gene complexes) such as those manifested in locally adapted landraces of domesticated animals. Traditional varieties of a wide range of crops with specifically adapted gene complexes to local environments of rice, small millets, and pseudocereals have disappeared from several agro-ecosystems. It is believed that around 5,000 landraces of rice collected from North East region of India are no more under cultivation in the region. Loss of breeds and genetic diversity within breeds due to universal use of a very few improved breeds, limited number of certain superior individuals, often from exotic sources, to dominate as parents of future generations, has become a threat to native livestock diversity. The aquatic biodiversity faces threats from overexploitation, habitat alterations, and constructions and reduced flow of water in rivers (Reddy, 2008).

FOOD SECURITY

Food security exists when all people, at all times, have physical and economic access to sufficient, safe and nutritious food to meet their dietary needs and food preferences for an active and healthy life (FAO). Over one billion people experience hardship that hunger imposes, a figure which continues to rise even amidst the riches of the 21st century. Food security depends not only on the availability of food but also on its access and distribution. Food security is built on three pillars:

- Food availability: sufficient quantities of food available on a consistent basis.
- Food access: having sufficient resources to obtain appropriate foods for a nutritious diet.
- Food absorption: appropriate use based on knowledge of basic nutrition and care, as well as adequate water and sanitation (www.cbd.int).

Indian Scenario

The malnutrition scenario in India, as is evident from the following data, is cause of concern and deep worry.

- On an average 13.2% Indian population gets less than the FAO recommended caloric requirement for Indians (1,890 K cal/cu/day). This percentage is least for the state of Jammu & Kashmir (2.4) and highest for Tamil Nadu (23.4).
- Nearly 26.8% rural households do not have access to safe drinking water. While this figure is only 3.0% for Punjab, it is 64.5% for Jharkhand.
- Almost 40% of children under three are underweight and 45% are stunted.
- Nearly 22 to 30% children are born with low birth weight and 36% adult women and 34% adult men suffer from chronic energy deficiency.
- The National Family Health Surveys show a marginal increase in anemia from 74% to 79% in the children under five and 52% to 56% in young women.
- Iodine deficiency disorders, vitamin A and vitamin B deficiency are fairly rampant (State of Indian Agriculture, 2009).

- Food security in the national context needs to be assessed at three levels, national or macro food security, household food security, and individual food security. Today, there is marketable surplus of rice and wheat in India. The prevalence of widespread hunger is not due to the non availability of these food grains in the market, but due to lack of adequate purchasing power amongst the rural and urban poor and also lack of other food items- pulses, cooking oil, fruits, vegetables, milk, meat, eggs etc, demand for which is increasing (State of Indian Agriculture, 2009).

The Indian food basket has changed drastically since 1973-74. Both the expenditure on and the quantity of food consumed indicate that the share of cereals in the food basket has decreased in rural and urban areas, while the share of other items has increased. Energy derived from food has also decreased considerably. The National Council of Economic and Applied Research's (NCEAR) estimate of nutrition for rural and urban areas is less than NSSO's estimates by 53 k Cal. As one moves from poorer to richer households, per capita consumption of cereals reaches a plateau whereas in rural areas as one move to better-off classes, consumption increases. Consumption is dominated by rice and wheat. Other cereals such as pearl millet, sorghum, maize, barley and ragi are consumed only in rural areas and that too in small quantities.

Food demand and dietary pattern in the country depend predominantly on: 1) Increase in population; 2) Increase in per capita income; 3) Changes in preference due to change in tastes, life-styles and occupational structure; and 4) Increase in urbanization. Prices are other important factors affecting demand. Based on the likely changes in these factors, demand for food commodities has been projected (Chand, 2007). Total demand for the year 2020 is projected to grow to 262 million tonnes for cereals, 19.1 million tonnes for pulses, 54 million tonnes for oilseeds, 127 million tonnes of vegetables, 86 million tonnes of fresh fruits, 345 million tonnes of sugarcane, 141.5 million tonnes for milk products, 10.9 million tonnes for meat, 11 million tonnes of fish, and 11 billion eggs. Due to this increasing demand and potential for exports, the Government of India has introduced the National Food Security Bill 2010 (GOI, 2010) in the parliament. Several steps have been taken by Central and state governments for strengthening food and nutrition safety nets for the economically and socially underprivileged sections of our society. Some of these are:

- National Horticulture Mission, which can provide horticultural remedies to nutritional maladies.
- National Food Security Mission designed to increase the production and availability of rice, wheat and pulses.
- Rashtriya Krishi Vikas Yojana designed to help in bridging the gap between potential and actual yields in the fields of small farmers.
- National Rural health Mission
- National Rural Employment Guarantee Act
- Strengthened Integrated Child Development Services (ICDS)
- Enlarged Mid Day Meal Programme in Schools

BIODIVERSITY MANAGEMENT FOR FOOD SECURITY

The management of biodiversity requires coordinated approach from several agencies/organizations and stakeholders. Appropriate policy and institutional changes are, thus, needed for developing a national action plan for its management. The Indian Council of Agricultural Research has come up with Vision 2025 document to achieve targets in this direction for plant, animal, fish and microbial genetic resources. Several institutions are already actively involved, and coordinated effort is needed at all levels to ensure effective reforms and genetic resources conservation policies that benefit the public. Policy changes that attack the root of the problems and ensure people's rights are needed. Further attention is required in the development of the agricultural and resource use policies ensuring public participation, support and incentives for effective agro-ecological methods that make sustainable intensification possible. Building complementarities between agriculture and biodiversity will also require changes in agricultural research and development, land use and breeding approaches. Integrated conservation of genomic resources is required to collect, validate and facilitate use of useful genes and gene-constructs generated in the country. This venture will enhance utilization of genomic resources and their availability to the researchers across the country. Indigenous resources possessing genes and combination of genes for desirable traits provide a buffer output in times of drought, heat tolerance, floods and physiological responses like photo insensitivity, high photosynthetic rates, low respirations, flood and other biotic stresses. Current economic decisions are largely based on the direct use of the values, although other uses may be of equal or greater importance. By focusing exclusively on direct use values, biodiversity and genetic resource are likely to be consistently undervalued.

In India, there is need for exploring and promoting underutilized genetic stocks, developing and creating genetic diversity to reduce vulnerability of production systems to climate changes, and to overcome other yield and/or quality limiting characters. Indian agriculture must continuously evolve to remain ever responsive to manage change and to meet growing and diversified needs of different stakeholders in the entire production to consumption chain. Approaches of better management of genetic resources need to focus looking for species and genes to provide bio-alternatives and combining both traditional breeding approaches with modern technologies.

Biodiversity affords the development of plant varieties with novel genetic combinations, which will be required to meet the challenges arising from adverse alterations in temperature, precipitation, sea level and the frequency of droughts and floods – all of which are anticipated from human-induced climate change. The loss of each gene and species therefore, limits the options for the future. Millets, tubers and grain legumes are rich in micronutrients but require less water than the major crops. These plants and others are also sources of genes that confer tolerance to drought, floods and the increased salinity of soils.

Recent estimates indicate that some 7,000 species of higher plants have been cultivated or collected for food at one time or another and have the potential to contribute to our food security needs. Nonetheless, only 30 of these 7,000 species have performed the function of feeding the world, because they contribute 95% of our dietary energy needs. Evidence also exists that three of those 30 crops- wheat, rice and maize – which have received the greatest attention in terms of conservation and improvement – provide more than 50% of the global plant-derived energy intake, even though some other crops such as cassava, sorghum, millet, beans, and plantain which serve as staples in particular regions or countries, also contribute, but to a lesser extent, to the global plant-derived energy intake. A similar trend of concentrating on and using only a limited number of available biodiversity is true of the genetic resources of livestock, aquatic and forest products. According to information from FAO, only 30 of the 50, 000 known mammalian and bird species have been used extensively in producing food and only 15 species currently account for over 90% of global livestock production.

Given the strong and well-known linkages between biodiversity, food and nutrition, we cannot continue over-exploiting a limited portion of our biodiversity endowment neglecting a larger portion. Paradoxically, going either way – over-exploitation or neglect – leads eventually to the same outcome: loss of biodiversity. We, therefore, need to develop strategies and activities that will better balance our resource use. These includes (State of Indian Agriculture,2009):

- Enhancing public, private and civil society partnership in research in genetic resource handling and use and confidence building to increase of private sector in exchange and collaborative use of genetic resources for food and agriculture.
- Creating variability, conservation and commercialization of genes and gene-constructs. New initiatives in processing, product development and value-addition of suitable crops and varieties by appropriate exploitation of their genetic resource.
- Bio-prospecting and domestication of unexploited medicinal plants based on indigenous knowledge unravel new molecules having neutraceutical and pharmaceutical potential. India with its huge wealth of medicinal plant diversity can focus on bioprospecting molecules of antioxidants, therapeutic and fragrance value using latest techniques including cell culture, biotransformation, cell permabilisation, cell immobilization, elicitation and hairy root cultures.
- Encouraging networking for exchange, evaluation and sustainable utilization of genetic resources. enabling comprehensive bio-security system of India, protecting the nature diversity from alien invasive species, exotic pests and unwanted genes/ genotypes from entry, spread and establishment.
- Developing ability, appropriate institutional arrangements and policy framework for handling intellectual property rights related issues. Need for safeguarding the national interests in a harmonized manner to meet the challenges and opportunities of the new legal regimes of CBD, WTO & TRIPS, UPOV and ITPGRFA.

INSTITUTIONAL FRAMEWORK FOR BIODIVERSITY CONSERVATION AND USE

Conservation and sustainable use of biodiversity have been an integral part of Indian ethos. The CBD is a comprehensive international agreement that addresses all aspects of biodiversity in a holistic manner. Reaffirming sovereign rights of nations over their biological resources, the Convention has set three objectives: (i) conservation of biological diversity;

(ii) sustainable use of its components; and (iii) fair and equitable sharing of benefits arising out of the use of genetic resources (India's 4th National Report to CBD,2009).

India has participated in major international events on environment and biodiversity conservation since 1972. It has contributed significantly to developing the agreed texts, ratified, and complied with the commitments in various international conventions relating to biodiversity. These agreements are: Convention on Migratory Species, Convention on International Trade in Endangered Species (CITES), Ramsar Convention on Wetlands, World Heritage Convention, and the Bonn Guidelines on Access to Genetic Resources and Fair and Equitable sharing of the benefits arising out of their utilization, 2002 and Convention on Migratory Species (CMS). Some other international agreements which have bearing on biodiversity to which India is a Party include UNFCCC, United Nations Convention to Combat Desertification, UN Commission on Sustainable Development (UNCSD), World Trade Organization (WTO), International Treaty on Plant Genetic Resources (ITPGR) for food and agriculture, and UN Law of the Seas (India's 4th National Report to CBD, 2009).

Biological Diversity Act, 2002

The Government of India has enacted the Biological Diversity Act in 2002 after intense consultation with different stakeholders, for the conservation and protection of India's biological resources. The Act provides for reaffirming sovereign rights over the biodiversity, regulation of conservation and sustainable use of biodiversity and associated knowledge. In order to implement the provisions of the Biodiversity Act, the National Biodiversity Authority (NBA) has been established in 2003 at Chennai. It is a corporate body established by the Ministry of Environment and Forests (MoEF), Government of India to perform regulatory, advisory and monitoring roles for implementation of the Act. The implementation of Biological Diversity Act is operated through decentralized regulation of activities through local management committees, state biodiversity boards and the national biodiversity authority, each with well-defined functions within their respective jurisdiction. Accordingly, it is being implemented at national, state and local level, as a three tier system. It places emphasis for meeting its objectives by involving different stakeholders in public, private and NGO sectors. To accelerate the process of implementation of the act, NBA is working closely with taxonomists, researchers, stakeholders and other Government departments for the conservation of the biological resources. The NBA thus operates through consultative process involving expert committees and stakeholders (Gautam et al. 2010). The SBBs have been established in 24 states in India and the Biodiversity Management Committees have been established in 14 states and 419 People's Biodiversity Registers have been documented containing information on bio-resources and/or associated traditional knowledge. The NBA is also involved in the capacity building, awareness generation & creation and strengthening institutional structure to implement the act. The MoEF, Government of India, and NBA are taking initiatives for strengthening the institutional mechanisms for implementing the said Act at the national, state and local levels (GOI, 2010).

CONCLUSION

The sustainable food security, conservation of biodiversity along with mitigation and adaptation to climate change are crucial, in our journey to future. This calls for the involvement of large number of stakeholders at national, state and local levels. Research on conservation, cataloguing and maintaining material and information on biodiversity assumes utmost importance for future food security. Efforts need to be intensified to identify, characterize and conserve the rich agro-biodiversity hotspots with special emphasis on some of the important crops, trees and animals at given locations. The community based agro-biodiversity practices like establishment of genetic heritage gardens, conservation of plant, animal and fish genetic resources including underutilized crops like millets, medicinal plants, population of pollinators and predators and raising public awareness about agro-biodiversity are to be implemented to develop a hunger free India and to sustain food security. Biodiversity Act encompasses many of these issues to be implemented at national, state and local levels, but the efforts need strengthening.

REFERENCES

Arora, R.K., Roshini Nayar, E. and Pandey, Anjula (2006). Indian centre of floristic and economic plant diversity. In: *Hundred Years of Plant Genetic Resources Management in India* (review).

CBD website, (www.cbd.int).

Chand, Ramesh (2007). Demand for foodgrains. *Economic and Political Weekly,* 42(52): 10–13.

ENVIS, Centre for Ecological Science (www.ces.iisc.ernet.in/ energywater/ paper/ fishdiversity/diversity1.html).

Gautam P.L., Reddy, C.A., Thomson, C.J. and Raghuram, K.P. (2010). Implementation of biological diversity act in India. Proc. Indian Science Congress. *Biodiversity and Sustainable Development,* 69–81, NBA and MoEF.

GOI, (2010). Draft National Food Security Bill, 2010, Government of India (http://www.indiaenvironmentalportal.org. in/content/draft-national-food-security-bill).

India's Fourth National Report to the CBD, Ministry of Environment and Forests, Government of India, GoI, 2009.

Jenkins, P.T. (1999). Trade and exotic species introductions. In: *Invasive Species and Biodiversity Management.* 229–235. Sandlund, O.T., Schei, P.J., Viken, A., Ignacimuthu, S. and Jairaj, S. (Eds.). Kluwer Academic Publishers, London.

Khoshoo, T. N. (1995). Census of India's biodiversity: Task ahead. *Current Science,* 69(1): 14–17.

National Bureau of Agriculturally Important Insects (www.pdbc.res.in).

NBFGR, (2007). *Perspective Plan Vision, 2005,* National Bureau of Fish Genetic Resources, Canal Ring Road, P.O Dilkusha, Telibagh, Lucknow, U.P, India (www.nbfgr.res.in).

Protection of Plant Varieties & Farmers' Rights Authority, Ministry of Agriculture, Government of India, 2010.

Reddy, C.S. (2008). Biological invasion: Global terror. *Current Science,* 94(10): 1235.

State of Indian Agriculture, (2009). Published by National Academy of Agricultural Sciences, New Delhi.

UNESCO, (2010). Statement and Recommendations from the UNESCO International Year of Biodiversity Science Economic and Political Weekly, 42(52): 10–13. Policy Conference, UNESCO Headquarters, Paris, 25–29 January, 2010.

16

Biodiversity of the Sundarbans: Status and Challenges

L.K. Banerjee

INTRODUCTION

The Sundarbans is the Sea-land interphase regions formed by the World's largest deltaic complex from the sediments deposited by three great rivers, the Ganges, the Brahmaputra and the Meghana. The name 'Sundar' is probably chosen for beautiful looking of the forests or due to predominance of the tree 'Sundari' (*Heritiera fomes*). This wonderful forest, with unique biodiversity extends as a largest single and continuous block of World's mangrove land and tiger habitat over the two countries, India and Bangladesh. Out of 26000 sq. km cover areas, about 62% is situated in Bangladesh and 38% of its western part lies in India.

Indian Sundarbans: The Sundarbans which is the largest and remarkable sheltered part of Indian Wet coastal ecosystem lies between 21°32′-22°40′ N. and between 88°85′-89°E. Within 24-parganas district of the State, West Bengal covering an area of 9630 sq. km. It is bounded by the Hooghly river to the West, the Bay of Bengal to the South, the Harinbhanga-Raimangal river along the international boundary with the Bangladesh to the east and the Dampire-Hodges line to the north. This great deltaic complex gradually or step by step is developed by the enormous silt and sediment, deposition carried down by the bordering rivers, the Hooghly, Muriganga, Sapta Mukhi, Thakuran, Gosaba, Bidya, Matla, and Haringbhanga. Due to presence of Bay of Bengal at the apex, the areas receive saline tidal thrust twice a day and also intermix with the fresh water flow from the up steam of the Ganga-Brahmaputra system. Since the areas are covered with several estuaries, many distributaries, innumerable meandering creeks and canals interconnected each others, it remains under the ecological stress such as continuous saline inter tidal flow, unstable and premature substrates of silt, sand and clay, regular erosion and accretion of the river banks and frequent storms and cyclones. There are about 102 islands of which 54 islands are inhabited and rest of the islands are supported by mangroves. Present population of the Sundarbans is about 4.2 million during 2005. These intertidal dynamic ecosystem, along with the intermixture of saline and fresh water flow create a very specialized habitats in this ecologically stressed environment and some of the specially adopted flora and faunal biodiversity are habituated to survive in this condition.

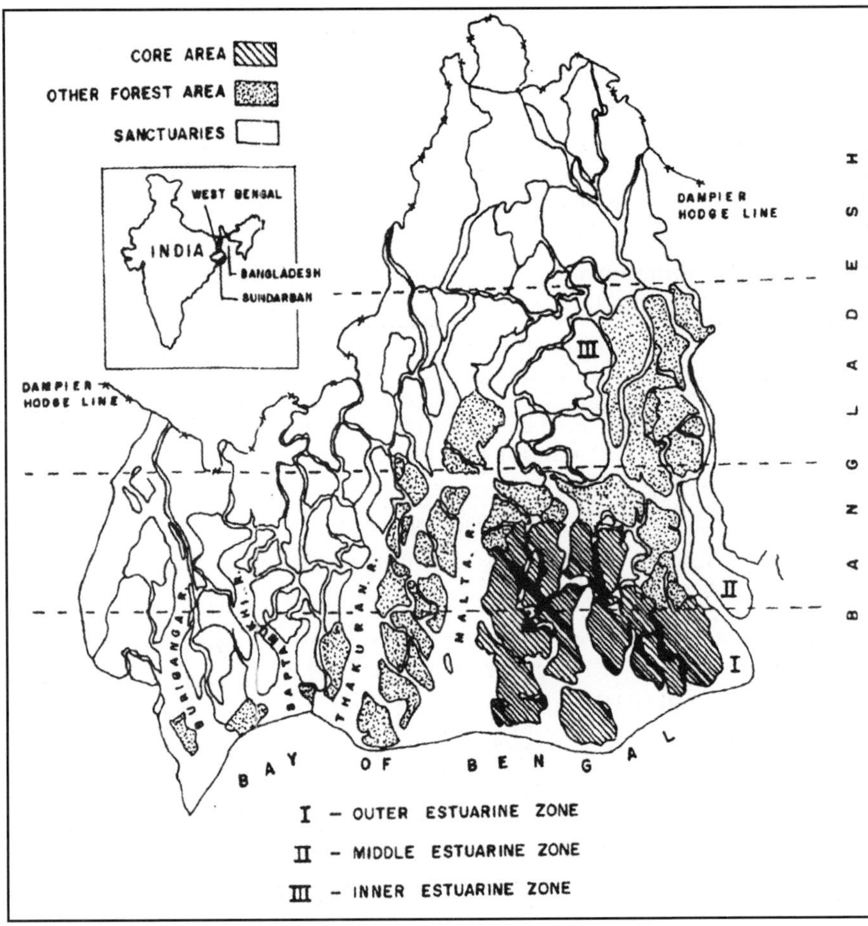

Map. Sundarban Biosphere Reserve

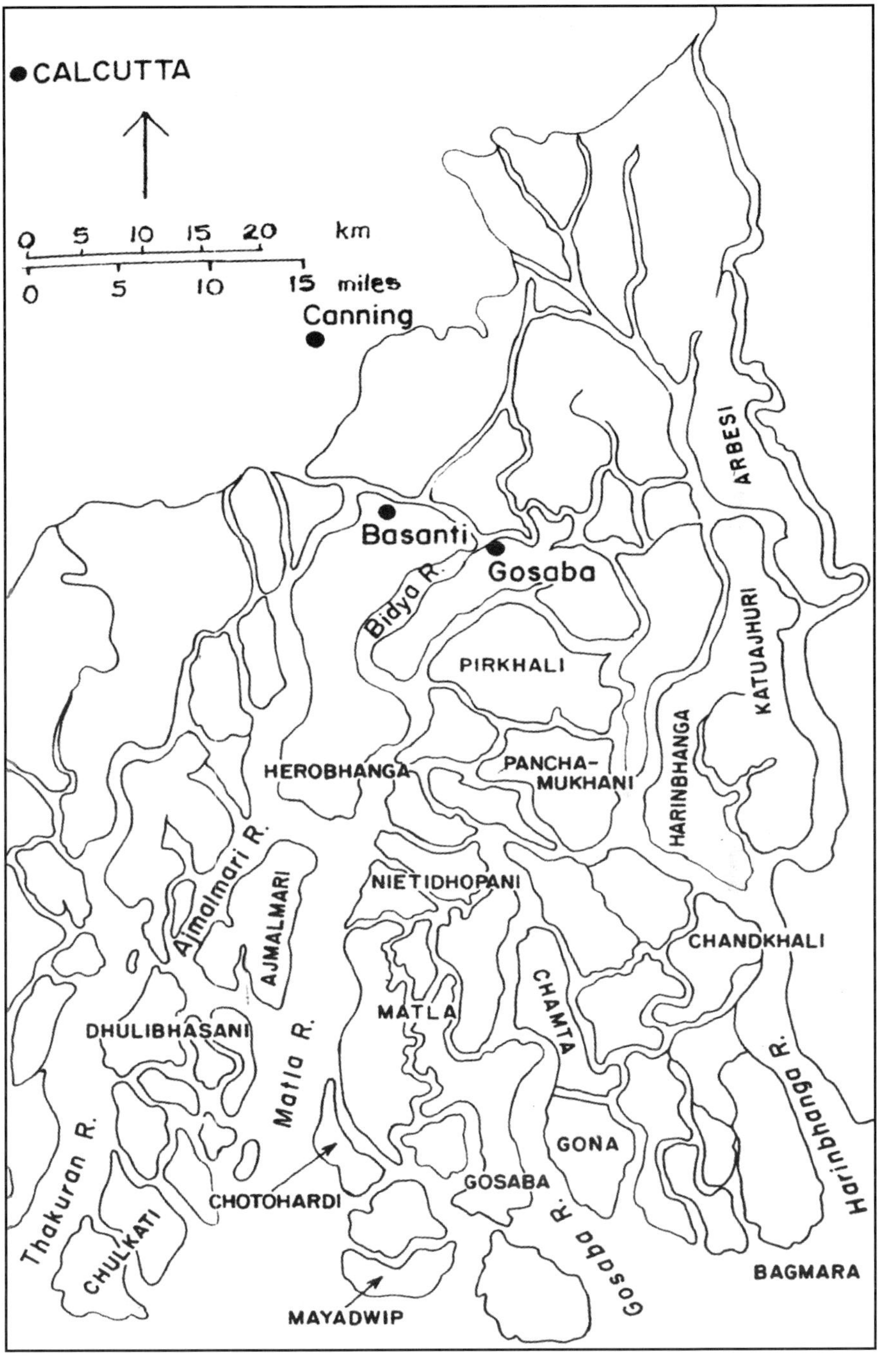

Forest blocks of the Indian Sundarbans

Considering the management of the largest India's wet coastal ecosystem and its specialized adapted biodiversity components, on 29th March 1989 it was established as a Biosphere Reserve (S.B.R.) under the National Man and Biosphere Programme to conserve the unique biodiversity of the area. Out of the total geographic area, an area of 4330 sq. km. forms the Reserve forest area and the rest is the reclaimed area for human habitation. The Reserve forests consist of many protected Zones namely National Park in Tiger Reserve, Wildlife Sanctuaries at Sajnekhali, Holiday island and Lothian island. For easy management the entire Biosphere Reserve has been divided into Core Zone (1700 sq. km), Manipulated Zone (2400 sq. km), Restoration Zone (230 sq. km) and Developmental Zone (5300 sq. km). The Sundarbans Tiger Reserve is situated within the Core Zone and Mangrove forests occupy about 4316 sq. km area of these Core, Manipulation and Restorations Zones. Remarkable Mangrove habitats are surrounded by Hooghly-Matla estuarine Complex consisting the major rivers like the Hooghly (Ganges and Bhagirathi), Sapta mukhi, Thakuran, Matla, Bidya, Ajmalmari, Bidyadhari, Gosaba, Kalindi and Royimangal. All the rivers flow from north to south forming large estuaries and are influenced by tides from the Bay of Bengal. Various silt-laden distributaries, innumerable meandering creeks, several criss-cross canals and rivulets with various grades of tides, water currents and water salinity cover the entire area (178 sq. km water and 4316 sq. km forest land) and form a composite net work of large saline tidal Swamp or coastal wet land.

This deltaic tidal swamp ecosystem is the largest unique area of the World for threatened Royal Bengal Tiger and third largest Mangrove formation after Indonesia and Australia. It is well known for its greater degree of specialization of its Biodiversity Structure, function and valuation and commands highest importance by virtue of its biological productivity and adaptive biodiversity. On the other hand, it is one of the most backward regions of the State where more than 50% of the inhabitants live much below the poverty line.

Since the regions are covered with dense mangrove forests, constant flow of saline tide water, unstable silty, sandy and muddy saline soil and greater depth of fresh ground water sources, the region is befitted for any major Agriculture. Majority of the fringed inhabitants of these regions are fully dependent on its biological resources directly or indirectly for fuel, charcoal, tannin, timber, thatching, house and boat building materials, fishes, shell fish, tiger prawn seeds, honey, crabs and various other materials for their liveli hood. Environmentally the region is very rich and play most significant role for protection of frequent soil erosion, cyclones, storms and floods, conservation of accreting shore and stabilization of new lands.

TOPOGRAPHY & GEOLOGY

Sundarbans, the World's largest deltaic complex has been formed through the deposition of silt carried down by the rivers Brahmaputra and Ganges and a number of distributaries. The area is characterized by a group of islands, many estuaries, dense network of rivers, intertidal creeks, canals, swamps, tidal flats, coastal beaches, dunes, back dunes, few agricultural fields and dense forest covers. It has been suggested that, in the past, the tidal swamps of the Sundarbans extended land ward to the base of the Rajmahal hills and filled up in the latter part of the Tertiary period (Furgusson, 1863). Geologists believe that there was a general south-eastern slope of the Bengal Basin during the Tertiary period. The Ganges started flowing down this slope and created a new delta through silt deposition over thousands of years. Gradually, the Western part of Bengal below the Rajmahal hills rose in level and became suitable for human settlement. Later evidence suggested that, the Bengal Basin tilted eastward during 12th Century because of neo-tectonic movements (Morgan and McLntire 1959).

The rising of the Western part of the delta separated the ancient branches of the river Ganges from the area which today comprises the Indian Sundarbans. This alteration of the river course resulted in a reduction of fresh water flow into the western sides of the delta and finally led to accretion at the river growth and increase of river water salinity into the Western part. This is very much significant on the absence of fresh water loving arborescent tree, *Heritiera fomes*, the famous 'Sundari tree' along the Western part of the delta. Today, the actual delta of the river Ganges lies in the Eastern part of the Sundarbans due to all changes of the rivers.

Depending upon the characteristic of different deltaic heads and influence of fresh and salt water flow, entire Sundarbans can be divided into 6 main habitat types:

1. **Sagar-Mahisani-Ghoramara Sand head group of Islands:** These Western part of the Sundarbans with an area of about 90 sq. km are formed at the estuarine mouth of the river Hooghly. Here due to elevation of Western part and construction of Farakka barrage, water salinity has been increased and except a few old mangrove stand all the intertidal flats and river banks are devoid of good,

healthy mangroves. Plantation of mangroves and cultivation of water melon and chilli are now significant over these areas.

2. **Area between Mahisani island and the river Thakuran:** This 900 sq. km area is the second deltaic lobe. Due to lack of fresh water flow and as a typical sheltered part of the delta, along with mangrove vegetation these regions show rich communities of Zooplankton and Phytoplankton and thus the areas are commonly used for breeding and hatching of tiger prawns, crabs and crocodile farms and Wild life sanctuary at Lothion island.
3. **Area between the river Thakuran and Matla River:** This 1600 sq. km area is the 3rd lobe of the Gangetic delta. It is totally detached from any upstream flow and mostly influenced by backwater tidal flash. Erosion and deposition are both taking place at different parts of the river banks but the rapid erosion is predominant in some localities the. Though the tidal flats are densely covered with mangroves and palm swamps but frequent subsidence of some part of forests, causing complete disappearance of chunks of forest land, are very prominent due to large scale erosion in these regions.
4. **Area lying between the river Matla on the West and the river Harinbhanga on the east:** This is the Core Zone of Sundarbans Tiger Reserve. It is bounded by the river Netidhopani and Gosaba in the north and Bay of Bengal in the South. This 1692 sq. km area is influenced by both salt and fresh water flow and due to this intermixture, the area is recognized as an ideal site for the growth of maximum mangrove diversity and good habitat of Royal Bengal Tiger. The area is more or less free from interference.
5. **Northern part of the Sundarbans Core area surrounded by the river Ichhamati:** This 893 sq. km area is supposed to be the buffer zone of the Sundarbans Biosphere Reserve where many activities are manipulated for conservation and sustainable development to the forest areas. Southern part of this region receives sweet water flow from the river Ichhamati.
6. **Extreme northern area lying to the east of the river Matla:** The easterly flowing effect of the river Ganges through tributaries of the river Padma influence the areas in an ideal agricultural manipulated region. Due to sweet water flow of the river Ichhamati these 4455 sq. km areas are more advance in Agriculture, aquaculture and Agroforestry.

There are two most prominent geological phenomenon need to be remembered for understanding of physiographic structure of the Sundarbans. The first one is formation of several new islands continuously in the prograding delta and second one is the frequent subsidence and disappearance of chunks of forest lands and settlement areas.

Oldham (1893) pointed out that area south of the river Raimangal and Malancha estuary between 21°00′-21°22′ latitude in the Bay of Bengal the occurs in almost all the estuaries of Sundarbans delta. Fergusson (1863) opined that this feature used to form by the eastward and Westward current interaction in the Sundarbans and this phenomenon do not allow river carried silt to settle in the area. These sits are being diverted further south or east causing gradual deposition and formation of new islands in the Sundarbans delta.

Second geological feature of Subsidence and disappearance of chunks of forest land is probably due to large scale erosion effect. In general, large volume sediments carried by tidal water in different rivers of the Sundarbans delta are mostly deposited on the river beds due to constricting effect of river embankments. The sediments are raising the level of river beds with respect to the surrounding land and thereby increasing the rate of erosion and chances of flood. Erosion and accretion are both common phenomenon at different parts of the delta or locally alternating. However, as indicated from the stratigraphy and aerial photographs, erosion is the principal process shaping the modern surface of the Sundarbans delta. It is reflected in numerous small and large scale geomorphic features, the precise form of which depend on location, type and cohesion of substrate, current flow, slope of surface and others. There are three types of erosion such as sheet erosion, river bank erosion and creek or gullies erosion. All these three types of erosion appear to bear direct effect on tidal-flats and mangrove forests and settlement areas on these tidal flats are slowly subsidies and disappear due to the effect of these erosion.

Physiography

The rivers flowing within the Sundarbans form different estuaries. Physiographically these estuaries can be divided into outer part, middle part and inner part. Mayadwip, Gona blocks are situated in the outer part and Pirkhali, Jilla blocks in the inner estuarine part. Interaction of different ecological factors such as soil, tidal amplitude, water salinity wind and salt water spray and inter mixture of fresh and salt water differ consequently due to distance from the sea and approach of fresh water from the upstream; moreover, salinity of river water changes seasonally. These different physiographic units such as outer, middle and inner estuarine parts are very significant for zonation of different mangrove plant communities with their different morphological adaptive characters for withstanding the different ecological parameters.

The coastal processes of the Sundarbans are influenced by tidal waves, micro and macro-tidal cycles, and long shore currents along the coastal tract and vary during pre and post monsoon periods. Coast line is affected by tropical cyclones and is changing in the reclaimed areas due to high rate of cliff erosion. Biodiversity plays a significant role in the development of coastal evolution of the Sundarbans. A variety of habitats are unique in the Sundarbans formation including beaches, estuaries, mangrove swamps, tidal flats, saline mud flats, tidal creeks, innumerable meandering canals, coastal dunes and back dunes. Mangrove vegetation in the intertidal flat has an important role in the development of Swamp morphology and along the accreting site assist in creating new land formation. The baffle effect of mangrove roots and trunks substantially lower tidal currents and causes deposition of suspended sedimentary materials, while roots ramifying through and binding sediments also aid in stabilizing the substrate. Activities of mangrove fauna on the intertidal mud flats and beaches are very significant for sediment trapping and germination of mangrove seeds. The supratidal sand bars and dunes morphology are controlled by specially adapted sand binders like some xerophytes, creepers, sedges and grasses. The saline mud flats, due to hypersaline conditions as controlled by Samphires (salt marshes).

Geomorphology

The broad forest belt of the entire Sundarbans mangroves extends about 480 km. along the coast line and form a wild spectacular environmental unit of the area in transverse profile from the main land to the sea. The major geomorphic units in the Sundarbans delta may be mentioned as: **1.** Distributary channels with lag, sub-aerial and sub-aqueous levees, splays and tidal flats. **2.** marginal marshy areas above mean tide level. **3.** tidal sandbars and islands with network of tidal channels **4.** subaqueous distal sand bars and pro-deltaic clays and silts and **5.** Coast line, which varies during pre and post monsoon periods. Geomorphic feature of the Hooghly-Matla estuary is tide dominated funnel shaped deltaic head. From the month of February, the sea level start rising at the mouth of the rivers and reach maximum during September and again fall during winter months. This phenomenon is due to the change of wind velocity in the Bay of Bengal. There are some tidal islands on which tidal channels often flow in sinous shape and due to this shape those islands are very active sites for erosion. As discussed earlier, each outer, middle and inner parts of the river-estuaries has its different geomorphic conditions and different pattern of biodiversity distribution. Among the various interacting ecological factors, the salt water tide is the keystone factor, which play a vital role for distribution of different biodiversity components and shape different vegetation pattern. Thus, all the tidal flats and sand bars on the outer, middle and inner estuaries of the Sundarbans can be divided into 5 distinct units based on the approach of tidal flow, elevation, substrate types and pattern of vegetation zones such as: Intertidal flats, midtidal flats, supratidal flats, saline mud flats and riverine and estuarine sand stands.

1. **Intertidal flats:** These are low tidal flats along the river banks, creeks and channels and are colonized by saline grasses and mangroves. Due to erosion, several geomorphic modifications may occur and in many parts, entire intertidal geomorphic zone may be missing where mangrove flat may abut the midtidal flats.
2. **Midtidal flats:** These are little bit elevated part behind the intertidal flats and colonized mainly by well developed mangrove communities. In some parts, receeding tide may stay for longer period in its depressed zone and form saline flats.
3. **Supratidal flats:** These are more elevated and gently sloping surface, bordering hinter land behind. They are not inundated by regular high tides but sufficiently inundated at the highest tides. These zone is mainly dominated by mangrove species which need less saline tidal flow associated with other coastal plants of transitional nature.
4. **Saline mud flats or salt flats:** Towards the outer estuarine regions, there are some muddy inclined slopes followed by muddy flats exposed only at low spring tide known as saline mud flats. Also a broad, subhorizontal, salt crusted muddy surface within the depressed regions of the midtidal mangrove belt is known as slat flats. They are inundated when tides are higher than mean spring level. Due to hyper saline conditions these zones are colonized by samphires.

5. **Riverine and Estuarine sand bars:** Tendering north-south nearer to the river mouth sands are deposited from the rivers and form estuarine sand bars. Sometimes larger sand particles carried by the river currents are deposited in the supratidal mangrove stand and form the riverine sand bars. Along with many sand binding plants, some interesting coastal trees are found dominant in these regions.

Climate

The area falls within the subtropical humid climate with pronounced hot and wet season. Rainy season extends from May to September with annual rain fall ranging from 2500-3000 mm. Maximum temperature ranges from 25°c to 35°c and minimum from 12°c to 24°c with relative humidity 75 to 90%, winter months December to February remain dry. Tide level varies from 3.5 to 6.5 m depending upon the seasonal changes. River water salinity ranges from 0.7 to 3.1 g/lit. Due to fall of sudden pressure in the Bay of Bengal during May to November the areas are frequently affected by large storms, cyclones, floods and earth quake. History reveals that from the year 1582 to 2009, more than 95 storms, cyclones, super cyclones, hailstorm, floods and earth quake used to developed over these regions. These natural calamities with huge tidal waves, violent storms, cyclones and floods had caused wide spread devastation of many human settlements, mangrove forests and many lives. Recent cyclonic storm 'Aila' on 25th May 2009 with 90-100 km/hr wind speed, 10-12 m high tidal waves and superstorm damaged 6 lakh household 20 lakh people and 6 km area of dam in the Sundarbans.

Table 1. River water salinity ranges and tidal level of the Sundarbans delta

Areas	*Salinity in EC (dsm^{-1})*				*Tidal level in meter*			
Sundarbans	*Jan.-May*	*Apr.-June*	*July-Sept.*	*Oct.-Dec.*	*Jan.-May*	*Apr.-June*	*July-Sept.*	*Oct.-Dec.*
Outer estuaries	32.5	34.4	38.8	41.1	3.5	4.5	5.9	6.2
Middle estuaries	17.6	18.3	21.2	29.5	3.1	3.8	4.1	5.9
Inner estuaries	12.2	13.2	14.5	21.7	1.8	2.9	3.1	3.5

Soil

Deltaic soil of the Sundarbans comprises mainly with saline alluvium type consisting of clay, silt, fine sand and coarse sand particles. It falls under the category of saline coastal soil type. The outer estuarine soil consist of move sand particles than the clay and silt. The soil of the middle and inner estuarine mangrove forests is anaerobic and dominated by clay and silt particles with small percentage of organic matter, nitrogen, Phosphate, calcium oxide and carbon.

Table 2. Soil of the Sundarbans delta

Areas	*Particle size distribution (depth 30 cm)*				*pH*	*Organic matter (%)*	*EC (dsm^{-1})*
Sundarbans	*Clay (%)*	*Silt (%)*	*Fine Sand (%)*	*Coarse Sand (%)*			
Outer estuaries	12.5	23.5	33.5	30.5	7.8	0.79	38.4
Middle estuaries	46.5	23.5	20.0	10.0	8.02	1.49	32.5
Inner estuaries	33.5	30.5	23.5	12.5	8.0	1.5	22.8

The data represent average value of ten (10) samples from each type

BIODIVERSITY

Characteristics of the Sundarbans Ecosystem Biodiversity

Ecosystem Biodiversity of the Sundarbans is a unique character which falls within a dynamic deltaic system representing fragile and ecologically sensitive areas. Innumerable ramification of the river system, creeks and channels and several islands in the saline intertidal land along with the intermixing of sweet water flow from the upstream provide an ideal habitat for best growth and maximum diversity of mangrove species. These large mangrove forests re preserved with its wilderness under the protective shelter of the project tiger. Almost all the mangrove species in India except one or two are found in this Sundarbans forest ecosystem. It is a home for many rare and endangered species like *Heritiera fomes, Acanthus volubilis, Merope angulata, Kandelia candle, Sarcolobous globosus, Dischidia nummularia, Cynometra iripa, Sonneratia griffithii* along with the Royal Bengal Tiger, Estuarine crocodiles, Salvador lizard, Gray water monitor, Batagor terrapin etc. This ecosystem serves the nesting and breeding ground for many fishes, shell fishes, marine olive Ridley turtle, Green and Hawksbill turtle, Dolphin and many migratory birds like Curlew, Golden plover, Sea Eagle, Giant heron and many others.

During the high tide 70% of the land area is submerged with high degree of saline water, silt and sediments. Biodiversity in these regions is very adaptive and specialized with attractive characters. Vegetation represents a serial type depending on the gradation of salinity, tidal flow and soil morphology. Biodiversity is uniquely adapted to withstand high salinity conditions, great tidal velocity and muddy anaerobic soil. Majority of plant species dominated over the regions produce stilt roots, salt glands, vivipary, pneumatophores and lenticels to bring a remarkable balance for surviving in these specialized environmental conditions. Here both fish and crabs are found creeping on the tree, tigers are roaming easily on muddy substrate infested with strong woody peg like breathing roots and capable to swim large rivers and creeks. Tigers sustain in this ecosystem on both aquatic and terrestrial biodiversity with a unique habit of drinking even salt water. All the estuarine complex formed by several rivers in this ecosystem are highly productive due to constant mixing of brackish water with the sea water. In this ecosystem both marine and riverine microbial decomposer convert the mangrove litters into consumable protein, a primary source of energy which enters into the food chain of fish and other invertebrates and finally ends up with tigers and crocodiles.

The biomass productivity in mangrove forests is very high and mangroves have a staggering ability to sequester carbon from the atmosphere, about 150 m tons of CO_2 may store in the mangrove biomass per hector. In the accreting shore gradual increase of land level through the ramification of root system serve as a repository of sediments for other in shore marine habitats, such as many new islands seagrass beds and coral reefs. Under ground roots, bulbs, fruits and leaves of many species serve as good fodder for deer, monkeys and wild boars. Besides the primary energy, mangrove forests in their ecosystem play a vital role in the protection of soil erosion, storms, floods and cyclones as well as to develop a sustainable socio-economic structure to the livelihood of the surrounding people providing timbers, fuel, tannins, fodder, thatching materials, honey, fish, crabs, tiger prawn seeds and many others.

Characteristics of Mangrove Biodiversity and Importance in the Sundarbans Ecosystem

Mangroves: Group of plants isolated from their taxonomic family and successfully adapted in their morphology, physiology and biochemistry for colonizing in saline intertidal zone of ecologically stressed habitat at the interface of land and sea within the tropical and subtropical sheltered coast line known as Mangroves.

They are characterized by dark green foliage, negatively geotropic roots and absence of tap root system. They develop some superficial morphological characters like pneumatophores, stilt roots, vivipary, salt excretory glands and lenticels for withstanding every six hour submerged saline water tide and anaerobic soil conditions. To survive in the high speed wind and tidal current, they develop very dense growth without keeping any space in between the individual trunk. Their roots without taproots, produce numerous finger like lateral rootlets which in all corners anchor the unstable substrate. The baffle effect of mangrove roots and trunks substantially lower tidal current and causes deposition of suspended sediments while roots ramifying through and start functioning as land-building to stabilize the substrate. Best diversity of mangrove formation is found where regular mixing of sea water with the sediment loaded fresh river water occurs. Since the river Ganges carry maximum volume of sediment loaded fresh water compare to other Indian rivers, the maximum mangrove diversity and species richness is found in the Sundarbans throughout India.

Though species diversity in the mangrove ecosystem is usually low due to dominance of large Physical forces of saline tide water and lack of stable substrate but the biodiversity commands highest importance by virtue of biological productivity, ecological complexity and development of livelihood resources. Total 6,740 sq. km area of mangrove forests in India are represented by 66 mangrove vascular plant species belonging to 31 families, 59 genera and 2 varieties. Out of these 66 species in Indian mangroves, the Sundarbans represents 63 species belonging to 26 families, 40 genera and one variety, compare to 62 species in the Mahanadi delta, 32 species in the Godavai-Krishna delta, 16 species in the Cauvery delta, 33 species in the West coastal mangroves and 43 species in the Andaman and Nicobar islands. Along the mangroves about 11 species of salt marshes are found in the Sundarbans.

Table 3. Species diversity and distribution of mangroves and salt marshes in Mangrove area

Botanical name	*East Coast*				*West Coast*	*Andaman & Nicobar Is.*
	Sundarban delta	*Mahanadi delta*	*Krishna delta*	*Godavari delta*		
Acanthus ebracteatus	–	–	–	–	–	+
Acanthus ilicifolius	+	+	+	+	+	+
Acanthus volubilis	+	–	–	–	–	+
Acrostichum aureum	+	+	–	–	+	+
Acrostichum speciosum	–	–	–	–	–	+
Aegialitis rotundifolia	+	+	+	–	–	–
Aegiceras corniculatum	+	+	+	–	+	+
Aglaia cucullata	+	+	–	–	–	–

Avicennia alba	+	+	+	+	+	–
Avicennia marina	+	+	+	+	+	+
Avicennia marina var. *acutissima*	+	+	–	–	+	–
Avicennia officinalis	+	+	+	+	+	+
Amoora cucullata	+	+	+	+	+	+
Brownlowia tersa	+	+	–	–	–	–
Bruguiera cylindrica	+	+	+	+	+	+
Bruguiera gymnorrhiza	+	+	+	+	+	+
Bruguiera parviflora	+	+	–	–	–	+
Bruguiera sexangula	+	+	–	–	–	–
Caesalpinia bonduc	+	+	+	+	+	+
Caesalpinia crista	+	+	+	+	+	–
Cerbera manghas	+	+	–	–	+	+
Ceriops decandra	+	+	+	+	+	–
Ceriops tagal	+	+	–	–	+	+
Clerodendrum inerme	+	+	+	+	+	+
Cynometra iripa	+	+	–	–	–	+
Dalbergia spinosa	+	+	+	+	–	–
Derris scandens	+	+	+	+	+	+
Derris heterophylla	+	+	+	–	+	+
Dolichandrone spathacea	+	+	+	+	+	+
Excoecaria agallocha	+	+	+	+	+	+
Fimbristylis ferruginea	+	+	+	+	+	+
Finlaysonia obovata	+	+	–	–	–	+
Heritiera fomes	+	+	–	–	–	–
Heritiera litloralis	–	+	–	–	+	+
Heriteria kanikensis	–	+	–	–	–	–
Hibiscus tiliaceus	+	+	+	+	+	+
Hoya parasitica	+	+	+	+	–	+
Intsia bijuga	+	+	–	–	–	+
Ipomoea tuba	+	+	+	+	+	–
Kandelia candel	+	+	–	–	+	–
Lumnitzera littorea						
Lumnitzera racemosa	+	+	+	+	+	+
Merope angulata	+	+	–	–	–	–
Mucuna gigantea	+	+	–	–	+	+
Myriostachya wightiana	+	+	+	+	–	+
Nypa fruticans	+	–	–	–	–	+
Phoenix paludosa	+	+	–	–	–	+
Porteresia coarctata	+	+	+	+	+	–

Rhizophora apiculata	+	+	+	+	+	+
Rhizophora mucronata	+	+	+	+	+	+
Rhizophora stylosa	–	+	–	–	–	–
Sarcolobus carinatus	+	+	+	+	–	–
Sarcolobus globosus	+	+	+	+	–	+
Scirpus littoralis	+	+	+	+	–	–
Scyphiphora hydrophyllacea	–	–	–	–	–	+
Salvadora persica	+	+	+	+	+	–
Sonneratia alba	–	+	–	–	+	–
Sonneratia apelala	+	+	+	+	–	–
Sonneratia caseolaris	+	+	–	–	+	+
Sonneratia griffithii	+	+	–	–	–	+
Thespesia populnea	+	+	+	+	+	+
Thespesia populneoides	+	+	–	–	–	–
Tylophora tenuis	+	+	–	–	–	–
Xylocarpus granatum	+	+	–	–	–	+
Xylocarpus mekongensis	+	+	–	–	–	+
Xylocarpus moluccensis	–	–	–	–	–	+

Salt marshes						
Botanical name	*East Coast*				*West Coast*	*Andaman & Nicobar Is.*
	Sundarban delta	*Mahanadi delta*	*Krishna delta*	*Godavari delta*		
Aeluropus lagopoides	+	+	+	+	+	–
Arthrocnemum indicum	+	+	–	–	+	–
Cressa cretica	+	+	–	–	+	–
Heliotropium curassavicum	+	+	+	+	+	–
Salicornia brachiata	+	+	+	+	+	–
Sesuvium portulacastrum	+	+	+	+	+	–
Suaeda fruticosa	–	–	+	+	+	–
Suaeda maritima	+	+	+	+	+	–
Suaeda monoica	+	+	+	+	+	–
Suaeda nudiflora	+	+	+	+	+	–
Tamarix troupii	+	+	–	+	+	–

Importance of Mangrove diversity

Mangrove ecosystem commands highest importance for colonizing special type of biological diversity and maximum biological productivity. Majority of the fringe inhabitants in this ecosystem are fully dependent on its biological resources directly or indirectly in multiple ways for their livelihood. It is estimated that the Mangrove forests Provide at least US$ 1.6 billion each year in ecosystem service. Loss of mangrove diversity will follow devastating economic and environmental consequences. Mangrove diversity is acting as cost free, self-repairing static border security force for saving inland vegetation, crops, households and lives from storm surge, floods and cyclones. They are not only a vital component in efforts to sequester atmospheric carbon and fight against climate change but also protect some of the world's most vulnerable people from extreme weather and provide them income, food and shelter. Total importance of use and nonuse value of mangrove diversity in the Sundarbans can be expressed as follows.

Table 4. Total Economic Value of Mangroves

USE VALUE		
Direct Use Value	Direct Consumed use value	Fodder, fuel wood, timber, tannin, thatching, food, medicine, charcoal, industrial input, onsite and offsite fishery
	Direct Non-Consumed use value	Recreation, tourism, etc.
Indirect Use Value	Functional Benefit	Control of flood, storm surge, cyclone, biodiversity, stabilization of river and creek banks, carbon sequestration and transplantation of genome
NON USE VALUE		
Option value	Future Direct and Indirect values	Conserved habitat, Biodiversity
Existence value	Value from knowledge of continued existence, cognitive value	
Bequest value	Preservation of mangroves for future generation	

Mangrove forests play a significant role for sequestering carbon dioxide and store in their biomass as carbon sink. It is found that about 150 m tons of CO_2 may store in the mangrove biomass per hector. The baffle effect of mangrove roots and trunks substantially lower tidal current and soil erosion. Growth of mangroves on accreting shore, bind the sediments through their roots gradually increase the level of land. Both these processes ultimately fight climate change and sea level rise.

FLORAL DIVERSITY IN DIFFERENT HABITAT FORMATION OF THE SUNDARBANS FORESTS

The Sundanbans, the largest coastal wetland in India and 3rd largest mangrove formation of the World, ecologically exhibit different habitat conditions within a short distance due to impact of various gradations of water salinity, fresh water flow, seasonal tide level, edaphic conditions and elevation of land. It is found that mangrove species richness gradually taper from the Sundarbans and the Mahanadi delta to the South wards of the east coast of Andhra Pradesh and Tamil Nadu and then tapering again to the West coast and finally form mostly a single species dominated stand in the Gujarat regions.

Classification of Floral diversities

Depending on the various land forms and interaction of environmental factors, plant diversity of the Sundarban forests can be divided as follows:

1. Diversity of Plant communities in the Mangrove forest habitat:

(a) **Outer Estuarine Mangrove diversity:** All the estuarine regions of the Sundarban Biosphere Reserve (SBR) having less than 3-4 m elevation with maximum influence of water salinity, tidal inundation, silt and sand along with high wind speed, salt spray and wave action as it is near to the open coast are vegetated with special type of adapted mangroves. Diversity in this zone is dominated by *Avicennia marina* associated with *Sonneratia griffithii, Bruguiera cylindrica, Bruguiera parviflora, Ceriops tagal, Aegialitis rotundifolia* and sometimes *Avicennia alba, Phoenix paludosa* and *Excoecaria agallocha* are also found commonly exposed under eroded conditions of the river banks. Presence of salt excretory glands in the leaves and petioles of *Avicennia marina* and water storage mechanism in the leaves of the species of *Sonneratia, Ceriops* and *Aegialitis* are the adaptive mechanism for withstanding high salinity conditions. All the above plants are found specially along the forest block of Chulkati, Chotohardi, Mayadwip, Baghmara and Gona river mouths.

(b) **Middle Estuarine Mangrove diversity:** Best development of mangroves are found in the Middle estuarine or Riverine mangrove habitat and flora becomes rich and diversified due to sheltered situation and availability of more fresh water. This zone covers the lowermost part of the river system, just away from the estuarine mouth where tidal flats are associated with maximum numbers of creeks and channels. Here the salinity is lower than the river mouth but tidal velocity is higher as the flow enters into the creeks. Diversity is mainly dominated by *Rhizophora apiculata, R. mucronata, Kandelia candle, Aegiceras corniculatum.* Common associates in this type are: *Excoecaria agallocha, Bruguiera gymnorrhiza, Xylocarpus granatum, X. mekongensis, Avicennia officinalis, A. alba, Ceriops decandra, Dalbergia spinosa* and *Phoenix paludosa. Finlaysonia obovata, Derris scandens, Tylophora tenuis, Sarcolobus globosus, Acanthus volubilis* and *Hoya parasitica* are found as common climbers in this zone. Most of the species in this typical mangrove formation are by formation of stilt roots and vivipary for survival in the high speed tidal condition and easy regeneration. All the above mentioned plants are associated in Dhuli Bhasani, Ajmalmari, Matla, Netidhopani, Chamta, Chandkhali, Durgaduani, Pancharnukhani and other forest blocks.

(c) **Inner Estuarine or Riverine Mangrove diversity:** This habitat occurs in the middle part of the riverine system which is slightly elevated from the former type and is associated with less number of tidal creeks. As this part is further away from the sea and more nearer to the fresh water source, the salinity is less and sticky, muddy condition of the soil become very prominent. This habitat is mainly dominated with *Sonneratia apetala, Heritiera fomes, Nypa fruticans, Excoecaria agallocha* in association with *Brownlowia tersa, Sonneratia caseolaris, Xylocarpus mekongenis, Bruguiera gymnorrhiza, Avicennia officinalis, Aglaia cuculata, Cerbera manghas, Intsia bijuga, Cynometra iripa* and *Phoenix paludosa.* Some common shrubs, climbers, grasses and sedges in this formation are *Acanthus ilicifolius, Clerodendrum inerme, Caesalpinia bonduc, Sarcolobus carinatus, Derris trifoliata, Pentatropis capensis, Solanum trilobatum, Flagellaria indica, Myrostachia wightiana, Porteresia coarctata, Crinum asiaticum, Cyperus exaltatus, Fimbristylis ferruginea, Scirpus articulatus* etc. Most of the tree species in this zone produce knee roots, pneumatophores, peg-like woody root suckers and buttresses to withstand the less aerated condition of the muddy soils and to support their massive boles. These adaptive features give a distinct stamp to the overlying topography, distinguishing this zone as less pronounced or semi mangrove formation from the previous zone. All the above mentioned plants are found in Harinbhanga, Pirkhali, Katuajhuri, Arbesi and other forest blocks.

(d) **Hinterland mangrove diversity:** This habitat lies towards the hinter part of the river system where tidal flats are more elevated and creeks and channels are suddenly shut off due to formation of natural sand bars. These areas are usually devoid of regular tidal flow. This zone is mostly dominated by *Clerodendrum inerme, Hibiscus tiliaceous, Avicennia officinalis, Derris scandens, Thespesia populnea, Pongamia pinnata, Acanthus ilicifolius* and *Acrostichum aureum.* Further increase in elevation, deposition of sand and accumulation of more organic matter towards the high limit are found associated with fresh water elements like *Salvadora persica, Syzygium ruseifolium, Carissa spinarum, Ochna obtusata, Manilkara hexandra, Diospyros cordifolia, D. buxifolia, Trewia nudiflora, Kleinhovia hospita, Salacia chinensis, Barringtonia racemosa, Aristolochia indica* etc. Most of the species in this habitat do not show any adaptive features like stilt roots, Pneumatophores and vivipary. This formation may transit successfully towards non-saline zone. The plants are associated towards the most inner part of the islands and found in Basanti, Canning, Gosaba, Morij-jhapi, Sajnakhali and other forest blocks.

2. Diversity of plant communities in Salt Marsh Habitat

Some areas within the mangrove forest blocks become dried up and slightly elevated along its margins due to deposition of sands and other debris. Due to these microtopographic changes, the rate of outflow of saline tide water decreases gradually and eventually the outflow comes to a halt on the shallower places. Accumulation of saline water and high evaporation rate on these areas make the soil hypersaline conditions than the adjacent mangrove swamps. Thus the areas suffer from regular tidal inundation and remain physiologically dry with high concentration of salts. Common plant species growing on this habitat are *Sesuvium portulacastrum, Heliotropium curassavicum, Suaeda maritima, Suaeda nudiflora, Suaeda monoica, Salicornia brachiata, Arthrocnemum indicum, Tamarix troupii, Aleuropus lagopoides, Scirpus littoralis* and other halophytes.

3. Diversity of Plant Communities in Seashore Sandy habitat

(a) **Diversity on Open beach habitat:** Open beach habitats are mainly divided into two major types i.e. areas under the influence of regular inter-tidal sea waves and areas above the influence of regular sea waves. Inter-tidal areas of the open beaches are devoid of any diversity due to unstable nature of sandy breaches. In the supra-tidal regions where the areas are more or less semi-stabilized or stabilized with loose sandy grains, the following sand binders are common such as: *Ipomoea pes-caprae, Hemathria compressa, Canavalia gladiata, Vigna marina, Cyperus arenarius, Aristolochia indica, Fimbristylis dichotoma, Gisekia pharnaceoides, Hybanthus enneasperma, Imperata cylindrica, Launaea sarmentosa, Rivea hypocrateriformis, Rothia indica, Oldenlandia biflora, Saccharum spontaneum, Vetiveria zizanoides, Wedellia biflora* and others.

(b) Diversity on Sandy dunes/back dunes/sandbar habitat: Just after the open and flat sand beaches in the supra-tidal habitat there are the formation of small sand dunes, sand bars and backside of the sand dunes which are commonly covered with the following plant species - *Alternanthera paronychoides, Alysicarpus vaginalis, Ammannia baccifera, Argimone mexicana, Aristida setacea, Asparagus domosus, Atylosia scarabaeoides, Azima tetracantha, Bacopa monnieri, Bergia capensis, Bidens pilosa, Biophytum sensitivum, Blepharis indica, Blumea lacera, Calotropis procera, Calotropis gigantea, Canavalia gladiata, Catharanthus rosea, Cayratia pedata, Cayratia trifoliata, Cenchrus cilliaris, Centella asiatica, Chloris barbata, Chrozophora plicata, Cyperus arenarius, Cyperus haspan, Abutilon indicum, Aerva lanata, Allophylus cobbe, Antidesma acidum, Barringtonia racemosa, Bridelia retusa, Calophyllum inophyllum, Capparis decidua, Capparis sepiaria, Carissa spinarum, Cassia tora, Cassia sophera, Cissampelos pareira, Cissus vitiginea, Cleome aspera, Convolvulons arvensis, Crateva adansonii* sub.sp. *odora, Crotalaria verrucosa, Datura metel, Desmodium gangeticum, Diospyros cordifolia, Dodonea viscosa, Ehretia acuminata, Euphorbia nivulia, Ficus religiosa, Glycosmis pentaphylla, Hyptis suaveolens, Cyperus pangorei, Eragrostis gangetica, Evolvulus alsinoides, Gisekia pharnaceoides, Hybanthus enneaspermus, Imperata cylindrica, Indigofera linifolia, Ipomoea pes-caprae, Kyllinga brevifolia, Leptadenia pyrotechnica, Leucas stricta, Lippia alba, Ludwigia perennis, Malchra capitata, Merremia gangetica, Mollugo disticha, Mollugo pentaphylla, Murdania vaginata, Opuntia dillenii, Oxalis corniculata, Rotala verticillaris, Rungia pectinata, Sida cordifolia, Sida rhombifolia, Solanum* *surattense, Spinifex littoreus, Sporobolus marginatus, Tephrosia purpurea, Trianthema portulacastrum, Tribulus terrestris, Tylophora indica, Vernonia cinera, Xanthium indicum, Jatropha curcus, Lannea coromandelica, Litsea glutinosa, Malinkara hexandra, Ochna obtusata, Pandanus tectorius, Passiflora foetida, Paveta indica, Pentatropis capensis, Phoenix sylvestris, Phyllanthus fraternus, Physalis minima, Pongamia pinnata, Perma corymbosa, Prosopis juliflora, Solanum trilobatum, Sphenoclea zeylanica, Streblus asper, Strychnos nuxvomica, Syzygium ruscifolium, Tiliacora acuminata, Tinospora cordifolia, Vigna trilobata, Viscum ovalifolium, Vitex negundo, Vitex trifoliata, Zizyphus oenoplia, Zornia diphylla* and others. These are found in Bok-khali, Sagar island, Jamboodweep, Shaimari, Kalas, Mayadweep, Thakuran beach, Bijawara beach and other sandy beaches.

(c) **Diversity of beach and dunes plantation:** For protection of seashore wind following species are planted along the open coast sands: *Casuarina equisetifolia, Prosopis juliflora, Thespasia populnea, Acacia auriculiformis, Eucalyptus tereticornis, Lannea coromandelica* and others.

4. Diversity of Plant Communities in Non-Mangrove habitat

(a) **Diversity in close mosaic village habitat:** Typical village with hutments, adjacent vegetable gardens, orchards, bamboo grooves, ponds and road sides are covered with following plant species - common trees along the roadside, bank of ponds, canals and around the hutments are: *Syzygium jambos, Casuarina equisetifolia, Eucalyptus tereticornis, Acacia auriculiformis, Ceiba pentandra, Acacia nilotica* subsp. *indica, Pithocellobium dulce, Borassus flabellifer, Phoenix sylvestris, Azadirachta indica, Lannea coromandelica, Dendrocalamus strictus, Thespesia populnea, Cocos nucifera, Samania saman, Leucaena leucocephala, Tamarindus indica, Albizia lebbeck, Albizia procera, Zizyphus mauritiana, Ficus beghalensis, Erythrina variegata, Cordia dichotoma, Delonix regia, Peltophorum pterocarpum, Limonia acidissima, Bombax ceiba, Ficus hispida, Ficus religiosa, Trema orientalis, Areca catechu, Anthocephalous chinensis, Cicca acida* (*Phyllanthus acida*), *Dillenia indica, Pongamia pinnata, Ailanthus excelsa, Sapindus pinnata, Sapindus laurifolia, Cassia fistula, Sesbania grandiflora, Dalbergia sissoo, Bauhinia purpurea, Prosopis chinensis, Terminalia arjuna, Barringtonia acutangula,*

Mimusops elengi, Alstonia scholaris, Plumeria alba and others. Common herbs and shrubs in the same habitat are – *Pluchea indica, Ipomoea fistula, Abutilon indicum, Euphorbia tirucalli, Adhatoda vasica, Sida cordifolia, Sida rhombifolia, Sida cordata, Urena lobata, Malacra capitata, Crataeva roxburghii, Cissampelos pareira, Tinospora tomentosa, Tinospora cordifolia, Tiliacora acuminata, Capparis sepiaria, Cleome viscose, Flacourtia indica, Portulaca oleracea, Hibiscus tetraphyllus, Abroma augusta, Waltheria indica, Triumfetta rhomboides, Cissus repanda, Cissus trifoliata, Leea indica, Cardiospermum helicacabum, Crotalaria verrucosa, Crotalaria pallida, Desmodium gangeticum, Indigofera linifolia, Indigofera tinctoria, Jatropha curcas, Jatropha gossypifolia, Dioscorea bulbifera, Coccinia grandis.* Plants on aquatic and marshy habitat are: *Nymphaea stellata, Nymphaea nouchali, Nelumbo nucifera, Bergia ammanniodes, Bergia capensis, Aeschynomene indica, Medicago polymorpha, Melilotus alba, Sesbania cannabina, Neptunia oleracea, Myrophyllum indicum, Ammannia baccifera, Rotala indica, Ludwigia adscendens, Trapa natans var. bispinosa, Dentella repens, Emilia sonchifolia, Enhydra fluctuans, Gnaphallium indicum, Vernonia cinerea, Sphenoclea zeylanica, Pergularia daemia, Hydrolea zeylania, Bacopa monnieri, Limnophylla heterophylla, Lindernia ciliata, Hemigraphis hirta, Hygrophila difformis, Justicia simplex, Phyla nodiflora, Alternanthera paronychioides, Alternanthera sessilis, Polygonum barbatum, Polygonum orientale, Rumex dentata, Hydrilla verticillata, Eichhornia crassipes, Monochoria hastata, Commelina benghalensis, Cyanotis cristata, Typha domingensis, Ipomoea aquatica, Colocasia esculenta, Scirpus articulatus, Sagittaria guyanensis, Najas graminea, Cyperus alopecuroides, Cyperus haspan, Fimbristylis cymosa, Scirpus supinus, Phragmites karka, Sporobolus tremulus, Paspalum distichum, Perotis indica, Panicum repens, Leersia hexandra* and others. In case of tidal water canals some species of *Excoecaria agallocha, Acanthus illicifolius, Avicennia alba, Flagellaria indica, Acrostichum aureum, Cryptocoryne ciliata* and Water fern, *Ceratopteris thalictroides* are common. Some of the common vegetables are *Pisum sativum, Solanum tuberosum, Momordica dioica, Amorphophallus campanulatus, Brassica oleracea* var. *gongylodes, Brassica oleracea* var. *botrytis, Brassica oleracea* var. *capitata, Colocasia esculenta, Momordica charantia, Momordica cochinchinensis, Cucurbita pepo, Daucus carrota, Benincasa hispida, Trichosanthes anguina, Dioscorea alata, Luffa acutangula, Lycopersicon esculentum, Abelmoschus esculentus, Luffa cylindrica, Spinacia oleracea, Allium cepa, Vigna sinensis, Solanum melongena, Beta vulgaris, Cucurbita moschata, Raphanus sativas, Capsicum frutescens* and others.

(b) **Diversity in dispersed village habitat:** Vegetation of dispersed habitat where village areas are mainly dominated by paddy fields and scattered hutments are: *Oryza sativa* and other varieties, *Leptochloa chinenisis, Panicum paludosum, Paspalidium punctatum, Sporobolus tremulus, Sacciolepis indica, Croton bonplandianum, Leersia hexandra, Echinochloa colonum, Cyperus iria, Cyperus exaltatus, Lemna perpusila, Pistia stratiotes, Murdania nudiflora, Utricularia aurea, Utricularia stellaris, Crinum asiaticum* and others in and around the paddy fields. Among the roadsides and surrounding the hutments are commonly dominated by dispersed trees either planted or natural such as: *Eucalyptus citriodora, Ficus hispida, Ficus benghalensis, Trema orientalis, Alstonia scholaris, Albizia lebbeck, Azadirachta indica, Lannea coromandelica, Pithocellobium dulci, Acacia auriculiformis, Moringa pterigosperma, Cocos nucifera, Phoenix sylvestris, Borassus flabellifera, Tamarindus indicus, Zizyphus mauritiana, Areca catechu* and others.

5. Diversity of Plant Communities in Semi Urban Towns

Diversity along the semi-urban town associated with roads, gardens, orchards, ponds, canals and markets are:

(a) **Gardens and Orchards:** *Thevetia nerifolia, Hibiscus rosa-sinensis, Tecoma grandis, Mimusops elengi, Tabernaemontana coronaria, Malvaviscus conzattii, Plumeria alba, Aegle marmelos, Annona reticulata, Annona squamosa, Achras zapota, Averrhoa carumbola, Mitsa ornata, Psidium guyjava, Carica papaya, Mangifera indica, Cocos nucifira, Borassus flabellifer, Syzygium jambos* and others.

(b) **Ponds, canals and ditches:** *Nymphea nouchali, Nymphea stellata, Nelumbo nucifira, Ipomoea aquatica, Marsilea minuta, Oxalis corniculata, Ludwigia perennis, Sphenoclea zeylanica, Bacopa monnieri, Utricularia stellaris, Hygrophila polysperma, Crinum asiaticum, Eichhornia crassipes, Blyxa octandra, Pistia stratiotes, Lemna perpusila, Spirodela polyrrhiza, Najas graminea, Hydrilla verticillata, Typha domingensis, Phragmites karka* and others.

(c) **Roads, houses and markets covering common trees are:** *Polyalthia longifolia, Ceiba pentandra, Delonix regia, Aegle marmelos, Limonia acidissima, Ailanthus excelsa, Azadirachta indica, Mangifera indica, Swietenia macrophylla, Lannea coromandelica, Dalbergia sissoo, Bauhinia varigata, Pithocellobium dulci, Albizia lebbeck, Prosopis chinensis, Samanea saman, Terminalia arjuna, Syzigium cumini, Alstonia scholaris, Eucalyptus citriodora, Ficus religiosa, Ficus benghalensis, Casuarina equisetifolia, Phoenix sylvestris, Cocos nucifera, Borassus flabellifer* and others.

6. Diversity of Plant Communities on Wetland habitat

(a) **Diversity of Fish Ponds (bheri etc.) without embankment plantation:** *Anabaena circinalis, Oscillatoria limosa, Nostoc punctiforme, Oedogonium* spp., *Hydrilla verticillata, Ceratophyllum demersum, Enteromorpha compressa, Najas graminea, Spirogyra* spp., *Nymphoides indica, Salvinia natans, Schoenoplectus maritimus, Cladophora glomerata, Chara brachipus, Eichhornia crassipes, Eragrostis nutans, Paspalum punctatum* and others.

(b) **Diversity of Fish Ponds (bheri etc) with embankment:** *Lemna perpusila, Hydrilla verticillata, Ceratophyllum demersum, Najas graminea, Salvinia cuculata, Pistia stratiotes, Eichhornia crassipes, Azolla pinnata, Spirodella polyrhiza, Nyphoides indica, Paspalum vaginatum, Phragmites karka, Marsilea minuta* along with *Chaetomorpha, Oedogonium, Spirogyra* species. Embankment are commonly planted with *Eucalyptus citridora, Acacia auriculiformis, Samanea Saman* and *Acacia nilotica* subsp. *indica* and others.

(c) **Diversity on the embankments of tidal rivers, rivulets or sea sides are:** *Casurina equisetifolia, Eucalyptus citridora, Derris indica, Thespesia populnea, Prosopis juliflora, Avicennia alba, Avicennia officinalis, Excoecaria agallocha, Acanthus ilicifolius, Caesalpinia major, Derris scandens, Clerodendrum inerme, Cryptocoryne ciliata, Salvadora persica, Sonneratia caseolaris, Aeluropus lagopoides, Heliotropium curassavicum, Sesuvium portulacastrum, Tamarix indica, Cocos nucifera, Phoenix sylvestris, Datura metel, Carissa spirarum, Calotropis procera, Chrozophora plicata, Crateva adansonii, Ficus religiosa, Opuntia dillenii, Premna corymbosa, Tinospora cordifolia, Vitex negundo* and others.

(d) **Diversity in fish ponds inside the village area with small settlement:** *Ipomoea aquatica, Ceratopteris thalictroides,*

Echinochloa colonna, Eclipta prostata, Eichhornia crassipes, Heliotropium indicum, Hydrolea zeylanica, Hygroryza aristata, Limnophila indica, Lemna perpusilla, Ludwigia perennis, Murdania nudiflora, Nymphaea nouchali, Nymphaea pubescens, Persicaria berbata, Phragmites karka, Pistia stratiotes, Polygonum plebeijum, Schoenoplectus articulatus, Salvinia cuculata, Vallisneria spiralis, Colocasia esculenta, Commelina benghalensis, Sphenoclia zylanica and others.

(e) **Diversity in other fresh water ponds ditches and canals:** *Acorus calamus, Aeschynomene aspera, Alloteropsis cimicina, Alternanthera sessilis, Ammania baccifera, Aponogeton appendiculatus, Arundo donax, Baccopa monnieri, Bergia ammannioides, Bergia capensis, Blyxa octandra, Canna indica, Centella asiatica, Centipeda minima, Coldenia procumbens, Commelina diffusa, Cryptocoryne retrospiralis, Cyperus iria, Cyperus rotundus, Dentella repens, Drosera peltata, Eichhornia crassipes, Enhydra fluctuans, Eriocaulon quinquangularae, Grangia maderaspatana, Centella asiatica, Hygrophila auriculata, Lindenbergia muraria, Lindernia ciliata, Lindernia crustacea, Lindernia oppositifolia, Ludwigia adscendens, Monochoria hastata, Monochoria vaginalis, Nelumbo nucifera, Neptunia oleracea, Nymphoides hydrophylla, Nymphoides indica, Ottelia alismoides, Oxystelma esculentum, Persicaria glabra, Persicaria hydropiper, Phragmites karka, Phylla nodiflora, Rotala pentandra, Sagittaria sagittifolia, Sphaeranthus indicus, Sphenoclea zeylanica, Spilanthes calva, Tonningia axilearis, Trapa natans* var. *bispinosa, Typha domingensis, Veronica anagalis-aquatica, Vetiveria zizanioides, Xyris indica* and others.

7. Diversity of Plant Communities on Rivers

(a) **Diversity on Open stretch of flowing water with tidal influence:** *Najas graminea, Potamogeton crispa, Potamogeton nodosus, Schoenoplectus maritimus, Hygrorhyza aristata, Nymphoides hydrophylla, Phragmitis karka, Ruppia maritima, Enteromorpha intestinalis, Cladophora glomerata* and others.

8. Diversity of Common Crops, Pulses and Vegetables

Hordeum vulgare, Sorghum vulgar, Zea mays, Eleusine coracana, Echinochloa colonum, Cajanas cajan, Dolichos biflorus, Lathyrus sativas, Cicer arietinum, Pisum sativum, Phaseolus aconitifolius, Phaseolus aureus, Lens culinaris,

Apium gravelens, Solanum tuberosum, Zingiber officinalis, Momordica dioica, Oxalis tuberosa, Amorphophallus campanulatus, Brassica oleracea var. *gongylodes, Brassica oleracea* var. *acephala, Ipomoea hederacea, Colocasia esculenta, Momordica charantia, Momordica cochinchinensis, Cucurbita pepo, Coccinia grandis, Artocarpus communis, Persea americana, Daucus carrota, Allium porrum, Glycine hispida, Benincasa hispida, Trichosanthes anguina, Dioscorea alata, Luffa acutangula, Rumex vesicarius, Lycopersicon esculentum, Manihot esculenta, Ficus carica, Abelmoschus esculentus, Luffa cylindrica, Amaranthus caudatus, Portulaca oleracea, Trichosanthus dioica, Spinacia oleracea, Allium cepa, Brassica oleracea* var. *botrytis, Amaranthus oleraceus, Vigna sinensis, Vicea faba, Brassica oleracea* var. *capitata, Chenopodium album, Solanum melongena, Beta vulgaris, Nasturtium officinale, Canavalia gladiata, Cucurbita moschata, Raphanus sativas, Momordica balsamina, Allium sativum, Ipomoea batatus, Capsicum frutescens.*

9. Diversity of Phytoplankton Communities

Abundance of different phytoplankton species in different locations in Sundarbans varies in monthly counts. Maximum number of species concentration are found during the pre monsoon and minimum during the post monsoon. Species richness are found maximum during monsoon and minimum during pre-monsoon months. In total 33 phytoplankton species are commonly recorded in the Sundarbans, *viz.*, *Skeletonema* sp., *Coscinodiscus excentricus, Coscinodiscus jonesiamus, Coscinodiscus perforatus, Coscinodiscus asteranphalus, Coscinodiscus coneinnus, Lauderia annulata, Rhizosolenia stotterfothii, Rhizosolenia robusta, Rhizosolenia setigera, Rhizosolenia alata, Bacteriastrum hyalinum, Bacteriastrum varians, Bacteriastrum cosmosum, Bacteriastrum princeps, Chaetoceros curvisetus, Chaetoceros iibenii, Chaetoceros peruvianus, Chaetoceros lorenzianus, Chaetoceros mindicus, Eucampia* sp., *Ditylum brightwell, Ditylum sol, Biddulphia sinensis, Biddulphia mobiliensis, Hemidiscus cuneiformis, Diotoma vulgare, Thallassionema mitzschioides, Thallassiothrix frauenfeldii, Pleurosigma elongtum, Pleurosigma normanii, Bacillaria paradoxa, Nitzschia seriata*

10. Diversity of Bacteria and Fungi Communities

A wide diversity of bacteria and fungi have been isolated from the Sundarbans substrate and from the decomposed mangrove litter (Bhaumik et al. 1985). About 48 bacterial strains have been isolated from the Sundarbans. Some of the common bacteria are – *Bacillus alvei, Brevibacterium lypolyticum,*

B. marinopiscosum, B. minutiferula, B. sociovivum, B. stationis, Clostridium carnis, C. pectinivorum, Kurthia besgonii, K. zopfi, Lactobacillus brevis, L. delbrueckii, L. fermentii, L. plantarum, Listiria monocytogenes, Micrococcus agilis, M. varians, Pseudomonus sp. and others.

Among the fungi the following genera are very common such as *Aspergillus, Collectotrichium, Fusarium* and *Helminthosporum*. The genus *Aspergillus* plays a major role in the degradation processes. It is found that the genus *Aspergillus* comprised about 75% of the total fungal population while the genus *Helminthosporium* comprises remaining 25%.

11. Diversity of Lichen Communities

Along the inner estuarine mangroves, towards supratidal regions, lichen species are found commonly on *Ceriops decandra, Excoecaria agallocha, Xylocarpus granatum* and many other mangroves species and on old branches of the coastal trees. Total 23 lichen species are reported from the Sundarbans as follows:

Arthonia antillarum, A. arnorme, A. cinnabarina, A. subgyrosa, A. subvelata, Buellia lauri-cassiae, Caloplaca encephalarii, Chiodecton micrographum, Dirinaria consimilosis, Graphina leuconephala, Laurera madreporiformis, Lecanactis premnea, Lecidea granifera, Melaspilia insitiva, Opegrapha bonplandiae, O.martii, O. valgata, Renodia colobina, R. intrusa, Schismatomma glaucomoides, S. kurzi, Trypethelium eluterae, T. tropicum.

12. Diversity of Seagrasses and Seaweeds

(a) **Diversity of Seagrass:** Seagrasses are marine submerged angiosperms, successfully adapted in saline water belonging to three monocotyledonous families, Hydrocharitaceae, Ruppiaceae and Potomogetonaceae. They grow on muddy, sandy, clayey, coral rubble bottoms and on rock-crevices under water, forming a thick and dense meadows which are very potential for various marine fauna, fishes, prawns, crabs etc. for food, shelter and breeding.

There are 14 species distributed in Coromandel coast, and Andaman and Nicobar group of Islands. In the Sundarbans diversity of sea grass community is very rare and intensive survey along the estuarine mouths was not possible due to presence of Royal Bengal Tigers. However, only two species such as *Ruppia maritima* and very rarely *Halodule uninervis* are reported.

(b) **Diversity of seaweeds and benthie algal communities in Sundarbans:** When the coast and intirtidal areas are happen to be flat and gradually sloped towards the sea and completely submerged during high tide and successive exposure during low tide, that offer favourable habitat for algal population. The best growth is influenced where clarity of sea water is high and penetration of sunlight is more. Diversity of algal communities in the Sundarbans is very low due to the flow of turbid water in comparison with Indian West coastal algal diversity. Distribution extends from muddy shore to mud flats and sloped intertidal creeks and inlets in the Sundarbans. Some of the algal diversities of the Sundarbans are *Boodleopsis sundarbanensis, Bostrychia tenella, B. radicans, Catenella nipae, C. repens. Chaetomorpha aurea, C. brachygona, C. gracilis, Cladophora echinus, C. sundarbanensis, Enteromorpha calthrata, E. compressa, E. intestinalis, E. prolifera, Galoglossa leprieurie, Rhizoclonium hookeri, R. riparium, R. grande, Ulva fasciata, U. lactuca, U. patengensis, Vaucheria prescotii, V. pronosperma* and others.

13. Zooplankton Diversity

The Sundarbans support a diverse array of Zooplankton communities. They comprises a heterogenous assemblage of animals covering many taxonomic groups such as Copepods, Myside, Lucifers, Gammarid, Amphipods, Cladocerae, Ostracods, Cumacea, Hydromedusae, Ctenophores, and Chactognaths. Spatial and temporal distribution of Zooplankton show definite seasonal trends associated with fluctuation in environmental parameters. They play an important role in the trophic structure and in energy transfer. Copepods show a major proportion among the zooplankton population. About 70-90% of copepods are found in the total count of zooplankton. Information on the distribution of zooplankton in different parts of the Sundarbans is available from the research of Kolkata University, Department of Marine Science. Recorded zooplankton from the Sundarbans are – Copepod – *Paracalnus* spp., *Acrocalnus* spp., *Acartia* spp., *Acartia spinicauda, Acartiella swwelli, Acartiella keralensis, Labidocera euchaeta, Labidocera minuta, Labidocera* spp., *Pontella andersoni, Pontellopsis herdmani, Eucalanui* spp., *Pseudodiaptomus annandalei, Pseudodiaptomus hickmeni, Pseudodiaptomus aurivilli, Pseudodiaptomus tollingeri, Pseudodiaptomus masoni, Pseudodiaptomus daughlishi, Pseudodiaptomus binghami, Pseudodiaptomus* spp., *Pseudodiaptomus strigilipes, Euchaeta marina, Euchaeta wolfendeni, Euchaeta* spp., *Centropages dorsispenatus, Centropages furacatus, Temora turbinata, Temora discaudata, Toranus gracinis, Tortanus forcipatus, Candacia bradyi, Canthocalanus pauper, Undimula darwini, Oithona* spp., *Saphirella indica, Corycasus danae, Corycasus agilis, Corycasus catus, Halicyclops tenuispina, Mesocyclops* spp., *Cyclops* spp., *Oncaea venusta, Microsetalla rosea, Laophonta* spp., *Euterpina acutifrons, Tachidius discipes, Cladorstrata brevipoda*; Caetognath – *Sagitta bedoti, Sagitta enflata*; Mysid - *Mesopodpsis orientalis*; Sergestid – *Lucifer hanseni, Gammarid amphipods* (unidentified); Cumacea – *Paradiastylis* sp.; Cladocera – *Evadne* spp.; Ostracoda – *Philomedes* sp., *Hydromedusae* (unidentified); Ctenophora – *Pleurobrachia globosa, Beroe* spp.

14. Diversity of Benthic fauna

Benthic fauna is present in different mud flats and sand bars of the mangrove forests in the Sundarbans. They produce millions of larvae which support fish population and also make healthy condition of the ecosystem by making burrows for air and water exchange to the soil. It is found that majority of intertidal macrofauna prefer brackish water conditions and develop stenoeuryhaline adaptation for with standing the environmental tress.

Benthic macro fauna in the Sundarbans show marked variation in structure and composition due to seasonal interactions of different ecological factors such as soil, salinity, tide, erosion, accretion and temperature. Common macrofauna represented in these area are: Cnidaria, Nemertine, Annelida, Polychaetes, Isopods, Amphipods, Insects, Echiuroids, Shrimps, Stomatopods, Sea anemones, Decapods, Gastropods, Bivalves, Holothuroids, Gobioid fishes, Hemichordate, and Sipunculids.

(a) **Sea anemones** like *Pelocoetes exul, Phytocoeteopsis gangeticus, Actiniavas* sp. are found to the mangrove mud or on fallen logs.

(b) **Benthic crabs:** These constitute the largest group along the intertidal mud flatls or sandy seashore. There are seven species reported from the Sundarbans, such as *Dotilla blanfordi, Illyoplex gangetica, Macrophthalmus pectinipes, Metaplax crenulata, M. intermedia, Scylla serrata*, and *Uca dussumirri*. They are found along the lower intertidal zone of the mangrove forest. Along the mid-intertidal zone the species like *Sesarma chiromantes* and *S. taeniolatum* are dominated with *Uca acuta, Euricascinus grandiduri, Metapograpsus messor, M. maculata* and others. Population density show highest in the lower level of the intertidal belt and decreases towards the high water level.

(c) **Hermit crabs:** *Clibanarius padavensis, Diogenes avarus, Coenobita cavipes* are common hermit crabs found in the mangrove intertidal zone. They use to live for shelter on the shells of various molluscan species like *Telescopium telescopium*, *Cerithidea* sp., *Nerita* sp. and others.

(d) **Fiddler crabs:** The genus *Uca* belong to this group. Sediments in the mangrove forests or on sea shore with high sand particles are the most favourite place for distribution of this genus. There are four species such as *Uca acuta, U. lacteal, U. dussumuri* and *U. triangularis*. Population is highest during post-monsoon period and decreased from pre-monsoon to monsoon periods.

(e) **Estuarine edible crabs:** *Scylla serrrata* is the most favourite edible crab which not only provides a very delicious dishes but help the mangrove communities for Nitrogen fixation. The male crabs attain about 350 gm in weight and female crabs attain about 250 gm or less. Maturity in different depth of sediments shows that females mature earlier than the male.

(f) **Horseshoe crabs:** These group is very important for various medicinal values. In the sundarbans mud flats and muddy estuarine regions of the Bay of Bengal, three species of Horseshoe crabs, *Xiphosura polyphemus, Carcinoscorpius rotundicaudata* and *Tachypleus gigas* are commonly found during the late February to March. They are usually found breeding under the estuarine muddy layer with large sized female and smaller sized male on the back of the female.

(g) **Diversity of Benthic Gastropods and Bivalves:** Most gastropods and bivalves are primary consumers in the trophic levels of the mangrove ecosystem, being either grazers or suspension feeders. Filter feeders are also there among bivalves. They are exposed along the muddy substrate of the mangrove forests to a large number for the predators inhabiting in the Sundarbans. Some of the predators including fish, birds and mammals depend entirely upon this molluscs for their survival. They are also used for making poultry food and preparation of lime.

In total 61 species of gastropods and many number of bivalves are found in the Sundarbans ecosystem. Both numerically and in term of biomass, gastropods and bivalves constitute a predominant group in the mangrove communities. The species, *Cerithedia cingulata* and *Assiminea brencula* are found very dominant and common. Some species of *Neritina* and *Littorina* are found only in the mangrove forests. Among the 12 genera of bivalves, some of the commercially important species are *Dosinia trigona, Anadara granosa, Meretrix meretrix* and *Crossastrea gryphoides*. They are sedimentary epibenthos and found in submerged areas.

(h) **Diversity of Benthic polychaetes:** The deltaic Sundarbans appear to harbour many Polychaete but very little attention has been given on these group. Studies indicate that there are 30 species so far recorded from the Sundarbans and most common of which are Nereids, Eunicids, and Capitellids. The tubiferous polychactes, *Diapatra* sp. *Spio* sp. *Marphysa* sp. *Lumbrineris* sp. and *Laonome* sp. are found common along the mangroves and mangrove fringes of the Sundarbans.

(i) **Diversity of Arboreal and Aerial Invertibrates:** There are many arboreal invertebrates found common on the tree trunk, rotten logs, tree stump, old branches, woods, pneumatophores, stiltroots etc. *Littorina scabra* occupies 2-3 m length on the tree. Gastropods like *Onchidium tigrinum, Nerita articulata* are found on tree stump or on logs. *Onchidium* sp. Move on the tree with the movement of tide water. *Nerita semirugosa* are found attached to the stilt root of *Rhizophora* and Pneumatophores of *Brugueria*,. The bivalves, *Crassostrea cucullata*, and *C. gryphoides*, are found attached to the woods. Other arboreal mollusks in the Sundarbans are *Cerithidua ablusa, Littorina melanostoma, Ellobium arisjudae Nerita violacea*, etc. The *Sesarma* crabs are commonly found on trees. Among bivalves, *Crassostrea* sp. and *Teredo* sp. are very common in the Sundarbans. Among the aerial invertebrate, Butterflies, Moths, Dragonflies, Damselflies, Bees, Wasps, Fire flies, Ants, Beetles and Mosquites are very common in the Sundarbans.

The honey bee, *Apis dorsata,* migrates to the vast tidal forests of the Sundarbans during the months of March to June every year. Numerous swarms of bees start collecting nector from the flowers of different mangrove species of the genera, *Aegiceras, Acanthus, Ceriops, Phoenix, Sonneratia, Excoecaria* etc. and build huge hives, which are the source of commercial honey and wax in the Sundarbans. Total production of honey in the Sundarbans is about 400 to 700 quintals per year and wax is 500 to 600 quintals.

15. Diversity of vertebrate Fauna

Mangrove ecosystem of the Sundarbans forests with swamps, water ways, tidal creeks, estuaries, various muddy and sandy flats provide important habitat, nursery and breeding grounds and migratory places for rich terrestrial, aquatic and avifaunal species. It is estimated that vertebrate diversity of the Sundarbans includes about 8 species of amphibians, 250 to 6000 species of fishes, 58 species of reptiles, 196 to 300 species of birds and 40 species of mammals.

(a) **Diversity of Reptiles in the Sundarbans:** Among the 57 species of reptiles, 12 species belong to the order *Chilonia* (Tortoises), 44 species belong to *Squamata* (snakes and water monitor) and one species belong to the order *Crocodilia* (Crocodile).

In the *Squamata* group, some representative species are: *Chamaeleon zeylanicus* – the Indian Chamaeleon, *Varanus salvator* – Water monitor, *V. flavescens* – yellow monitor and many terrestrial and aquatic snakes like *Dendrelaphis tristis* – Bronze back or Tree snakes, *Ahaetulla nasuta* – Green whip snake, *Lycodon aulicus* – Wolf snake, *Acrochordus granulatus* – File snake, *Cerberus rhynchops* – sea snake or Dog-faced water snakes and Indian Python, Russell's sand boa, Rat snake, Keel back, Krait, Cobra and Vipers are common.

Among the *Chilonia* group, some representative species in the Sundarbans are *Batagur baska* – the endangered river terrapin, *Lepidochelys olivacea* – Olive Ridley turtle, *Trionyx gangetica* – Ganges soft shelled turtle, *T. hurum* – Peacock marked soft shelled turtle, Lissemys punctat – Peacock marked soft shelled turtle.

In the *Crododilia* group only one species known as Estuarine or salt water crocodile, *Crododylus porosus* is dominated through out the 2500 sq. km of the Sundarbans and found common along the Saptamukhi river, Thakuran river, Gona, Naubanki and Mechua Khal river areas. The estuarine crocodile is so dangerous that in many places an average of 4-6 people have been killed annually by this animal in the Sundarbans.

(b) **Diversity of Amphibians:** Among the 8 species of Amphibians, the common occurrence of Frogs and toads in the Sundarbans mangrove forests are *Bufo melanostictus* – common toad, *Rana cyanophlyctis* – skipper frog, *R. limnocharis* – cricket frog, *R. tigerina* – Indian bull frog, R. *hexadactylus* – pond green frog, *Rhacophorus maculatus* – Tree frog, *Microhyla ornata* – Narrow mouth frog.

(c) **Diversity of Fishes:** Many rivers, estuaries, several creeks, channels, and Bay of Bengal along with largest mangrove forests creat an ideal environment for large number of fish species in the Sundarbans.

It provides a good nursery ground for salt water, brackish water and also some fresh water fishes. Total value of fish catch in the Sundarbans exceeds more than 10 cores per year.

About 250 to 6000 species of fish have been recorded in the estuaries, rivers, creeks, ponds and Bay of Bengal of the Indian Sundarbans. Some of the fishes have high degree of Physiological adaptivity expressed in morphological character – such as

Gobid fishes are *Boleopthalmus boddeerti, B. histophorus, Periopthalums pearsei, P. schlosseri.*

Some estuariane fishes are *Gibus chuno, Brachygobius nunus, Gobius alegans, Gobiades rudicaudus* and others.

Among the eel fish, common species are *Anguilla bicolor, Pissodonophis boro, Muraenesox cinereus, Pangasius pangasius* and other species of the genera *Arius, Mystus, Osteogeniosus, Polydactylus* etc. The species of *Plotosus canius, Liza tade, L. parsia, Lates calcarifer, Harpodon nehereu, Eleutheronema tetradactrylum* and others are also common.

Some marine fishes that migrate upriver through the estuaries to breed in fresh water are *Hilsa hilsha, Sillaginopsis panijus, Pama pama, Polynemus paradiseus* and others.

Some species of fresh water migrate to estuaries for breeding are *Pangasius pangasius, Anquilla bicolor, Moringus raitaborua, Muraenesox cinereus* and *Pisodonophis boro.*

Several species of Sharks and rays are also found in the estuaries such as – *Scoliodon* sp. *Pristis* sp., *Sphyrna blochii, Rhinobatus* Sp., *Dasyatis zagei* and *Aetobatis* sp.

(d) Diversity of Birds in the Sundarbans: Extensive water ways, large mangrove forests and many mud flats create a beautiful habitat of many avian species in the Sundarbans. Sajnakhali in the Sundarbans is an important wild life sanctuary for birds. About 196 to 300 species are recorded from the Sundarbans including many migratory birds. During the onset of winter, November-December, more than 100 migratory birds visit the sundarbans and take Shelter in the mangrove forests.

Common resident birds like herons, cormorants, egrets, king fishers, storks, darters, openbill stork comprise more than 50% of the total avifauna.

A number of species of eagles and owls such as: Sea Eagle, Crested serpent Eagle, Palla's fishing Eagle, Pariah kite, Shikra, Barn owl, Scops owl, Spotted owlet and Brown fish owl are commonly found in this region. Different species of plovers and sand pipers are common along the intertidal areas. In marshy places herons, egrets, bitterns, storks and rails are seen mostly throughout the year. During winter, migratory flocks of Sand pipers, Red shanks, Whimlocks and others are common. The Swamp parridge and Jungle fowl, the ground dwelling birds are found in the dense forests.

The commonly found water birds of Sundarbans are listed below:

Podiceps ruficollis – Little grebe
Phalacrocorax niger – little cormorant
Anhinga melanogaster – oriental darter
Ardea cinerea – Grey heron.
A. perpurea – Purple heron.
Ardeola grayii - Indian Pond heron.
Butorides striatus – Little heron.
Nycticorax nycticorax – Night heron, black-crowned.
Ixobrychus cinnamomeus – Chestnut bittern.
Bubulcus ibis – Cattle egret.
Egretta alba – large egret.
E. intermedia – Intermediate egret.
E. garzetta – Little egret.
Anastomus oscitans – open bill stork.
Nettapus coromandelians – Pygmy cotton teal.
Gallinula cholorpus – Eurasian moorhen.
Metopidius indicus – Bronze-Winged jacana.
Vanellus indicus – Red-Wattled Lapwing.
Tringa glareola – Wood sand piper.
Calidris minuta – Little stint.
Chlidonia hybridus – Whiskered tern.
Ceryle rudis – Pied kingfisher.
Haleyon smyrnensis – White-brested kingfisher.

(e) **Diversity of Mammals in the Sundarbans:** Mammals of the Sundarbans are very uniquely adapted for survival to this ecologically stressed environment where the Royal Bengal tigers are the largest single surviving population on the wet land mangrove forests in the World. Numerous species of spotted deer, wild boar and rhesus macaque are abundant. Leopard cat, Fishing cat, Mongoose and Rodents are common but crested porcupine and Chinese pangolin are rarely seen. Indian flying fox, fruit bats, greater false vampire, Indian pigmy pipistrelle, lesser yellow bat, Cave bat and squirrels are found common. Among the aquatic marine mammals Plumbeous dolphin, Malay dolphin, Irrawaddy dolphin, Gangetic dolphin and black finless porpoise are more or less common along the bay, estuaries, rivers and creeks.

Total 40 species of Mammal diversity reported from the Indian Sundarbans are as follows:

Panthera tigris – Tiger
Felis viverrinus – Fishing cat
F. chaus – Jungle cat
F. bengalensis – Leopard cat
Canis aureus – Golden jackal
Vulpes bengalensis – Indian fox
Lutra lutra – Estuarine otter.
Viverra zibetha – Large Indian civet
Paradoxurus hermaphroditus – Common palm civet.
Herpestes edwardsii – Indian Grey mongoose
H. javanicus – small Indian mongoose
H. palustris – Marsh mongoose
Neophocaena phocaenoides – Black finless porpoise
Orcaella brevirostris – Irrawaddy dolphin
Stenella attenuata – Pantropical spotted dolphin
Platanista gangetica – Gangetic dolphin
Macaca mullata – Rhesus macaque
Sus scrofa – Wild pig
Axis axis – spotted deer
Funambulus pennanti – Northern palm squirrel
Bandicota indica – Greater bandicoot rat.
B. bengalensis – Lesser bandicoot rat.
Mus musculus – House mouse
M. booduga – Indian field mouse
Rattus rattus – House rat

Hystrix indica – Indian porcupine
Suncus murinus – House shrew
Manis pentadactyla – Chinese pangolin
Cynopterus sphinx – Short-nosed fruit bat.
Pteropus giganteus – Indian flying fox
Rousettus leschenaulti – Leschenault's rousette
Megaderma lyra – Greater false vampire
Rhinopoma hardwickei – Lesser mouse-tailed
Coelops frithi – Tailless leaf-nosed bat
Hipposideros lankadiva – Cave bat
Hesperoptenus tickelli – Tickell's bat
Pipistrellus coromandra – Indian pipistrelle
P. tenuis – Indian pygmy pipistrelle
Scotophilus kuhlii – Asiatic leser yellow bat

16. Rare, threatened and extinct Biodiversity of the Sundarbans

Vertebrate: A number of species have become extinct, and others are in danger of extinction in the Sudnarbans due to habitat degradation, coastal development and other factors including climate change, logging and agriculture. There is inadequate information regarding the fate of some smaller invertebrate. Further research is urgently needed to investigate the future status of the vertebrate species for faunal diversities of the Indian Sundarbans. Some of the vertebrate species which are in danger for extinction are as follows:

REPTILES	*BIRDS*
Crocodilus porosus	*Pelecanus phillipinensis*
Varanus bengalensis	*Theskiornis melanocephalus*
Varanus salvator	*Leptoptilos javanicus*
Varanus flavessens	*Serkiodornis melanotus*
Eretmochelys imbricata	*Carina scutulata*
Caretta caretta	*Platanista gangetica*
Dermochelys coriacea	
Troonyx gangeticus	
T. hurum	
Batagur baska	
Python molurus	

Mammals

Panthera tigris – Not extinct but in danger.
Muntiacus muntjac - Now not found
Bubalus bubalis - Now not found
Rhinoceros sondaicus - Now not found
Cervus deruchea - Now not found
Axis porcinus - Now not found

17. Rare, extinct and endemic plant species in the Sundarbans

Endemic Taxa: Global distribution of mangrove species indicates that there are 10 species such as *Rhizophora mangle, R. racemosa. R. samoensis, Avicennia rumphiana A. integra, A. schaueriana, Conocarpus erectus, Laguncularia racemosa, Osbornia octodonta* and *Pelliciera rhizophora* are restricted along the New World group including North, South and Central America and West Africa. None of these 10 species are found in the Old World. Within the old world, more than 65 mangrove species distribution originated from Persian Gulf-Madagascar region, Indo-Malaysian region and Australian region. This Indo-Malaysian group includes mangroves of Pakistan, Bangladesh, India, Myanmar, Malayasia, Indonesia, N. Australia, Celebes and Papua New Guinea.

Endemism of Indian sub group in the Sundarbans mangrove forest is very low. *Sonneratia apetala, Heritiera fomes* and *Phoenix paludosa* are found major endemic species of the Sundarbans mangroves. Their distribution towards the other mangrove region of India is very much restricted.

Rare, threatened, endangered and extinct mangroves: Recently IUCN, in a Press Release on 9th April 2010, reported that 'More than one in six mangrove species World wide are in danger of extinction due to coastal development and other factors, including climatechange, logging and agriculture.' As a result, 11 out of 70 mangrove species (16%) which were assessed will be placed in the IUCN Red list. In the Sundarbans where the habitat itself is very vulnerable due to sensitive and fragile ecosystem, majority of the mangrove species are rare and threatened. However, the following mangrove species must be considered in danger for extinction:

Dalbergia spinosa *Amoora cuculata*
Finlaysonea obovata *Cynometra iripa*
Heritiera fomes *Sarcolobus globosus*
Intsia bijuga *Thespesia populneoides*

Merope angulata *Tylophora tenuis*
Kendelia candel *Sonneratia griffithii*
Mucuna gigantia and *Acanthus volubilis*

Some of the orchids which were once reported by Prain (1903) from the Sundarbans are now extinct such as - *Acampe rigida, Dendrobium anceps, Luisia brachystachys, Oberonia gammiei, Saccolabium ochraceum, Sarcanthus insectifor*, and *Trias oblonga*

CHALLENGES

Various field studies reveals that 35% of the total floral diversities and 45% of the fanual diversities in the Sundarbans are in danger for extinction due to wide range of influences of many factors such as over exploitation, demographic pressure, loss of habitat due to coastal development, frequent natural calamities, massive erosion of the river bank, global climate changes, increase of water salinity and reduce inundation due to irregular construction of embankment, collection of shrimps and development of aquaculture ponds, formation of saline blanks, oil pollution due to inter national water ways, offshore drilling and large number of mechanized boats regularly passing through the forest areas. All these above factors are the real challenges and how to restrict those for conservation and management of the vast mangrove forests are the problems. The potential loss of flora and fanual species in an ecosystem is a symptom of wide spread destruction and exploitation of the mangrove forests. Biodiversity of mangrove forests form one of the most important tropical and sub tropical habitats that support many species and their loss can affect marine and terrestrial biodiversity much more widely. Some of the most important challenges and problems of mangrove degradation with possible restoration measure are discussed as follows:

Demographic pressure: Increased population and land-need pressure for resettlement, agriculture, fishing harbours, construction of bheries, development of new tourist camps etc. have led to the serious exploitation of mangrove forests specially along the Sagar island, Namkhana, Gosaba, Basanti, Canning, Dhamakhali and many other places. Population has been increased 6-8 times and their requirements of fuel woods, house building and thatching materials, timber etc. have also been increased within the limited forest areas. It is estimated that more than 80% mangrove areas have been cleared up during 15 to 20 years. As these areas are influenced by large tidal surges, high wind speed and cyclones; sudden flooding and loss of lives have become almost a seasonal event.

***Restoration*:** With the help of administrative control and useful planning these may be restored.

Degradation of future fish yield: For quick economic benefit large number of people are engaged in shrimp collection and during this time, they killed and throw billions of other fish seeds causing a serious damage of future fish yield in the Sundarbans.

Development of aquaculture ponds: Construction of aquacultural ponds for brackish water shrimp culture damages many mangroves and also creates a serious impact upon the existing ecology.

***Restoration*:** Both the shrimp collection and construction of aquaculture ponds should be stopped from the administration.

Degradation of mangroves due to erosion: The Gangetic delta seems to be tilted from the west towards the eastern region due to Neo-tectonic movement and that has created enormous deposition of sediment load to the river beds of the two large rivers Matla and Thakuran. The western part of the Sundarbans is therefore, suffering with severe erosion, resulting degradation of mangrove vegetation in different forest blocks. Along the forest blocks of Herobhanga, Ajmalbari, Chulkati, Dhulivasani and other places where erosion dominates, chunk of tidal flats are subsidizing with the mangrove vegetation and settlements. Here the intertidal flats are disappearing rapidly by cliff erosion at the edge of mangrove lands, by sheet erosion across the entire tidal surface and by creek erosion from the sides. As erosion proceeds, mangrove communities face vertical and horizontal fall and move progressively land wards where ecological conditions have been changed due to high elevation and lack of regular tidal flow. These polyvalent ecological nature create such a constrain and unfavourable conditions that all the mangrove tree species survive in a dwarf and thickly populated pattern. They survive there spending most of their energy just to keep them alive without much being left for growth and development.

***Restoration*:** With the help of artificial barrier, speed of erosion can be checked and massive plantation along the eroded sites can restore the situation.

Formation of saline blanks/Saltmarshes: Salt marshes usually found in changed ecological situation where the areas become hyper saline and suffer from regular drainage pattern. In the Sundarbans, saltmarshes are found on the inner elevated edges of the mangrove forests, within the middle elevated parts of the Swamps or as a secondary formation on cut-over or degraded places of the mangrove forests. Some of the back ward part of the old mangrove forest blocks become cut-off from regular inundation where due to high evaporation, soil become hyper saline and the mangroves slowly die away, ultimately entire areas are transformed into salt marshes. Here chemical characteristics of soil conductivity is more than 4 mhos/cm, exchangeable sodium ion is more than 15% and pH is more than 8, compare to that of the mangrove habitat.

***Restoration*:** arrangement of proper drainage system from the rivers or canals to these areas may restore the situation.

Natural calamities: The delta enjoys tropical humid climate with moderate temperature through out the year. Due to sudden fall of pressure in the Bay of Bengal, the area is suffering from frequent high wind speed, storm, earth quake, floods and cyclones almost every year or every alternate year during the months of May-June and again on November December. Last 10 years record reveals that average 12 devastating calamities and damages of mangrove forests, human lives with many properties occurred in the Sundarbans areas.

***Restoration*:** Massive plantation of mangrove species and protection of existing mangrove forests may restore these damages.

Global climate change: Global threat regarding the rise of temperature due to green house drift and sea level rise are well known. Last 40 years data on climate change reveals that glaciers have degraded double the amount, level of sea water rises from 0.14 mm/year to 0.35 mm/year. Increase of average temperature goes from 0.2 to 0.7°C. In these ways, if temperature rises and level of sea water increases 1 meter, then half of the Sundarbans in the lower Gangetic plain would be sub merged under water. Though land building activity of mangrove communities may increase the level of land but is would not be possible to save the mangrove lands and its biodiversities from the impact of climate change in the future.

***Restoration*:** Plantation and protection of mangroves along the wet coast and genetic research for molecular mapping and transplantation of mangrove gene to the crop plants to develop a capability to withstand and resist coastal salinity may restore the conditions to some extent.

Increase of water salinity: Tilting of lower Bengal basin from the West to the East has disturbed the fresh water flow to different distributaries, creeks and canals, more over, volume of fresh water has also been reduced due to construction of Farakka barrage. The Sundarbans areas are now influenced by the range of 3-3.5 gm/lit. Salinity instead of 2.01-2.8 gm/lit. This changes have a tremendous impact on the mangrove species distribution. The dominant species of *Heritiera fomes* and *Nypa fruticans* are now found more fresh water influenced zone. The medium salt loving species, *Rhizophora apiculata* and *R. mucronata* have been found in reduced population but high salt tolerant species, *Avicennia alba* and *A. marina* are now dominating.

Restoration: The arrangement of adequate freshwater flow from the river Ganges and padma can restore the problem.

Awareness and understanding: This part is very important for providing knowledge to the villagers who are using regularly the direct and indirect value of the mangroves for their livelihood. If possible every quarter or every year a workshop on mangroves should be organized in the remote villages for understanding the Sundarbans biodiversities, status and Challenges.

REFERENCES

Ahmed, E. (1972). *Coastal Geomorphology of India*. Orient Longman, New Delhi.

Anonymous, (1983). Global status of mangrove ecosystem. *Commission of Ecology, Paper 3*. IUCN, Switzerland.

Banerjee, L.K. (1992). Mangal formation of Mahanadi Delta: Exploitation and management. *Tropical Ecosystem*. Singh, K.P. and Singh, J.S. (Eds.). Wiley Eastern Limited Publisher, printed in India.

________. (1993). *Influence of Salinity on Mangrove Zonation. Towards Rational use of High Salinity Tolerant Plants*, Vol. 1: 181–186. Lieth, H. and Masoom, A. (Eds.). in Kluwer Academic Publishers, printed in Netherlands.

________. (1994). Conservation of coastal plant communities in India. *Bull. Bot. Surv. India*, 26(1–4): 160–165.

________. (2002) Sundarbans Biosphere Reserve. In: *Floristic Diversity and Conservation Strategies in India*. Vol. 5. *in-situ* and *ex-situ* conservation. Singh, N.P. and Singh, K.P. (Eds.). Botanical Survey of India.

Banerjee, L.K. and Rao, T.A. (1989). *Mangroves of Orissa Coast and their Ecology*. Bishen Singh, Mahendra Pal Singh, Dehradun.

Banerjee, L.K., Sastry, A.R.K. and Nayar, M.P. (1989). *Mangroves in India, Identification Manual*. Botanical Survey of India, Kolkata.

Banerjee, L.K., Ghosh, D. and Sastry, A.R.K. (1998). *Mangroves, Associates and Saltmarshes of Godavari and Krishna Delta–Andhra Pradesh*. ENVIS Centre, Botanical Survey of India.

Banerjee, L.K., Rao, T.A. Sastry, A.R.K. and Ghosh, D. (2002). *Diversity of Coastal Plant Communities of India.* ENVIS Centre, Botanical Survey of India.

Bhaumik, et al. (1985). A list of 24 strains of bacteria from decomposed litter of mangrove area of Sundarbans submitted in Symposium on Biology, p. 38.

Blasco, F. (1977). Outlines of ecology, botany and forestry of the mangals of the Indian subcontinent. *Ecosystem of the World 1: Wet Coastal Ecosystems,* 241–260. Chapman, V.J. (Ed.). New York.

Chapman, V.J. (1976). *Mangrove Vegetation.* J. Cramer. Lehre Fl. 9490. Vsaduz.

Chaudhuri, A.B. and Chaudhuri, A. (1994). *Mangroves of the Sundarbans.* Vol. 1: India, IUCN, Bangkok, Thailand.

Davis, J.H. (1940). The ecology and geologic role of mangroves in Florida. *Carnegie Institution of Washington Publication.* 517: 303–412.

Dey, Partha. (2009). *Changing Climate: Red Alert in Sundarbans.* NEWS, Kolkata.

Fergusson. (1863). Recent changes in the delta of the Ganges. *Qr. J. Geo. Soc.,* 19: 321–354.

Lugo, A.E. and Sneclaker, S.C. (1974). The ecology of mangroves. *Annual Review of Ecology and Systematics,* 5: 39–64.

Mukherjee, B.B. and Mukherjee, J. (1978). Mangroves of Sundarbans, India. *Phytomorphology,* 28(2): 217–229.

Naskar, K.R. and Guhabakshi, D.N. (1983) Sundarbans – the world famous mangrove forest of the District 24 Parganas. *J. Econ. Taxon. Bot.,* 3: 1–883.

_________. (1987). *Mangrove Swamps of the Sundarbans.* Naya Prokash Publication, Kolkata.

Oldham, R.D. (1893). *A Manual of Geology of India.* Calcutta.

Prain, D. (1903). Flora of Sundarbans. *Rec. Bot. Surv. India,* 2: 31–70.

Scoffin, T.P. (1970). The trapping and binding of subtidal carbonate sediments by marine vegetation in Bimini Lagoon, Bahamas. *Journal of Sedimentary Petrology,* 40: 249–273.

Mangrove community in Low Tide

Mangrove community in High Tide

Aegiceras corniculatum (L.) Blanco. in troop – a good source of Honey

Dense population of *Aegialitis rotundifolia* Roxb.

Porteresia coarcata Tateoka – A wild rice in mangrove habitat

Ipomoea pes-caprae (L.) R.Br. – a coastal sandbinder

Nypa fruticans Wurmb. – An endangered palm

Dense population of *Phoenix paludosa* Roxb. – A den of Royal Bengal Tiger

Salicornia brachiata Roxb. – A salt mass in mangrove habitat

UNDARBAN – A habitat for salt water Crocodile

Mangrove species – struggle under stress

Vivipary of *Bruguiera gymnorrhiza* (L.) Savigny

Pneumatophore of *Xylocarpus granatum* Koenig

17

Biodiversity of Oceanic Islands – Andaman and Nicobar: Status and Challenges

K. Venkataraman

INTRODUCTION

The Andaman and Nicobar Islands are the summit of a submarine mountain range lying on the great tectonic suture zone that extends from the eastern Himalayas to the Arakan along the Myanmar border and finally to Sumatra and Lesser Sunda Islands. This archipelago consists of a group of 572 islands, islets and rock outcrops, but there are a total of 352 important islands comprising the main chain of Andaman and Nicobar, Ritchie's Archipelago and the out laying volcanic islands of Narcondam and Barren. The former lies at about 96 km east of Port Cornwallis. It is a cone-like island attaining a height of about 710 m above sea level. The Barren Island still shows signs of volcanic activity and the most recent ones being in 1991 and 1993 is about 128 km northeast of Port Blair. The islands are spread over an area of 8,249 sq km of which 6,408 sq km are occupied by the Andaman group and 1,841 by the Nicobar group of islands. The Andaman group extending from Land Fall Islands in the north to the southernmost point of Little Andaman stretches over a length of about 467 km. The Andaman group has a maximum width of about 52 km extending from the west coast of the Great Andaman to the East in the Ritchie's Archipelago. However, in middle Andaman the maximum width is about 30 km. The average width of the Andaman group is about 24 km. The Nicobar group of islands stretches over to a total length of 259 km and a maximum width of 58 km.

The North Bay of Andaman

The Andaman Sea is a semi-enclosed basin connected to the Bay of Bengal on one side and to the Pacific on the other. The sea has an average depth of 1096 m with the deepest point stretching up to 4500 m. Sea bottom varies with depth for the major part it had hard rock. The shallower depths of < 200 m are dominated by coral reef, coralline debris and coralline sand. The depths between 200 and 1000 m contain sandy mud or clay, rarely interspersed with rock formations (Anon., 1995). The depths beyond 1000 m consisted of mostly clay or hard rocks. The continental shelf is narrow and dips steeply down to a depth of 400 m on the eastern side and 500-600 m on the western side. Surface temperature of the sea water varies between 27°C and 28.5°C. Southern parts of the Andaman Sea have generally warmer waters. Surface salinity varied between 31.2 $\%_0$ to 33 $\%_0$. It also changes with different season. It is reported that there is limited mixing between surface, sub surface and bottom waters. The Andaman Sea is oligotrophic in nature with low primary and secondary productivity. The low productivity of living matter is offset by the production of large quantities of detritus.

The Andaman group consists of 324 islands of which 24 are inhabited while the Nicobar group includes 28 islands of which 12 are inhabited. The physiography of this Archipelago is characterised by undulating topography and intervening valleys. There are however some flat islands like Car Nicobar and Trinket. The principal ranges of hills are in the Andaman group and situated closer to East coast than to the west coast. The highest peak among these hills is Saddle Peak in North Andaman raising to about 732 m above sea level. There are several other smaller peaks in the hilly ranges; Mount Diavolo (503 m) and Mount Angelica (460 m) in the Middle Andaman; Mount Koyob (450 m) in South Andaman. Mount Harriet overlooking the Port Blair harbour raises to about 365 m. There is another peak Mount Hext in the same Mount Harriet range raising to 425 m. Mount Ford in Rutland raises to about 430 m. In comparison, west coast has lesser hilly ranges than the east coast. Mount Oldham in the Middle Andaman is reported with a height of about 310 m. Two peaks on the west coast of South Andaman, namely Mount Cadele and Mount Cholunga raise to about 320 and 325 m respectively. In Nicobar hills over 405 m occur on Tillanchang, Great and Little Nicobar islands Mount Thullier, the highest is in Great Nicobar (852 m). Mount De oban (578 m) and Express Peak (575 m) in Little Nicobar are some important peaks. Car Nicobar is a flat coral covered island. Trinket and Chowra are also flat but the latter has a small table hill in the south.

Andaman and Nicobar Islands have typical tropical temperatures with very little monthly or seasonal variation. The mean annual maximum remained around 30°C with an average mean minimum of 23.6°C. There has been a slight fall in the mean minimum, which was about 24°C in the earlier years. The difference between night and day temperatures is between 8-9°C.

These islands are under the influence of both southwest and northeast monsoons. The former is active from the month of May to October, while the latter is in November – December. Normal rainfall is measured to be 3180mm. The rainfall may vary substantially from year to year by 30% at the maximum levels. During monsoon months, June to September, Port Blair, Long Island and Mayabander (10°N Lat) get more rainfall than those in south. Rainfall in general decreases from north to south. Rainfall during the transitional period, October, November and May is more or less uniform over all the islands. But during December to April it decreases from south to north. During this period southern parts get more rainfall.

Perennial freshwater rivers are not available in these islands, with the exception of Great Nicobar. There is however, several rain fed streams which dry up mostly during summer. All the major islands support a luxuriant growth of evergreen, semi ever green, moist deciduous and littoral forests from water edge to the mountain top depending on the topography and nature of the soil.

These islands have a good forest cover, which extends over 92.3% of the geographical area. Among inhabited islands Little Andaman has 98.8%

forest cover followed by Middle Andaman with 87.57% cover. The forest is categorised into reserved (2929 sq km) and protected (4242 sq km). Tribal reserve constitutes about 38.5% percent of the total. About 14% of the forest area is under revenue settlement. Forest land was cleared for agriculture, plantation and allied purposes and also for human settlement. About 48,000 ha have been cleared for agriculture and allied purposes. It was estimated that about 10,000 ha of mangroove forest has been lost. There are conflicting figures on the agriculture and plantation land use. But the official figures show that 19,910 ha are under cultivation and 24,410 ha under coconut plantation. Paddy is the important crop and cultivated over an area of 12,570 ha. The agricultural holdings in the islands are 10,414 numbers with 10,313 in Andaman group. The population of cattle, buffalo and goats had shown increase. This puts a demand for the increased fodder production. Fodder and feed scarcity are the constraints in livestock productivity.

ETHNIC POPULATION

There were once 12 tribes roaming in the virgin forests of the islands. The establishment of penal settlement in 1858 and subsequent inflow of settlers from the mainland have their impact on the twelve aboriginal tribes. Six have become extinct, five have their population dwindled, only one tribe has grown in population. The demographic profile of tribes portray miserable picture. Five tribes namely Great Andamanese, Onges, Jarawas, Sentinelese and Shompen are represented by a few hundreds and less. The Nicobarese have admirably adapted to changed circumstances and showed great survival instinct. The tribal population constituted 39.2% of the total population of the islands in 1951. The latest census figures of 2001 indicate that the percentage has come down to less than 9.5%.

The settler population had increased between 1951 and 2001, showing an average decadal annual rate of 92.76%. For the same period Nicobar had an average of 34.85%; which reflects the growth in nicobarese population. The Nicobar is still free from anthropogenic pressures, except in pockets where there are settlers and defense establishments. Rural population constitutes about 69% and distributed in 547 small villages. The average size of the settlement is about 408. There is an overall increase in the number of households and villages. Car Nicobar has the highest population density and household size.

For administrative purposes the Islands are divided into two districts, namely Andaman and Nicobar. The former consists of two sub divisors,

Mayabunder which is again divided into three tahsils namely Mayabunder, Diglipur and Rangat, and South Andaman division is further divided into Port Blair, Ferrarganj and Little Andaman. The Nicobar district is divided into Car Nicobar and Nancowry tahsils. There are a total of 204 revenue villages of which 197 are in the Andaman District.

BIODIVERSITY

Flora

The flora in Andaman and Nicobar Islands has immense potential both in terms of ecology and economy. The coastal vegetation plays a major role in maintaining the ecology of the coastal zone. The analysis of plant diversity in the islands revealed the presence of about 3552 species forming 76.5% of estimated species. Indigenous angiosperm constitutes about 2000 species of which 14% are endemic (Fig. 1). The flora in general has affinities to the Indo-Malayan region (Ellis, 1987).

Littoral beach forests include about 400 species of trees, shrubs, herbs and climbers forming impenetrable thickets along the shore. The broad natural zonation is often disturbed by human activities in populated islands. The settlement species of the littoral forest include *Ipomoea pescaprae, Vigna marina, Scaevola sericea, Vitex trifolia etc.* The inner littoral species are *Hernandia peltata*, the giant *Manilkara littoralis*, the massive *Calophyllum inophyllum etc.* Among the species thriving in the coastal zone about 100 species have economic value and are subjected to human pressure. Important coastal plant species, about 76 in number and their economic utility is listed. Of these coastal plant species 7 are endemic while 60 species are common. There are 14 rare species and 2 species namely *Neisosperma oppositifolium* and *Xylocarpus moluccensis* are threatened with exploitation.

Tidal swamp forests in the islands are constituted by about 30 species. Large scale exploitation of germplasm *i.e.* seeds, fruits and associated macrofungi may endanger the natural regeneration capacity of these species. The most important component of coastal vegetation is the occurrence of fringing mangroves in several islands. Occupying about 11.6% of the total geographical area, the mangroves of Andaman and Nicobar Islands are the third largest in India after Sunderbans and Gujarat on the mainland. There are about 50 species of mangroves, semi-mangroves and associate mangroves occurring in these mangrove forests. There has been a decrease area in mangrove forests, which were cleared for human settlement, and for plantation of red oil palm in Little Andaman, besides extraction of timber and fuel wood. Between the years 1981-82 and 1984-85 a total of 73,668 m^3 of wood was extracted from mangroves for charcoal. This and other exploitative activities have reduced the mangrove cover from 1826 sq km in 1957 to 966 sq km in 1998. Recent management practices such as cessation of wood extraction from mangroves, the process of regeneration in some areas and declaration of Wild Life Sanctuaries, National Parks have prevented mangroves from further degradation. The mangroves in the islands are still in a better condition than those on the mainland.

Marine macro-algae (sea weeds) and sea-grasses are the prominent plant communities of coral reef habitats, which contribute a major share to the primary productivity of reef ecosystem. The reef flats and sandy and muddy areas of the 352 islands of Andaman and Nicobar are the important substrata for the growth of algae and sea-grasses respectively. A total of 184 algae and nine species of seagrasses are reported from Andaman and Nicobar Islands. The sea weeds have a great deal of optional and indirect use values other than the direct or commercial use values as sources of food, fodder, fertilizer and chemicals. About 65.3% of the estimated algal diversity has been enumerated up to species level. But there are still gaps in our knowledge on their spatial distribution. Coralline algae are the least known group. Altogether, 16 species of crustose corallines and 8 species of articulated corallines are known to occur in the islands. The scanty ecological information revealed that intertidal flats and shallow areas with sand and mud are ideal for maximum cover of sea grasses. The cover of *Thalassia hemprichii, Cymodocea rotundata* and *Enhalus acoroides* was reported to be more than the other species. The sea grass beds have good fishery potential and habitat for endangered Dugong and Turtle.

Fauna

Andaman and Nicobar Islands have two important tropical ecosystems, forest and marine ecosystems, which are strongly interlinked with an inter-phase of mangrove ecosystems. The forest ecosystem is smaller in comparison with the marine one. The former extends over an area of 8249 sq km, while the latter has 6 lakh sq km of EEZ and 16000 to 35000 sq km of continental shelf. Considerable data has been generated on the forest ecosystem and its biodiversity, but our understanding of marine ecosystem and its biodiversity is still based on preliminary data. The complexity of habitats offered by the tropical insular ecosystem has provided opportunities for the growth of rich biological diversity.

The islands occupy zoo-geographically an important position. Their isolation from the mainland India has lead to the evolution of distinct terrestrial flora and fauna which exhibit high degree of endemism varying between 10% and 70%. The faunal affinities can be traced to Indo-Chinese, Myanmar and Indo-Malayan regions. The proximity of islands to the 'Marine Faunistic Centre' or 'Centre of Dispersal' facilitated recruitment of several species, which do not occur in other parts of the Indian Ocean or along the mainland coast. However, endemism in marine fauna is not very significant as many of the species enjoy wide spread distribution in the Indo-Pacific. The meiofauna, which are fettered to the bottom soil, are reported to have 21% endemism.

The islands are considered as biodiversity hot spots and the terrestrial and marine ecosystems figure in over 200 sites identified as global eco-regions. About 6540 species of fauna and 2500 species of flora were reported from these islands. But the spatial and temporal distributions are known only for a few vertebrate species. The available data indicates that the marine biodiversity constitutes about 63.7% of the total reported species. During the last three and half decades many species, including some new species, were added to the faunal diversity. Ten new subspecies of birds, five new species of mammal and many species of insects were reported after 1964. New species and new records were also reported from the marine fauna.

The view of coral reefs of Andaman

These findings indicate that there may be many more species awaiting the attention of taxonomists.

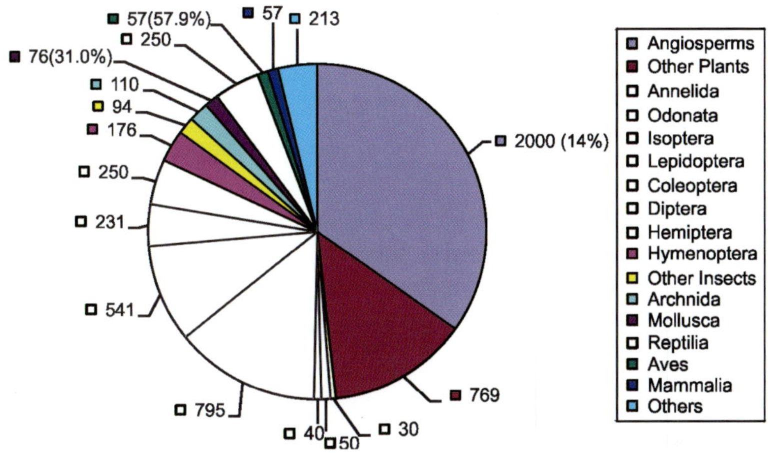

Fig. 1. Biodiversity of terrestrial ecosystem of Andaman and Nicobar Islands.

The insular fauna has some special characteristics. Mammals, which have 68.5 % endemism, are characterised by the absence of large mammals (Table 1). Twenty six species of bats and 14 species of rats constitute about 76.9% of terrestrial mammals and account for 44.23% of endemism. The Crab-eating Macaque and the two subspecies of Nicobar Tree Shrew are restricted 'endemic to Nicobar group', while Andaman Masked Palm Civet and Andaman Wild Pig are restricted endemic to Andaman group. Largest number of introduced species, about eight, is from mammals. Seven species of marine mammals were sighted in the near shore and oceanic waters. The sea cow, dugong, is sighted rarely.

Avifauna of these islands is far richer than any other vertebrate fauna. A total of 270 species and subspecies with about 39.6% endemism were reported from these islands. Vijayan (1999) reported 214 species, of which 62 are endemic. These include 100 species and subspecies of migrants, 95 residents and endemics, 43 residents and 4 introduced species. Andaman and Nicobar Islands are identified as Important Bird Area (IBA) by Birdlife International. The islands have 22 restricted range bird species, which include six species listed in IUCN Red Data Book. These are Andaman Crake (*Rallina canningri*), Narcondum Hornbill (*Aceros narcondum*), Nicobar Bulbul (*Hypsipetes nicobariensis*), Nicobar Scrub foud (*Megapodicus nicobariensis*), Nicobar Serpent – eagle (*Spilornis minimus*) and Nicobar Sparrow hawk (*Accipiter butlen*). Reptiles are represented by 76 species on land and

12 species in the marine environment. Majority of the species belong to snakes and lizards. There are three endemic species of pit vipers. Four species of turtles, and two amphibious snakes, *Lauticauda lauticauda* and *Lauticauda colubrina* come to the shore to lay their eggs. Andaman water monitor (*Varanus salvator andamanicus*) is the largest lizard and grows to a length of 2.5 m. Andaman day gecko is endemic in the islands. The burrowing lizard (*Dibamus novaeguineae*) occurs only in Nicobar. Amphibians constitute a small group. *Rana cancrivora*, a mangrove frog, which does not occur on the main land was reported from these islands. During the last one decade six species are added to the amphibian fauna of islands. There is a large diversity in fish fauna. The freshwater fishes constitute only the cultured species introduced from the mainland. Marine fish diversity is very rich and represented by about 1200 species.

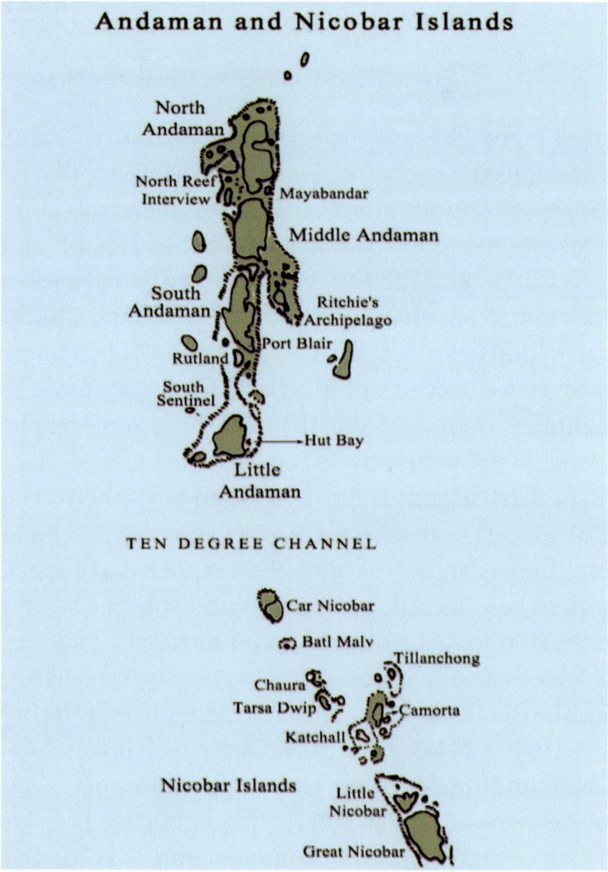

Map of the Andaman and Nicobar

MARINE BIODIVERSITY

These islands, besides having typical insular habitats in its rich tropical evergreen forests, can boast of important marine ecosystems, such as smooth, undisturbed sandy beaches, rocky coasts, mangroves, coral reefs, sea grass beds and algal communities.

Sandy beaches are the nesting sites for turtles and birds, like tern. The sandy coasts have their typical meiofauna with high rate of endemism. Rocky coasts offer ideal habitats for mollusks, seamats, crustaceans and other cryptofauna. Coral reefs and associated marine ecosystems such as, mangroves, seagrasses are considered as marine priority ecoregions.

Many tidal estuaries, small rivers and lagoons in the islands support luxuriant mangrove growth. Mangroves in the islands, spread over an area of 966 km^2 are the third largest in India (20%). Mangroves constitute about 12% of the total forest area of the islands. Luxuriant mangrove growth occurs on the sea ward fringe and on the sides of creeks in North Andaman (Mayabunder, Betapur, Rangat, etc.), Ritchie's Archipelago, South Andaman (Navy Bay, Wandoor, Chiriatapu etc.) and Nicobar group of islands (West Katchal Bay, Nancowry Island; Galathia river mouth, Dogmar and Alexandra in Great Nicobar). Out of a total of 65 exclusive species of mangroves known from the world, 37 species comprising 27 trees including 2 palms, 5 shrubs, 2 shrubby trees and 3 climbers were recorded from the islands (Balakrishnan, 1989).

Sea Grass Ecosystems

Out of 14 species reported from India nine occur in the islands spread over an area of 100-200 hectares (Das, 1996). Many areas in these islands are still either unexplored or underexplored. Consequently, there are gaps in our knowledge on biological diversity, more particularly terrestrial invertebrates and reef inhabiting animals. In view of this and habitat complexity in the islands, it can safely be inferred that more intensive surveys by specialists of the group concerned would reveal many more species hitherto unknown.

Marine algae of the coral reef habitats of Andaman and Nicobar Islands are less investigated, when compared with the work done on algae of the Gulf of Mannar Islands and Lakshadweep. Venkataraman and Wafar (2005) reported the total number of algae from Andaman and Nicobar Islands.

Mangrove Ecosystems

The fauna of mangrove ecosystem consists of terrestrial and aquatic components. An analysis of the faunal diversity is attempted after collating all the available data. Hitherto, 838 species of macrofauna and 53 species of meiofauna were reported to occur in the mangals of the Andaman and Nicobar Islands. At the phyletic level 9 are represented in the mangroves as against 21 in the coral reef ecosystem (Table 1). However, several species of shrimps, prawns, hermit crabs as well as insects already collected from the mangrove areas of these islands are yet to be identified and named up to species level. It is estimated that when thoroughly sampled the biodiversity of mangrove ecosystem in the islands may encompass about 1000 species, the richest for any mangrove area in India.

A total of about 311 species of insects belonging to 10 orders were reported from the mangals of the islands (Kathiresan, 1998). Of the total known insects 232 are herbivores, 43 parasitoids and 36 predators. Mangals also give shelter to 16 species belonging to 4 orders, namely, Homoptera, Coleoptera, Lepidoptera and Hymenoptera, which are pests on plants of economic importance. Among herbivores, Lepidoptera, Coleoptera and Hemiptera comprise majority of the species in that order. Over 85% of the herbivorous species were found to be monophagous, while only 25 species were observed to feed on more than one species (Das and Dev Roy, 1989). Most of the mangrove species are reported to be host to many species of insects. A total of about 55 and 33 species of insects were recorded respectively from the mangroves belonging to *Rhizophora* and *Avicennia*. The former occupies largest area in the mangals. The other common genus *Bruguiera* was found to host at least 30 species of phytophagous insects. The other important mangrove genera, *Sonneratia* and *Excoecaria* were found to harbour 22 species and 20 species of insects respectively. Parasitoids belong to one order Hymenoptera, while predators were classified into six orders. Insects in the mangals of Andaman Islands appear to be more diverse than those of the mainland. The folivorous and frugivorous Coleoptera, Diptera and Lepidoptera have better representation on the islands. The herbivorous insects were reported to affect leaves, buds, flowers, fruits, inflorescence etc. Some bore into seedlings, terminal shoots, propagule, fruits, buds, flowers etc. The species of Diptera, Lepidoptera and Acarina form galls on the leaves and stems. But, overall it was observed that insects do not cause much damage to the mangrove plants (Veenakumari et al. 1991).

Table 1. Faunal Diversity and Endemism in Andaman and Nicobar Islands

Group	*Species*	*Endemics*	*Endemism %*
Terrestrial ecosystem:			
Annelida	30	9	30.0
Insecta			
Odonata	50	4	11.0
Isoptera	40	19	47.5
Lepidoptera	795	52	12.5
Coleoptera	541	92	10.5
Diptera	231	24	11.2
Hemiptera	250	22	15.0
Hymentopora	176		
Others	94		
Arachnida	94	38	40.4
Centipedes	17	–	–
Millepedes	5	–	–
Mollusca	110	77	70.0
Amphibia	16	3	18.8
Reptilia	76	24	31.0
Aves	270	97	39.0
Mammalia	52	33	68.5
Total	**2410**	**494**	**20.5**
Marine ecosystem			
Meiofauna	486	102	–
Sponges	72	–	21.0
Coral	196	–	–
Other Cnidarians	147	2	–
Polychaeta	184	4	0.6
Sipuncula	25	–	2.0
Crustacea	829	6	–
Mollusca	1000	18	1.0
Pycnogonida	8	–	1.9
Echinodermata	336	4	–
Pisces	1200	2	1.2
Reptilia	12	–	0.2
Mammalia	7	–	–
Total	**4502**	**220**	**4.9**

Mangrove ecosystem provides a subsidiary habitat for many of the terrestrial fauna of these islands, which mainly invade mangroves as visitors. Only a few are residents. Mammals, many birds and reptiles stray into the mangroves for foraging. A number of insects, bats and birds visit mangroves, take shelter in foliage and derive food from leaves, flowers and fruits. But the importance of the ecosystem as a refuge for several of the animals cannot be overlooked. More over many of the vertebrate species reported from the mangals are endemic in the islands. Out of the eight species of mammals reported from mangals five are endemic (62.25%) either to Andaman or Nicobar Islands. Among birds 30 species (56.6%) are endemic, while 10 are resident, 12 are winter visitors and one is an introduced species. There are some species of birds, which generally prefer mangroves or tidal swamps. The Andaman Teal, Serpent Eagle, Night Heron, Kingfishers, etc. prefer mangrove ecosystem. Some birds have also been reported to build nests on mangrove trees.

Coral Reef Ecosystems

Coral Reefs are estimated to extend over an area of about 1000–2000 sq km. Hitherto, about 180+ species of corals were reported to occur (Pillai, 1978; Venkataraman et al. 2003). In the Indo-Pacific the Western Pacific Arc from Philippines to Western Indonesia contains rich biodiversity including coral diversity. Southeast Asia has highest coral diversity represented by about 400 species belonging to 78 genera (Turner et al. 2009). At the other end Western Indian Ocean has 200 species. Thailand on the east coast of Andaman Sea has 250 species. It is presumed that the islands may have at least 250 species of corals, if not more.

The reefs built by the scleractinian corals harbour a rich diversity of other plant and animal phyla constituting a community of their own. The members include representatives from almost all the animal phyla and marine phytal groups. The phytal communities include Corallinacean red algae, Lithothamnion which help in cementing and binding of the detritus green algae, Halimeda contribute to sedimentation in lagoons. The green, brown and red algae are the primary producers of the ecosystem. The association of zooxanthellae and coral is very intricately woven.

The hydrozoans *Millepora*, *Heliopora* and *Tubipora* form calcareous skeletons and help in the reef building process. Gorgonians and pennatulids which possess spicules in fleshy bodies contribute to mass of detritus. Sea anemones, the flower like animals, have symbiotic relationships with clown fishes and anemone fishes. Sponges are common in reefs and give shelter to fishes, crabs, shrimps and other small animals. Living in crevices or between coral branches are shrimps, crabs, and lobsters. These include predatory coral crab cleaner shrimps which play a major role in maintaining the hygiene of reef ecosystem. The most dominant invertebrate group in the coral reef ecosystem is Mollusca. Species of *Tridacna, Turbo, Trochus, Lambis, Conus, Cypraea*, and *Drupa* abound in variety, abundance and importance. Shallow water species of octopus thrive in the crevices of reefs. Triton's Trumpet, *Charonia tritonis* which feeds on the tube feet of Crown of Thorns starfish, is a rare and endangered mollusca recorded from Nicobars (Subba Rao and Dey, 1991).

Several species of Echinodermata inhabit the coral reefs. Among these are commercially important sea cucumbers. Starfishes *Acanthaster, Culcita* and *Linckia* are chief predators on the live coral polyps. The brittle stars *Ophiocoma, Ophiolepis, Macrophiothrix* inhabit crevices or undersurfaces of lead corals while many euryaliids are epizoic on gorgonaceans. The unstalked crinoids commonly known as feather stars are richly represented in the reef communities. The chief herbivores of coral reefs are sea urchins like *Stomopneustes, Diadema, Echinothrix* etc. These feed on the algae and maintain the balance in the ecosystem by keeping the algae under check. Fishes play a vital role in the reef ecology and its food web acting as both predators and prey. Thus, the reef ecosystem has a rich gene pool and biodiversity.

Andaman and Nicobar Islands are the richest coral reefs in India. According to the satellite imagery data the coral reefs and associated habitats are spread over an area of 950 km^2 (Muley et al. 2000). The local estimates based on random ground truth data project 2000 km^2, which form 6% of the continental shelf.

The reefs found in the shallow regions of the coast are mainly fringing type. These are characterized by reef platforms or benches which extend to about 500 m in width from the shore in Andaman group and up to 1000 m width in Nicobar group. The latter are characterized by 1m deep and wide surge channels. Besides the fringing reefs, coral banks were reported on the west coast a few km away from shore. Important of these are west coral bank, which is formed out of three separate banks, a middle bank, South Bank and Darlymphe Bank, off Little Andaman. On the east coast an Invisible Bank lying east of Rutland Island, was reported.

The reefs in the islands are distinguished into recent or sub recent reefs, and modern reefs (Reddaiah, 1977). Recent or sub recent reefs, raising

2 m above the present sea level occur on the shores of Car Nicobar. The modern reefs are further distinguished into windward reefs, channel reefs, bay reefs, knolls and patch reefs. Windward reefs grow in the direction of monsoonal winds, while channel reefs are less influenced by winds (hence called leeward reefs also) and grow in channels between the islands in South Andaman and Camorta-Nancowry area. Bay reefs grow in bays (Ritchie's Archiepelago Chidiatapu) and are protected from strong winds as in the case of channel reefs. Knolls are circular and compact growths with flat tops having two hollows and two rims. These were reported from the Peile Island in Ritchie's Archipelago. These reefs may rise from about 20 m depth. Patch reefs are mushram – shaped growths which develop on the inner side of the main reefs. These grow into coral knolls. The smaller patch reefs are also termed coral pinnacles.

Coral heads, which may be dangerous for navigation were noticed between Landafall Island and West Island, north-east of Paget Island, South of Lewis inlet, West of Great Nicobar, West of South Brother Island, North of Tarmugli Island and a few other islands continuing coral growths were noticed near Little Andaman, in MGM National Park, and Camorta – Nancowry complex and were recommended for conservation. Degraded reefs were noticed north – east of John Lawrence Island, Wilson Island, Nicholson Island, West of Henry Lawrence Island, eastern side of main near Port Blair and Flat Bay, north of Kyd Island, Smith Island, near Gibb Creek in North Andaman of the Andaman group: In the Nicobar group degraded reefs were observed near Teressa Dwip, Bompoka, Tillangchang, North Bay in Katchal Island and Menchal Island (Nayak et al. 1991). The reefs near Car Nicobar are also very much degraded due to rummaging of the intertidal region by the tribal population.

Threats to Island Ecosystem

Our ignorance of the total economic value of coral reefs has made us to develop a careless approach to reefs and utilise the resources without realising the future consequences. Many of the shallow reefs are damaged due to human pressures. In Andaman Islands attempts were made to identify the threats to coral reefs but no critical analysis is made of the various stresses which lead to coral reef destruction. The dominant sources of threats to coral reefs may be either climate induced or due to human pressures. Environmental variables like temperature, light, CO_2 are influenced by climate change and also human pressures. Reefs in the islands are also subjected to fishing pressure, legal and illegal. These climate and human induced stresses are delivered through different pathways. Some factors are transmitted through atmospheric pathway via air-sea influence. Some factors are associated with physical oceanography and hence considered under hydrographic pathway. Sometimes stress may be delivered through hydrologic pathway, which include evaporation, precipitation, run off from land based sources.

Sedimentation

Sedimentation is a serious threat to coral reef ecosystem in the islands. It is caused by erosion, dredging, poor land use pattern, for example deforestation and agricultural practices and coastal construction. The currents, sea level, wave energy or storm patterns may change sediment accumulation patterns. In the islands, west coast is comparatively untouched by the developmental activities and the problem of sedimentation seems to be less than on the east coast, where there are more activities contributing to the sedimentation in near shore waters. Sediment transportation is controlled by wave and wind driven coastal currents.

Of the total terrestrial ecosystem of the islands 87% (or 92%) is covered by forest leaving 13% of land under revenue, which is about 1070 sq km. This revenue land is mostly confined to coastal belt, the theatre for all developmental activities. The need to protect this limited land from sea erosion has been realized and sea walls have been erected in all major erosion prone areas. Logging operations in some of the islands may lead to sedimentation in the near shore waters. Siltation due to poor land use pattern has been reported from Hut Bay, Little Andaman, and North Bay, Port Blair and also Ritchie's Archipelago Soil erosion is a serious problem in Little Andaman. The average silt deposition near Port Blair was found to be 1.46 mg/cm^2 in 1988. About 75% mortality of reef flat at Sisotris Bay was attributed to siltation from land run off from Aberdeen Bazar, Atlanta Point, Raj Niwas Hill, Gymkhana ground etc. Heavy deposition of silt was also noticed near New Wandoor. Clearing of littoral forests and nearby hills, for human settlement and agriculture had caused soil erosion increasing the sediment load in near shore waters. The runoff from adjoining agriculture fields enriched the nutrient content of the coastal water and phytoplankton blooms are considered favorable for rapid growth of Crown-of-thorns (Turner et al. 2009).

More than 90% of rain water flows into the sea carrying with it suspended silt from land. During heavy rains coastal waters become turbid and waters up to 300 to 500 m from the shore turn muddy. The ability of coral colonies to cleanse themselves of sediment depends on the coral morphology as well as sediment grain size. For some species sediment suspended in water may meet their energy demands.

Sand Mining

Sand mining in the coastal areas will have deleterious effects on the marine life. Removal of sand causes erosion of the coast and also increases sedimentation in the near shore waters. There is a lot of construction activity in the Andaman Islands. Thus there is a heavy demand for sand requirements. Sand collection at Dhani Nallah, Middle Andaman threatened the natural nesting grounds of turtles. It is also close to Cuthbert Bay Wildlife Sanctuary. The stored sand gets washed into the sea during monsoon changing the intertidal ecology and affecting intertidal fauna.

Pollution

The small size of many islands combined with the importance of their coastal and marine environments make them fragile ecosystems which are vulnerable to even small pollution incidents. Pollution may be caused by effluents from domestic sewage and industrial sources. Toxins may be introduced into the marine environment through oil, metals and pesticides. There are no major industries in the Islands and pollution due to industrial effluents is not known. But toxic discharges from saw mill at Chatham and Andaman Timber Industry at Bamboo Flat had caused damage to coral reefs.

Andaman and Nicobar Islands are near to the international tanker and major shipping routes. These islands are prone to operational oil pollution. Oil spills in the Islands are of rare occurrence. But thin oil film persists on the sea surface near harbours of Port Blair and Camorta. There was a major oil spill at the entrance to the Straits of Malacca on 21 January 1993 and the pollution of sea south of Great Nicobar was prevented by the timely action of Indian Coast Guard. The Islands have witnessed in the recent years a constant increase in the number of vessels plying the Mainland-Island and Inter-Island sectors and passing of large number of crude carriers through Andaman and Nicobar Waters. The harbour waters get polluted by the ships discharging bilge water/oil mixed with water. No specific studies are available on oil or toxic pollution in the waters around the Islands. Elsewhere ballast water is identified as an issue, especially in bulk material and energy export ports and ballast water management plans are developed. It is not known whether ballast water is an issue in Islands ports.

The greatest danger to the marine ecosystem is from domestic sewage. It is a source of nutrient enrichment in the coastal waters. Fertilizers and pesticides from agriculture trickle down through creeks scattered along the coast increasing the phosphorus and nitrogen levels in the inshore waters promoting the unnatural growth of plants. The increase in population may enhance the sewage load. Elevated nutrients encourage the growth of phytoplankton and benthic algae which gain competitive advantage over corals and may often overgrow at the cost of the latter. High levels of nutrient also encourage growth of bryozoans and barnacles, whose larvae become dominant preventing the recruitment of coral larvae (Smith and Buddemeir, 1992). It was proved experimentally that high levels of phosphate inhibited coral calcification. High levels of ammonium may cause mortality in corals. The effect of terrestrial run off on Islands inshore coral reefs is not known.

Poaching

Poaching of wildlife is common in the terrestrial ecosystem. Poaching in marine ecosystem is a peculiar and serious problem in the Islands. The rich and large fishing ground in the Islands' EEZ lies very close to Myanmar, Thailand, Malaysia and Indonesia. This richness and proximity have been sources of attraction to foreign fishing vessels, which have been intruding into India's territorial waters since 1940. Besides, marine resources have been scarce in the territorial waters of our neighbouring countries. Fishing operation in Andaman and Nicobar Islands waters may be very remunerative. Hence, many foreign nationals intrude into our territorial waters. These intruders sail in country crafts and on sensing danger they rush to creeks found all around the Islands.

The poachers come in search of shells, sea cucumbers and fishes. In their anxiety to collect as much as possible they cause irreparable damage to ecosystem. About 184 fishing boats were apprehended during the last one decade. The number of crew was 1750, on an average about 9.5 per boat. The poachers were mainly from Myanmar, followed by those from Thailand and Sri Lanka. Malaysians and Indonesian were apprehended during four and two years respectively in small numbers. The number of Sri Lankan poachers seems to have increased in recent years; all were apprehended after 1993 only. The available data from Indian Navy and Coast Guard do not indicate any trends in the poaching activities. It perhaps reflects the patrolling intensity of Coast Guard, Indian Navy and Police Departments.

Anti-poaching operations by the Indian Navy have started only in 1997. A greater co-ordination was suggested among Forest Department, Police Department, Fisheries Department, Marine Department, Coast Guard and the Indian Navy for effective control of poaching. Each department should nominate a nodal officer for taking up timely action.

Storms

Reefs in the open sea and wind ward reefs are subjected to physical damage due to high waves during storms. Cyclones, which may occur in November to December, uproot the corals breaking their branches. Branched, tree like corals are subjected to heavy damage. Reefs at Chidiatapu, South Andaman, Little Andaman and east coast of Great Nicobar have suffered physical damage.

Crown of thorn Star Fish

The crown-thorns star fish, *Acanthaster planci*, was identified as an important cause of coral mortality. In the Islands star fish was reported from several reefs but no authentic records exist to show that its population has reached menacing proportions. Star fish outbreak was observed in 1988 in Mahatma Gandhi Marine National Park, Wandoor. Feeding scars of the starfish were noticed in corals in the Marine National Park near islands such as Grub, the Twins, Chester etc. and also outside the protected area. However, after 1989 no such outbreak was noticed. The results of surveys in Middle Andaman, south Andaman, Ritchie's Archipelago, Little Andaman and Great Nicobar have revealed one or two star fish per sq km. Sedimentation and particular pattern of rainfall were considered responsible for the outbreak of star fish population. The poor condition of corals off the islands of Grub, Twin, Jolly Buoy and Chester were attributed to predation by crown-of-thorns (Wood, 1991, Venkataraman and Rajan, 1996).

Bleaching

Mass bleaching of corals was reported throughout length and breadth of the Andaman Islands. The phenomenon occurred in May 1998 in several coral reef sites in North, Middle and South Andaman. Bleaching was extensive in reef flat areas and on reef slopes it extended up to 10m depth. Bleaching was 100% in many branching types and in some massive forms of corals. In July 1998 bleaching was observed in 85% of the corals near Ross Island and Mahatma Gandhi Marine National Park. Ninety percent of the massive corals and more than 75% of the branched corals were observed to be affected by bleaching in Ross Island, North Bay and Marine National Park. All symbiotic reef taxa (*Millepora*, soft corals, zoanthids, giant clams) were reported bleached, but symbiotic sponges show fewer tendencies for bleaching. It may occur due to loss of pigmentation from the algae or loss of algae from the coral. Bleaching may cause mortality in corals or often they may recover from this stress (Venkataraman, 2002)

Over Exploitation

Over exploitation of corals, mangroves and their associated organisms may pose a significant threat to reef ecosystem. In these Islands coral reef ecosystems were not disturbed or damaged till 1970. However, from mid 80s impact of human pressure on the reefs was more discernable than ever before (Pillai and Venkataraman, 1999). Corals were collected as curios and were sold in the market. Fancy corals such as *A. humilis, Monitipora cocosensis, Pocillopora damicorins, Merulina ampliata, Tubipora musica* (organ-pipe coral), sea cucumbers, sea horses and ornamental fishes etc., were in great demand. It prompted the traders and tourists for indiscriminate collection. Commercial exploitation of corals and their associated organisms was banned in 1978, however, it continued clandestinely in remote places of the islands. The organ-pipe coral or red coral was once very common in the near shore waters of Great Nicobar. It was reported that the red coral was collected and smuggled out of the islands. Fishing hooks and lines, and synthetic nets get entangled in the reefs. In the islands, about 900+ units of hooks and lines, 1000+ units of gill nets and 600+ units of cast nets are operated in near shore coral reefs.

Although fishing regulations exist, there are difficulties in implementing those. Some of the invertebrates and fishes are fished so heavily and indiscriminately that some of the species have become 'commercially extinct'. Fishing can have secondary effects on the reef ecology. Over fishing of carnivorous species (shark) may increase the population of other fishes which may have adverse impact on the reef ecosystem. Investigations on the impact of fishing on the reef ecosystem in the islands are necessary to be studied.

Tourism

Unescorted and unguided tourism had its impact on the reefs. Scuba divers and snorkelers damage the fragile corals by trampling and walking on them.

Anchoring of boats also crush the corals (Mustafa, 1990). Thus heaps of dead corals are seen in Cinque and Jolly Buoy Islands. Only soft corals and boulder-type of horny corals are able to survive these impacts. Although only 36 islands are inhabited, the human activity spreads in some form or other to all the major islands in Andaman group. The fishermen from Mayabander go to Interview Island and even too far off island like Little Andaman in search of fishes especially sharks. As such, over exploitation of resource is not restricted only to thickly populated islands.

CONCLUSION

Many scientific conferences, congresses, symposia and seminars were held in recent times highlighting the need to conserve the vanishing biodiversity, including the island ecosystems, but very little was actually done to save them from destruction. In practice, priority was always accorded to human welfare and the consequent developmental activities at the cost of these wild resources. It is important to note that the Government of India has established the Nicobar Biosphere Reserve, Mahatma Gandhi Marine National Park and Rani Jhansi Marine National Park, to protect the endangered terrestrial and marine ecosystem. This area has been considered the least disturbed group of islands supporting myriad of biodiversity. It is also encouraging to receive recent reports that the creation protected area had shown improvement of its faunal resources.

Tourism has become a major trade and a luring business in Andaman and Nicobar Islands. These islands with exceptional scenic beauty offer a great potential for tourism and in this connection, the presence of luxuriant terrestrial and marine life would be an added attraction. To achieve the dual goals of biodiversity conservation and its exploitation for tourism, the zoning system may be introduced in these islands. Apart from marine aquaria, sea turtle, crocodile, dolphin and dugong sanctuaries may be established for propagation of these endangered species and promotion of tourism in the islands. If managed properly, these can form a great source of tourist attraction in this region.

Despite the Wildlife (Protection) Act, 1972 and Biological Diversity Act, 2002 and the great concern to conserve the natural ecosystems and their biodiversity in this region, they are only on the decline (Venkataraman, 2007). In the circumstances, unless effective conservation measures coupled with capacity building and training to the managers and researchers and awareness and education on biodiversity conservation at all levels, it would be intricate to save the rich heritage of animal life from irreparable damage and preserve the biological uniqueness of these islands. It is therefore, quite necessary to reserve sufficiently large virgin areas free from human interference as viable entities on the long run as biosphere reserves, national parks and sanctuaries for wildlife conservation, genetic diversity and scientific study.

Like any resource, exploitation of biodiversity by man shall be rational, sustainable and judicious. Otherwise, this paradise of rare plants and animals in no time share the same ill fate of many other denuded areas in the world. All destructive activities contributing to the degradation of the natural environment are to be arrested, which are not certainly as easy as the ways of their destruction. Exotic introductions should not be encouraged without a careful study of their consequences on the local biota. The littoral habitats are also to be protected from adverse effects of potential pollution.

Considering the fast depletion of biodiversity on these islands, the Government of India has established one biosphere reserve, besides the earlier established 9 national parks and 96 wildlife sanctuaries comprising an area of 1891.54 sq km on this union territory, of which in about 350 sq km pertained to the marine ecosystem. In addition, the keen interest evinced by the naturalists to protect our vanishing wild wealth, awareness and cooperation of common man, who comes in contact with these animals in day to day to life, is very necessary for their conservation. Involvement of local communities in resource management is found important for the successful implementation of our conservation strategies. Further, wide gaps always existed between law and its enforcement, although the legislation often proved better in halting the commercial exploitation of wild resources. Hence, the Wildlife (Protection) Act and Biological Diversity Act should be strictly enforced, all poaching arrested and deterrent punishment imposed in all cases of violation.

ACKNOWLEDGEMENT

I wish to express my sincere thanks to the West Bengal Biodiversity Board, Kolkata for giving me this opportunity to write this article.

References

Ahlawat, S.P.S., Rai, R.B. and Sharma, A.K. (1990). Livestock production in Andamans. *Islands on March,* October, 1990: 17–29.

Anon, (1995). *Bibliography on Zoology of Andaman and Nicobar Islands.* Zoological Survey of India, Calcutta, p. 284.

Balakrishnan, N.P. (1989). Andaman islands-vegetation and floristics. In: *Andaman, Nicobar and Lakshadweep – An Environmental Impact Assessment.* Saldanha, C.J. (Ed.). Oxford & IBH, New Delhi, pp. 55–68.

Das, H.S. (1996). *Status of Seagrass Habitats of the Andaman and Nicobar Coast.* SACON Technical Report 4, Coimbatore.

Das, A.K. and Dev Roy, M.K. (1989). A general account of the mangrove fauna of Andaman and Nicobar islands. *Fauna of Conservation Areas,* 4, p. 173. Zoological Survey of India.

Ellis. J.L. (1987). Floral composition of the nearshore vegetation of the Andamans, *Proc. Symp. Management of Coastal Ecosystems and Oceanic Resources of the Andamans.* Andaman Sci. Association. CARI, Port Blair, pp. 24–27.

Kathiresan, K. (1998). Distribution and status of mangroves in India. *Seshaiyana,* 6(2): 8–9.

Muley, E. V., Venkataraman, K., Alfred, J.R.B. and Wafar, M. (2000). Status of coral reefs of India. *Proceedings 9th International Coral Reef Symposium,* Bali, Indonesia. 23–27, October 2000, 2: 847–854.

Mustafa, A.M. (1990). Coral reef degradation, conservation and management in Andamans: An overview. *J. Sci. and Tech. Islands on March,* 1990: 104–106.

Nayak, S.R., Bahuguna, Anjali, Untawale, A.G., Jagtap, T. G. and China Raj, (1991). Coastal wetland mapping of the Andaman and Nicobar group of islands. Scientific Note. Space Applications Centre, Ahmedabad. RSAM/SAC/COM/SN/10/91. p. 9.

Pillai, C.S.G. (1978). *Stony Corals of A & N Islands.* CMFRI Cochin: 1–14.

Pillai, C.S.G. and Venkataraman, K. (1999). *Status, Conservation and Management of Corals and Coral Reefs of Andaman and Nicobar Islands.* UNDP, GEF, Project Report, p. 24.

Reddiah, K. (1977). The coral reefs of A & N islands. *Rec. Zool. Sur. India,* 72: 315–324.

Sewell, R.B.S., (1935). *Memoires Asiatic Society of Bengal*, 9(8): 461–540.

Smith, S.V. and Buddemeier, R.W. (1992). Global change and coral reef ecosystems. *Ann. Rev. Ecol. Syst.*, 23: 89–118.

Subba Rao, N.V. and Dey. A. (1991). Composition and distribution of marine mollusks of Andaman and Nicobar islands. *J. Andaman Sci. Assoc.*, 7(2): 50–55.

Turner, J.R., Vousden, D., Klaus, R. Satyanarayana, C., Fenner, D., Venkataraman, K., Rajan, P.T., Subba Rao, N.V., Ramakrishna and Raghunathan, C. (2009). Remote sensing and rapid site assessment survey. Report of Phase 1: April 2001. GOI/UNDP GEF. Coral reef ecosystems of the Andaman Islands, New Delhi. Occ. Paper, *Zool. Surv. India*, 301: p. 132.

Veenakumari, K., Mohanraj, P. and Bandyopadhyay, K. (1998). Insect diversity in mangrove canopies. *Seshaiyana*, 6(1): 12–14.

Venkataraman, K., Satyanarayana, C., Alfred, J.R.B. and Wolstenholme, J. (2003). Handbook on Hard Corals of India. (Published: Director *Zool. Surv. India*, Kolkata) 1–266.

Venkataraman, K. (2007). Conservation and management of coastal and marine ecosystems in India. *NBA Scientific Bulletin No 13*: National Biodiversity Authority, Chennai, p. 91.

Venkataraman, K. and Wafar, M. (2005). Coastal and marine biodiversity of India. *Indian J. Mar. Sci.*, 34(1): 57–75.

Venkataraman, K. and Rajan, P.T. (1996). Coral reefs of Mahatma Gandhi Marine National Park and crown of thorn phenomenon. *Proceedings of the Symposium on Island Ecosystem and Sustainable Development*. Andaman. 124–132.

Venkataraman, K. (2002). Status survey of the Gulf of Mannar coral reefs following the 1998 bleaching event, with implications for reserve management. *Proceedings 9th International Coral Reef Symposium*, Bali, Indonesia. 23–27, October, 2000, 2: 841–846.

Section – IV
Conservation Initiatives

18

An Appraisal of Global Initiatives in Protection of Biodiversity of Species and a Recommended Model

Promila Kapoor Vijay

INTRODUCTION

In the International Year of Biodiversity 2010 a shock has come to all involved that humanity is at crossroads faced with accelerated extinction of species and the concomitant rapid erosion of its gene pool diversity (Global biodiversity outlook 3, 94).While acknowledging our failure to meet overall target of reducing the loss of biodiversity, it is important to note that for halving poverty and hunger by 2015, meeting food, nutrition and health needs of the poorest of the poor a much higher level attention by world's scientists and intellectuals coupled with fast action on the ground is needed for stemming this loss. Countries are preparing for the forthcoming COP-11, and Rio+20 conferences and will have to carefully develop new strategy for conservation and sustainable use of biological diversity, genetic resources, and associated traditional knowledge in unique and diverse ecosystems of each country in the world.

Scientists have so far described between 1.5 to 1.6 million distinct species of plants and eukaryotic animals (Hammond & Heywood, 1995; May, 2005), while unexplored species can range from few million to 100 million (no verifiable reference available). About 10,000 species are being described per year (Hammond & Heywood, 1995). New discoveries related to genomic knowledge of single species are leading to a better understanding of their environmental relations. The importance of microbial diversity is providing insights into ecosystem function and stability (Bent et al. 2008). Their functional metabolic mechanisms can help to explain global evolutionary processes across diverse regions (Dinsdale et al. 2008). New scientific knowledge of population dynamics, conservation biology, biosystematics, and ethno-genomics are creating an understanding of species and their potentials for harnessing their distinct genetic, ecological traits for answering diverse and seemingly intractable problems caused by hunger, poverty, malnutrition, disease and ecosystem changes. Magnitude of the responses of a species, and its genetic and phenotypic diversity within and across environments is of great interest to scientists in evaluation and assessment of species (Schilchting & Pigilucci, 1998). The knowledge of biodiversity while anchored in advances in science requires basic research and assessments in understanding adaptive evolution and its constraints gained by using advanced molecular genetics. It involves scientific research over a long period of time (Schlichting, 2004). Every living organism discovered or undiscovered is important in functioning of life. Understanding structure and functional components of diversity of genes, species and ecosystems enables its conservation and sustainable use. Dazzling diversity of species is a vast reservoir of gene wealth and the associated traditional knowledge of species and its diversity can be leveraged to meet basic human needs to help us to cope with climate change and other global issues plaguing us.

Latest estimate (Global biodiversity Outlook 3,94) shows 23% of plant species on earth are threatened by extinction. Research shows corals will be lost from the Mediterranean Sea if commercial exploitation continues unchecked and measures need to be put in place. Scientists and experts have given (Global biodiversity Outlook 3,94) five principal pressures which contribute directly to the loss of biodiversity which are: habitat change, over exploitation, pollution, invasive alien species and climate change. Twenty one indicators have been used in the report (Global biodiversity Outlook 3,94) acknowledging species extinction rate has been accelerated. Asia has about half of world's population, though many countries of the region are rich in biodiversity, cultures, languages and ecosystems, and yet the poorest segment of the global population inhabits this region. Sustainable use of biological resources with proper species and ecosystem protection will ensure socioeconomic development with no loss of species, or their diversity and habitats.

Aim of this paper is to describe a conceptual model for developing study, conservation and sustainable use of biological diversity and genetic resources and to highlight various important initiatives in this area around the world. It gives an appraisal of the elements of programmes that made it successful.

Information, knowledge, specialized expertise of science related to biological diversity have been around since study on life began. Scientists started taking actions and attempted to develop proper definitions, meaning and importance in 1980s, on Biodiversity for global and human welfare (Kendrew, 1984; Wilson, 1985; Kapoor, 1988).

Several global initiatives on protection of species, their genetic diversity, associated traditional knowledge and ecosystems have been launched by International Organizations, both governmental and non-governmental, which have extensive networks of officials and experts. Two approaches used so far for the protection of species worldwide, have a fundamental difference. One is based on models where affluence and abundance provides the backdrop for building protection programmes for species, and ecosystem. The other is anchored in resource scarce settings. It derives its strength from contributions, inputs, participation and approval of local communities, organizations and leaders, who may represent urban or rural setting, and are from a background of scarcity and poverty. There are variations of both approaches. The Biological Diversity and genetic Resources (Kapoor, 1992) (BDGR) programme of the Commonwealth Science Council (CSC) was a hybrid being based on both approaches. It was based in UK. The peers and network of scientists, who worked in it and supported it, were from backgrounds of affluence and scarcity as the Commonwealth encompassed both the developed and developing countries, with the majority of its member countries being poor nations.

The conceptual model of this programme was designed (Figure 1) on the basis of identifying a six step action plan. This was followed by building networks of communities of scientists across 32 countries in seven geographic regions that constituted CSC. High level scientific expertise was brought to bear on the programme through national coordinators and members of national committees who helped searching priority issues in the study, sustainable use, and conservation of biodiversity (Kapoor, 1992; Kapoor & White, 1992; Kapoor & Usher, 1993; Kapoor et al. 1993; Harris & Kapoor, 1988; Paroda et al. 1988). Information and knowledge on many issues in biodiversity had been available in specialized disciplines across agriculture, ecology, genetics, population biology, forestry, molecular biology, and biotechnology. The challenge in the BDGR programme was to bring the science and scientists together and identify priority issues such as: (i) Which species needed to be given more attention, Life support species or Keystone species or both? (ii) What criteria should be used to study and conserve key species? (iii) Which techniques needed to be focussed on for the study, use and conservation of biological diversity work, and for developing key elements of the BDGR programme with local, national support?

Continued efforts were made to draw knowledge and wisdom from intellectuals, experts, peers, and young scientists from everywhere within limits and constraints (Kapoor, 1992). Members of various specialized organizations, academies, universities gave their inputs without any financial costs. This programme is exceptional as contributions made by a large number of learned scientists and experts between 1985 and 1994 were provided on a voluntary basis to build the Biodiversity and species conservation programmes across thirty two countries. Contributions of scientists was the primary motive force which allowed improving and adapting the programme at every step as it could be evaluated continuously by peers across the Commonwealth and others. The science and technology related to conservation and sustainable use of biological resources of the world evolved while senior and middle level scientists were being trained simultaneously in many areas of programme. This action learning led to building training programmes, materials and techniques at regional and national level. The BDGR programme is an illustration of how diverse inputs can be utilized drawing knowledge from many specialized disciplines of biology, anthropology, agriculture, and forestry. The programme at the operational level had four processes namely (1) Interactive dialogue, (2) adaptive transfer technology, (3) negotiations; and (4) field work at national level.

Activities undertaken in the programme resulted in indigenous capacity building and transfer of scientific knowledge and technology. A programme on Tropical Forest ecology was developed for Papua New Guinea jointly with UNESCO and Government of Australia; this subsequently led to and helped in the establishment of a Sustainable Tropical Forestry programme jointly with the Government of Guyana namely the Iwokorama Rainforest Programme of the Commonwealth Secretariat. The components of the BDGR programme and its evolution is given in Box 2 and Box 3 respectively.

A list summarizing key current and latest Global initiatives which have been launched by various organizations within the UN system and others to protect species, habitats and ecosystems are given in Box 1. Some also deal with Global warming and Climate change affecting biodiversity. The list does not include exclusive actions taken by member countries on protection and conservation of plant, animal and microbial, marine biological resources. The scope of this paper does not allow covering details of all initiatives taken for ex situ methods used for species protection such as Gene banks, Botanical gardens, Zoos, Museums, etc.

Box 1. Key initiatives listed here are from UN, International, Governmental non Governmental Organizations, and regional organizations who are giving support to protection and conservation of biological resources, biodiversity.

A. Legal initiatives for Biological diversity, Genetic resources and Traditional Knowledge

1. Convention on Biodiversity (CBD) is an International legally binding treaty, to which parties have committed to conservation of biological diversity; sustainable use of its components; and the equitable sharing of benefits arising out of utilization of genetic resources. It has convened nine Conferences of parties. COP-10 will develop a new strategy plan for the coming decades including a 2050 vision(www.cbd.org)
2. Food and Agricultural Organization (FAO) after lengthy negotiations in the Commission on Genetic Resources for Food and Agriculture (CGRFA) an International Treaty on Plant Genetic Resources for Food and Agriculture entered into force in 2004, first session of the Treaty's Governing Body was held in 2006.
3. United Nations Education and Scientific and Cultural Organization (UNESCO). Man and Biosphere programme started in 1971.www.unesco.org
4. World Conservation Union (IUCN) has been responsible for preparing "Red List of threatened species" jointly with Species survival commission of IUCN.www.iucn.org
5. The economics of ecosystems and biodiversity (TEEB) initiative of IUCN and the Ecological and Economic foundation, 2010 at international level seeking to attach a monetary value to species and ecosystems, cultures and languages at international level with responsibility to future generations. http://www.teebweb.org
6. World Wildlife Fund (WWF) is working on protection of wild life species. It has worldwide network of species and protected areas. programmes.www.wwf.org
7. Convention on international Trade in Endangered species of wild fauna and flora (CITES). About 5,000 animal species and 29,000 plant species are protected against overexploitation through international trade.www.cites.org
8. World Intellectual Property Organization (WIPO) of UN is involved in facilitating international negotiations through its Intergovernmental Committee (IGC) on Genetic resources, Traditional Knowledge and folk lore (GRTKF). It is working to an International binding for the protection of rights of access and benefit sharing, equitable sharing. www.wipo.int
9. International Union for the Protection of New Varieties of Plants (UPOV) protects breeder's rights and new varieties of crops.www.upov.int

10. Svalbard Global seed Vault-Seed Gene bank was operationalized in 2008 in Norway. The facility preserves a wide variety of plant seeds in an underground cavern. The seeds are duplicate samples, or "spare" copies, of seeds held in gene banks worldwide.

B. Initiatives for Conservation and sustainable use of biodiversity

11. United Nations Environment Programme (UNEP) is establishing an Intergovernmental Panel on Biodiversity and Ecosystem Services. www.unep.org. International council of scientific Unions(ICSU) has been working jointly with UNESCO on strengthening science in biodiversity work in CBD
12. World Conservation Monitoring Centre (WCMC) is collaboration between UNEP and WCMC 2000. It analyses and delivers information on animal and plant species in support of their conservation and sustainable use.www.wcmc.org
13. Species 2000, and Catalogue of life initiatives, is preparing a comprehensive catalogue of all known species of organisms on earth by 2011. http://www.sp2000.org/
14. Natura 2000 network is one of Europe's premier organizations with protected areas network. It focuses on wild birds and habitats to safeguard Europe's wildlife areas and species www.natura.org/
15. Biodiversity International has set up a new initiative on underutilized crops called Crops for Future was launched in 2010 with an International centre in Malaysia http://www.cropsforthefuture.org/
16. Global Mountain Biodiversity assessment 2010. Building inventory of habitats and species in mountains.http://gmba.unibas.ch/index/index.htm
17. Global Taxonomy Initiative to complete information on taxonomy of species of the world. http://www.wfcc.nig.ac.jphttp://www.nordgen.org/sgsv
18. UNCTAD has a development focus on trade and investment. It organized a global meeting of experts to give their views and recommendations to protect and promote traditional knowledge associated with biological diversity and genetic resources.

The design of the Commonwealth BDGR programme had its seeds in the work done in India (Box 2) on an underexploited plant species (Kapoor, 1978; Pratap & Kapoor,1985), *Chenopodium album* and its population. The species conservation studies within various Indian programmes gave a strong scientific and technical underpinning to the BDGR programme while the inputs from the developed countries helped in its evolution. A symbiotic relationship developed between the two different approaches as the programme grew and was implemented in 32 countries.

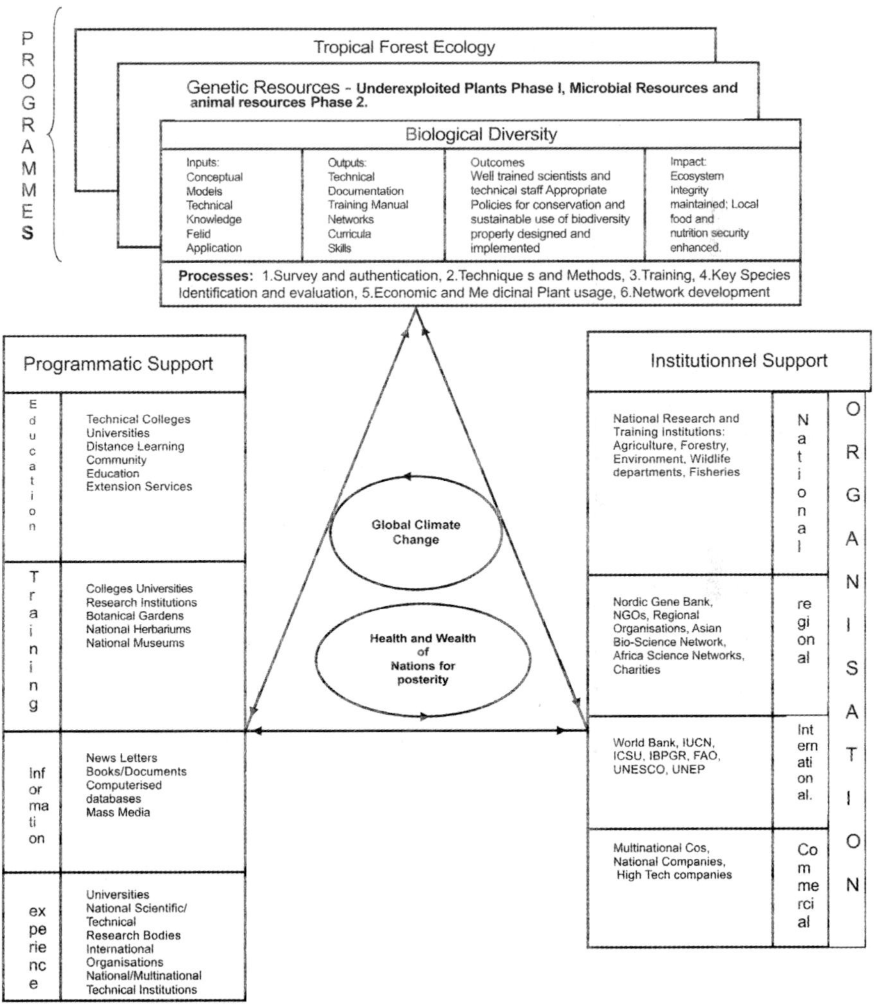

Adapted from: *Biological Diversity and Genetic Resources, The Programme of the Commonwealth Science Council, P. Kapoor-Vijay; The Commonwealth Secretariat London 1992; ISBN.0-85092-388-3*

Fig. 1. Conceptual Model

The conceptual design and implementation of the BDGR programme was defined with a clear chain of elements that were tightly coupled to get results as set out in Figure 2 below:

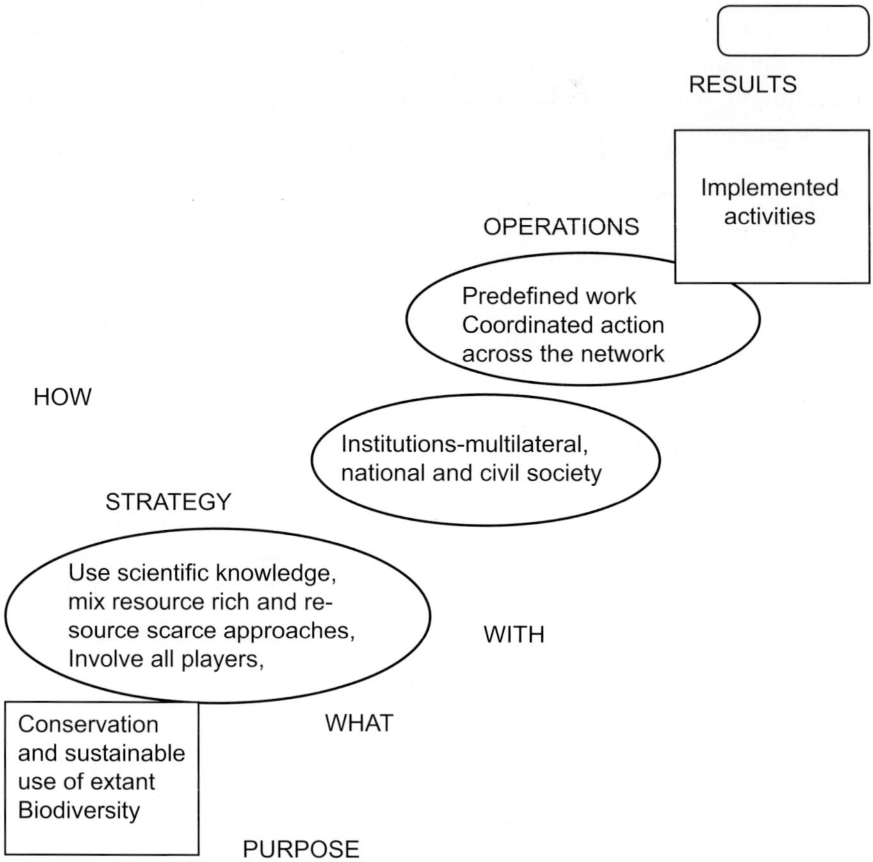

Fig. 2. Linkage among the Elements of the BDGR Programme

Box 2. A summary of India's key Initiatives for the protection of species and genetic diversity. India is party to the Convention on Biological Diversity since 1993. The country contains a great wealth of biological diversity in its forests, its wetlands and in its marine areas. Flowering plant species of country constitute about 6% of the world's flowering species. Bird species diversity and Fishes of India have high diversity and represent 11.7% and 12.6% of world's diversity of birds and fishes. Majority of the initiatives in India which have impact on the country and the international community relate to *in situ* and *ex situ* approaches to safeguard its national biological diversity and genetic resources. Following key initiatives summarizes species protection in force.

A. Legal Initiatives concerning Biological diversity

1. Biodiversity act 2002 and Rules 2004 have given it a distinct position enabling protection of unique gene, population, species and ecosystem wealth.
2. The National Biodiversity Authority created in 2003 with 25 state biodiversity boards in place till date. 20th meeting of National Biodiversity Authority India-held on June 2010. http://www.nbaindia.org/docs/20th_Proceedings_10-08-2011.pdf
3. The Protection of Plant Varieties and Farmers' Rights Act, 2001 has been enacted as a national legislation.
4. Traditional Knowledge Digital Library was prepared by Council of Scientific and Industrial bio-piracy.

B. Science, Technology, Environment, Agriculture, Forestry initiatives in Biodiversity

5. The National Bureau of Plant Genetic resources has set up national gene bank. Its status report and vision 2025 provides several other initiatives in place or planned to collect, characterize, evaluate, conserve and protect genetic resources of crops, medicinal and aromatic plants and underutilized plant species. Advances in basic and genomic knowledge are widely used http://www.nbpgr.ernet.in/
6. Many national governmental and non governmental organization working to inventories, characterize, and conserve species and genetic diversity of the country across diverse habitats and ecosystems on flora, wild species, forestry, wildlife, economical and industrial species are: Botanical Survey of India, Council for Scientific and Industrial Research (CSIR), Department of Biotechnology (DBT), Department of Environment, Forests and Wild Life along with bioinformatics cells/departments with scientific organization.
7. Wildlife Institute is a premier body protecting wild life species, and is working on environment impact assessment(Kapoor-Vijay, 2006).

8. National Bio-resource Development Board set up by Ministry of Science and Technology in 2000 http://dbtindia.nic.in/uniquepage.asp?Id_Pk=199&ParentID=55
9. Society for Research Initiatives for Sustainable Technologies and Institution (SRISTI),
10. National Innovation Foundation.

C. Traditional and ecological knowledge

11. All India Coordinated Research Project on Ethnobiology. 8,900 species used by tribal communities.
12. Mangrove Conservation Programme (launched by Ministry of Environment and Forests) since 1987.
13. Forest gene bank sites called Medicinal Plant Conservation Areas (MPCA) since 1993 have been set up in 54 areas.
14. Ashoka Trust for Research in Ecology and the Environment, and several civil society organizations in association with national and international agencies are engaged in situ conservation of targeted species. Foundation for Revitalization of Local Health Traditions focuses on medicinal plants of Ayurveda system.
15. Herbal, ayurvedic, yunani and other medicines are sourced from medicinal plants, and this sector is developing and needs support(Kala et al, 2006)· A vast numbers of species have not been studied for their population biology, on their extinction status. Protection programmes and initiatives are needed.

Increasing species extinction rates and continued loss of habitats and ecosystems has given a wakeup call to the world. A new strategy is needed which balances the use of scientific and economic principles. It should facilitate information flow for conserving and sustainably using biological resources which originate from biological diversity and ecosystems that support it and ensure equitable sharing of benefits(Global biodiversity Outlook 3, 94). Protection of habitats and areas and ecosystems is necessary, but approaching this with a well defined focus on species is more practical and achievable. Some successful programmes on biodiversity of species and their genetic diversity need to be examined. Earlier and current work on Life support species and their population diversity shows (Chenopods, Sorghum, Millets and many more species across unique ecosystems) many have in their biological system ability to occur and thrive in extremes of environments, and survive floods, drought, soil salinity etc, and are able to provide food, nutrition, medicine and other ecological benefits to people and marginal communities who live and protect these environments and species (Life Support Species in Asia-Pacific Region,1987; National Academy of Sciences, 1975). Promoting, and utilizing such positive results and using hybrid models can provide answers to call of the times to safeguard natural environments, cultures, biodiversity and gene wealth locked in the species. BDGR model described in this paper is possibly one such model. It has strength in science and technology applications and acceptance of local people and communities. An appraisal of work done in Box 3 sets out the historical context of the BDGR work and gives new developments in this area in two parts: one before the launch, and the other after the launch of the Convention of Biodiversity (CBD).

Box 3. An appraisal of various steps between 1976 and 2010. The key scientific programmes, publications, and reports and network which contributed immensely in developing the International Programme on BDGR provided inputs to negotiations of CBD in 1992. These steps and contributions added to the knowledge base for the locally used species, many of which while still underexploited, are source of food, nutrition, micro-nutrition, medicinal benefits, energy, ecological and soil restoration services, and socio-cultural value to poor people.

A. **International Efforts on Biodiversity 1970-1991.** Primary focus on Species and Protected areas, highlights as follows:

1. A programme on underexploited plants of Board of Science and Technology for Innovation and Development (BOSTID started in 1970s by the National Research Council of USA (see BOSTID reference)
2. National Seminar on Resources Development and Environment in Himalayas (1978) organized by Department of Science and Technology, Govt. of India, provided first forum to identify locally used species in Himalayas.
3. An All India coordinated project on Underexploited plants was started in 1976 by the Indian Council of Agricultural Research, New Delhi, India and is continuing (NBPGR Status report)
4. Report of the Expert Group of the Commonwealth led by Sir John Kendrew (Global Biodiversity Outlook 3,94) in 1984 identified need to study biodiversity and genetic resources with special focus on **underexploited food plants.**
5. First International workshop on Biodiversity and Genetic Resources held at Royal Botanic Gardens, Kew (Commonwealth Secretariat, 1986) in September 1986, launched International Biodiversity and Genetic Resources Programme (BDGR) on Underexploited plant species.

6. Life support plant species (Commonwealth Secretariat, 1987) and crops used by local people in extreme conditions such as soil salinity, flood, drought, famine and emergencies for Asia (in collaboration with National Bureau of Plant Genetic resources, New Delhi, India).
7. World Conservation Union (IUCN) and Commonwealth Science Council (CSC) (Secretariat convention on Biological Diversity 001) in 1988 identified for immediate protection, monitoring, and potential application KEY species which are of critical importance. It stated "some of the species (or species groups) which are under threat possess special economic or cultural significance, are essential for the survival of the ecosystems and landscapes in which they occur…there is increasing concern for species having particular ecological, economic and cultural significance…. although conservation priorities are recognized for species of obvious economic importance ....many other species of key ecological and socioeconomic importance will be lost unless action is taken urgently within the next decade. An unknown fraction of world's species used by people for food or for their culture, makes disproportionate contribution to people's social and economic well being. Such species are called service species or socioeconomic species; both ecological and socioeconomic species can be called Life Support Species" (Kapoor-Vijay & Usher, 1993).
8. Appropriate computerized databases of relevance to species and taxon of importance were initiated, e.g. International Legume database information system (ILDIS) in collaboration with National Botanical Research Institute, Lucknow, India (Kapoor-Vijay, 1992)
9. The Food and Agricultural Organization's (FAO) Food Policy and nutrition Division in 1980 started a programme on Underexploited Traditional Food Plants in 1980. Joint initiatives of World Health Organization (WHO) and FAO began joint work in 1986 to increase consumption of locally produced foods to increase diet diversity and prevent diseases form nutritional, micro nutritional and vitamin deficiency(www.fao.org)
10. World Bank's Biodiversity task Force meeting held in 1988 at which the Commonwealth's BDGR (Kapoor, 1988) programme invited to present and contribute towards the Global Biodiversity Action Plan.
11. Global Biodiversity Strategy is launched, Commonwealth's Training programmes initiated in: Ethnobotany (in Asia pacific), Conservation Biology (in Southern African countries), Inexpensive methods of Survey, authentication, characterization, in Africa; Herbarium Development at Royal Botanic gardens for pan-Commonwealth nations. Medicinal Plants work in Caribbean, Asia-pacific and Africa. Iwokorama Rain forest programme was initiated and designed for implementation in 1990-91.

B. After the launch of the Convention on Biodiversity (CBD) in 1992.
Salient landmarks are:

12. Second International workshop of BDGR and its Advisory Committee reviewed the work in June1992 immediately after Rio Summit, and recommended next phase BDGR to expand and include other species including microorganisms (Kapoor-Vijay, 1988).
13. Biodiversity International (formerly known as IPGRI) worked on conservation and use of neglected and underutilized plant species in the 1990s. Its Global Initiative on Neglected Crops funded by IFAD had several projects on underutilized plants. In 2008 an International centre called "Crops for Future" was recommended, is now in operation in Malaysia since 2010.
14. Global Forum on Agricultural Research (GFAR) in 2002 set up a Global Facilitation Unit for underutilized species (GFU) funded by Germany. An International Centre for Underutilized crops (ICUC) was created in 1989 in UK at Southampton University. Two workshops on Underutilized species organized in 2006 and 2008.
15. European Union has a multi-organizational regional programme in Africa, led by the Netherlands. In addition many NGOs, projects by many national agriculture research centres and International agriculture centres focus on underutilized crops.
16. USA has a Collaborative Crop Research Programme on *Chenopodium quinoa* (Quinoa) in Bolivia, on Millets in Africa and India, on Lupin in Ecuador.
17. UN's Millennium Development Goals in 2005 have recognized need to invigorate farming systems in developing countries where cultivation of local species using traditional methods of sustainable farming. The 2010 report, the target of MDG were not met (GBO-3.2010)
18. International Conferences organized on underutilized plants by MSSF foundation, Chennai, India(2002, 2006- www.msrf.org)
19. Calls made during e-conferences on Future of Andes and Himalayas facilitated by International centre of Mountain Development for supporting scientific programmes enabling protection, sustainable use and conservation of life support species and ecosystems (www.icimod.org)
20. In 2006, COP-8 of the CBD gave high priority attention to the use of local species by local communities. FAO, IPGRI, WHO and Standing Committee on Nutrition (SCN) were requested to implement the cross cutting initiatives on biodiversity for food and nutrition (www.cbd.org).
21. A review chapter written on Locally used species for food and nutrition in 2006 for a book (Kapoor-Vijay, 2006).

22. Scientific research, investigations and reviews on their use on locally used species, medicinal plants, grain crops, and pseudo cereals have provided a deeper understanding on its genetic and species diversity and variability. Work on ecological, ethno botanical, population biology information, and research on genetic relationships among cultivated and wild populations of Chenopodium species (Kapoor-Vijay, 2006; Kapoor, 1978; Pratap & Kapoor, 1985) (known as Quinoa in Bolivia, Peru, Chile and Latin American countries, and Bathu in Himachal Pradesh, India) has given new direction in their characterization and conservation. Chenopodium album populations are consumed for their nutrition and energy value in Himalayas. (Kapoor, 1978; Bhargava et al. 2007)

CONCLUSION

Tools and technologies are growing fast in the area of biological knowledge. Countries like India are now the leaders not only just in the study, sustainable use, but also conservation and protection of biodiversity. The focus of the countries with rich biodiversity should be on the: (i) survey, (ii) evaluation, (iii) use, and (iv) protection of their gene wealth embedded in populations, species and gene pool diversity.

Systematic inputs from traditional knowledge and modern, emerging scientific and technological know-how should increase. Heightened awareness should be created among policy makers of the importance of locally used and preferred species, the life support species, and their role in maintaining ecosystem integrity. They need to become cognisant of the fact that Scientific and Community based knowledge of biodiversity and genetic diversity of species is the key in saving humanity from hunger, malnutrition and health burdens, it can provide answers to all such problems. Existing initiatives based on the hybrid approach of using knowledge and field experience of both the affluence and scarcity based approaches related to biodiversity of species and their diversity and variability need to be integrated with initiatives on protected areas and ecosystems especially in extreme and changing global environments. Strengthening population biology, conservation biology, ethno-biology approaches, with conventional breeding methods and genomic technologies should become a central component of all programmes and actions that seek to conserve and use biodiversity in a sustainable manner.

Financial and human resources necessary for such work, underpinned by an appropriate legal framework, will need to be provided over a very long period of time for the work of life and its diversity must continue indefinitely to ensure future generations of humans can, in the first instance, survive and then benefit and enjoy prevailing biodiversity through well designed and synchronized actions across nations.

ACKNOWLEDGEMENTS

I would like to thank all the scientists, who gave their precious time and knowledge to BDGR programme. My special thanks to national coordinators who gave their valuable practical experience and various institutions involved in implementation of activities of the programme.

REFERENCES

Bent J. Stephen, Forney, J. Larry (2008). Tragedy of uncommons: Understanding limitations in the analysis of microbial diversity. *ISME Journal,* 2: 689–695.

Bhargava, A. Shukla, S., Kumar, R., Ohri, D. (2009). Metrogylph analysis of morphological variations in *Chenopodium spp. World J. Agric. Sci.,* 5: 117–120.

CSC (1986). *Biological Diversity and Genetic Resources.* Summary of an International Workshop held at Royal Botanic Gardens, Kew, London, UK, October, (1986). RNR-7, TP225.

Commonwealth Secretariat. (1987). *Life Support Species in Asia-Pacific Region.* CSC(87) RNR-9, TP232.

Dins dale, E., Dins dale, A., Edwards, R.A, Hall, Diana, Angle. et al. (2008). Functional met a genomic profiling of nine biomes. *Nature,* Vol. 452, No. 7187. pp. 629–632.

Hammond, P.M. (1995). *The Current Magnitude of Biodiversity in Global Biodiversity Assessment.* Heywood, V.H. (Ed.). Cambridge University Press.

Handbook of the Convention on Biological Diversity. Secretariat of the Convention on Biological Diversity (001) Earthscan Publications Ltd., ISBN 1-85383-737-7.

Harris. W. and Kapoor, P. (1988). *Nga MahiMoori O Wao Nui A Tane.,* DSIR New Zealand (*Contributions to an International Workshop on Ethno-botany* held in Christchurch, New Zealand). ISBN 0-477-02579-X.

Kala, Chander Prakash, Dhyani, Pitamabar P. and Sajwan, Bikram Singh. (2006). Developing the medicinal plant sector in northern India: Challenges and opportunities. *Journal of Ethno-biology and Ethno-medicine*, 2: 32.

Kapoor, P. (1978). Exploration of plant resources of Himachal Pradesh as food source. In: *Proceedings National Seminar Resources Development and Environment in Himalayas*. Deptt. Science and Technology, Govt. of India, pp. 153–163.

________. (1988). *Biological Diversity and Genetic Resources: Status Report*. Commonwealth Science Council, CSC(88)RNR-17, Commonwealth Secretariat, p. 23.

Kapoor, Vijay. (2006). *Locally used Species for Food and Nutrition*. Communicated (unpublished).

Kapoor-Vijay, P. and James, White. (1992). *Conservation Biology*: A training manual for biological diversity and genetic resources. Commonwealth Secretariat, London. (Ed.). ISBN 0-850-923-921, p. 248.

Kapoor-Vijay, P. and M.B. (1993). Usher key species: Identification of key species for conservation and socioeconomic development. Commonwealth Secretariat, London UK. (Ed.). ISBN 0-85092-403-0, p. 198.

Kapoor-Vijay, P. (1992). Biological diversity and genetic resources: The programme of the commonwealth science council, Commonwealth Secretariat, London. ISBN 0-85092-388-3, p. 145.

________. (2009). *Future of Underexploited Plants a Vision for 2030*. NCCR, Berne, Switzerland.

________. (2009). Mountain biodiversity, genetic resources and cultures. *Mountain Forum Bulletin*. Vol. 9. Issue 2, ISSN 1815–2139.

Kapoor-Vijay, P. Appanah, S. and Saulei, S.M. (1993). *Tropical Forest Ecology and Management in the Asia-Pacific Region*. Commonwealth Secretariat, London UK. (Ed.). ISBN 983-9592-03-3.

Kendrew, John. (1984). Science for technology for development: An expanded programme of scientific cooperation in the commonwealth. p. 89.

May, Robert. (2005). *Threats to Tomorrow's World*. Anniversary address, The Royal Society. London, UK, p. 28.

National Academy of Sciences (1975). *Underexploited Tropical Plants with Promising Economic Value*. Report of an Ad Hoc panel. National Academy of Sciences, Washington, D.C., p. 188.

Paroda, R.S., Kapoor, P., Arora R.K., Mal, Bhag. (1988). *Life Support Plant Species: Diversity and Conservation*. National Bureau of Plant Genetic Resources, ICAR, New Delhi, India. (Ed.). CSC (88) RNR-12 TP 24.

Pratap, T. and Kapoor. (1985). *Agriculture Ecosystems and Environment.* Vol. 14(3): 185–199.

Pratap, T. and Kapoor, P. (1985). The Himalayan grain chenopods. II. comparative morphology. *Agriculture, Ecosystems and Environment,* 14(4): 201–220.

Pratap, T. and Kapoor, P. (1987). The Himalayan grain chenopods. III. An underexploited food plant with promising potential. *Agriculture, Ecosystems and Environment,* 19(1): 71–79.

Schilchting and Pigilucci. (1998). *Phenotypic Evolution: A Reaction Norm Perspective.* Sinauer Associates. Inc. MA, USA. 387.

Schlichting C.D. (2004). *The Diversity of Complexity in Phenotypic Integeration.* Piggliucci and Preston. (Ed.). Oxford University Press, p. 443.

Secretariat of the Convention on Biological Diversity (2010). *Global Biodiversity Outlook* 3. Montreal, p. 94.

Slootweg, R., Rajwanshi, A., Mathur, V. Kolhoff, A. (2009). *Biodiversity in Environmental Assessment.* (Eds.). Cambridge University Press, p. 456.

Wilson, E.O. (1985). *Biodiversity.* (Ed.). National Academy Press, Washington. D.C., p. 521.

19

Strategies for Biodiversity Conservation in India

A.K. Ghosh

INTRODUCTION

'Biological Diversity' in India has been recognized unique in terms of richness of biota vis-à-vis limitation of geographical area. Nearly 1,37,000 species have been recorded from all major ecosystems; it is recognized as one of the eight Centres of Origin and Diversity of Crop Plants in the world; endowed with 300 wild ancestors and close relatives of cultivated plants and 140 native breeds of farm livestock, India has rightfully placed in the list of global megadiverse countries.

India has recognized Convention on Biological Diversity (CBD) and ratified the same in 18th February, 1994 and within next eight years enacted Biological Diversity Act, 2002 followed by Rules, 2004. The act was passed to ensure 'conservation, sustainable utilization and equitable sharing of the benefit arising out of utilization of genetic resources'.

India has announced 'National Forest Policy' in 1988, 'National Environment policy' in 1992 & again in 2006 and enacted 'Wildlife (Protection) Act', 1972, 'Forest (Conservation) Act', 1992, before the 'Biological Diversity Act', 2002. The 'Environment (Protection) Act' of India, 1986 laid the foundation of EIA Notification 1994, now amended.

The basis of India's strategy for Biodiversity Conservation can be traced back to the foregoing major policies and laws. 'National Biodiversity Action Plan' approved by the Union Cabinet on 6th November, 2008 provided a Roadmap for the future.

STRATEGIES

The National Action Plan proposed Nine Action Areas and spelt out 124 Action Points based on following 11 key points:

1. Strengthening and integrating *in-situ*, on farm and *ex-situ* conservation.
2. Augmentation of Natural Resource Base and its sustainable utilization, ensuring intra generational equity.
3. Regulation, introduction and managing invasive species.
4. Assessment of vulnerability and adaptation to Climate Change & Desertification.
5. Integration of Biodiversity concern in Economic and Social Development.
6. Impact of Pollution and initiate better control.
7. Developing and integrating biodiversity database.
8. Strengthening implementation of policy legislation and administrative measure.
9. Capacity Building for biodiversity conservation & use of new technologies.
10. Valuation of goods & services provided by Biodiversity and use of economic instrument in decision making.
11. International cooperation with UN & Regional bodies.

The on-going programmes to assess as far the key-points are being supported can now be briefly reviewed.

India has 96 National Parks and 509 wildlife sanctuaries distributed over 28 states and 06 union territories;

Table 1. State-wise Details of the Protected Area Network of the Country

S. N.	*States / UT's*	*Number of National Parks*	*Number of Wildlife Sanctuaries*
1.	Andhra Pradesh	4	22
2.	Arunachal Pradesh	2	11
3.	Assam	5	20
4.	Bihar	1	11
5.	Chhattisgarh	3	10
6.	Goa	1	07
7.	Gujarat	4	21
8.	Haryana	2	10

9.	Himachal Pradesh	2	32
10.	Jammu & Kashmir	4	16
11.	Jharkhand	1	10
12.	Karnataka	5	21
13.	Kerala	6	13
14.	Madhya Pradesh	9	25
15.	Maharashtra	6	35
16.	Manipur	2	05
17.	Meghalaya	2	03
18.	Mizoram	2	07
19.	Nagaland	1	03
20.	Orissa	2	18
21.	Punjab	Nil	10
22.	Rajasthan	5	23
23.	Sikkim	1	06
24.	Tamil Nadu	5	20
25.	Tripura	Nil	04
26.	Uttar Pradesh	1	23
27.	Uttaranchal	5	06
28.	West Bengal	6	15
29.	Andaman & Nicobar	9	96
30.	Chandigarh	Nil	02
31.	Dadar & Nager Haveli	Nil	01
32.	Lakshadweep	Nil	01
33.	Daman & Diu	Nil	01
34.	Delhi	Nil	01
Total		**96**	509
GRAND TOTAL		**605**	

[*Source:* NBAP, MoEF, GoI, 2008]

Besides, 38 tiger Reserves (Table 2), 11 Elephant reserves (Table 3) and 03 conservation reserves have been set up with keystone species in terms of ecosystems; 16 Biosphere Reserves (Table 4) have been notified and 14 more have been identified.

Table 2. Reserve-wise release for Centrally Sponsored Scheme of Project Tiger, During 2009-10 (as on 03.03.2010)

(Rs. In lakhs)

S. N.	*Tiger Reserve*	*State*	*Total Amount Released*
1.	Nagarjunsagar	Andhra Pradesh	94.9100
2.	Namdhapa	Arunachal Pradesh	–

3.	Pokke	Arunachal Pradesh	48.8600
4.	Kaziranga	Assam	165.0000
5.	Manas	Assam	–
6.	Nameri	Assam	19.7700
7.	Valmiki	Bihar	8.000
8.	Achanakmar	Chhattisgarh	1193.5000
9.	Indravati	Chhattisgarh	42.1500
10.	Udanti-Sitanadi	Chhattisgarh	103.0500
11.	Palamau	Jharkhand	110.7350
12.	Bandipur	Karnataka	161.6970
13.	Bhadra	Karnataka	128.0870
14.	Danheli Anshi	Karnataka	144.3700
15.	Nagerhole	Karnataka	210.8180
16.	Periyar	Kerala	151.8000
17.	Bandhavgarh	Madhya Pradesh	159.9600
18.	Kanha	Madhya Pradesh	277.0600
19.	Panna	Madhya Pradesh	175.8950
20.	Pench	Madhya Pradesh	158.3200
21.	Sanjay Dubri	Madhya Pradesh	145.8400
22.	Satpura	Madhya Pradesh	1502.3710
23.	Melghat	Maharashtra	155.1850
24.	Pench	Maharashtra	75.8720
25.	Tadoba-Andheri	Maharashtra	131.8200
26.	Dampa	Mizoram	2171.0000
27.	Satkoshia	Orissa	127.7300
28.	Simlipal	Orissa	42.3500
29.	Ranthambore	Rajasthan	10560.0000
30.	Sariska	Rajasthan	134.1700
31.	KMTR	Tamil Nadu	138.4550
32.	Mudumalai	Tamil Nadu	51.8540
33.	Corbett Tiger	Uttarakhand	241.7050
34.	Buxa	West Bengal	38.5800
35.	Sundarbans	West Bengal	259.9700
36.	(i) Dudhwa	Uttar Pradesh	128.4800
37.	(ii) Katerniaghat	Uttar Pradesh	50.2450
38.	Annamalai	Tamilnadu	129.3600
39.	Parambi Kulam	Kerala	129.3600
Total			**19724.93**

[*Source:* MoEF, Annual Report, 2009-2010]

Table 3. Elephant Reserves of India

S. No.	*Elephant Range*	*Elephant Reserve with date of notification*	*State*	*Total Area (Sq. Km.)*
I. II.	Eastern India (South West Bengal,	1. Mayurjharna ER (24.10.02)	West Bengal	414.00
III. IV.	Jharkhand & Orissa	2. Singhbhum ER (26.09.01)	Jharkhand	4530.00
V. VI.		3. Mayurbhanj ER (29.09.01)	Orissa	3214.00
VII.		4. Mahanadi ER (20.07.02)	Orissa	1038.00
VIII.		5. Sambalpur ER (27.03.02)	Orissa	427.00
IX.		6. Baitarani ER #	Orissa	1755.00
		7. South Orissa ER #	Orissa	4216.00
		8. Lemru #	Chattisgarh	450.00
		9. Badalkhol-Tamorpingla #	Chattisgarh	1048.30
	Total			**17092.30**
X.	North Brahamaputra	10. Kameng ER (19.06.02)	Arunachal	1892.00
XI.	(Arunachal – Assam)	11. Sonitpur ER (06.03.03)	Assam	1420.00
	Total			**3312.00**
XII. XIII.	South Brahamaputra (Assam –Arunachal)	12. Dihing-Patkai ER (17.04. 03)	Assam	937.00
XIV.		13. South Arunachal ER (29.02. 08)	Arunachal	1957.50
	Total			**2894.50**
XV. XVI.	Kagiranga (Assam – Nagaland)	14. Kagiranga-Karbi Anglong ER (17.04.03)	Assam	3270.00
XVII. XVIII.		15. Dhansiri-Lungding ER (19.04.03)	Assam	2740.00
		16. Intaki ER (20.02.05)	Nagaland	202.00
	Total			**6212.00**
XIX. XX.	Eastern Dooars (Assam – West	17. Chirang-Ripu ER (07.03.03)	Assam	2600.00
XXI.	Bengal)	18. Eastern Dooars ER (28.08.02)	West Bengal	978.00
	Total			**3578.00**
XXII.	Eastern Himalayas	19. Garo Hills ER (31.10.01)	Meghalaya	3500.00
XXIII.	(Meghalaya)	20. Khasi Hills ER #	Meghalaya	1331.00
XXIV.	**Total**			**4831.00**
XXV.	Nilgiri – Eastern	21. Mysore ER (25.11.02)	Karnataka	6724.00
XXVI.	Ghat	22. Wayanad ER (02.04.02)	Kerala	1200.00
XXVII.	(Karnataka – Kerala –	23. Nilgiri ER (19.09.03)	Tamilnadu	4663.00
XXVIII. XXIX.	Tamilnadu – Andhra Pradesh)	24. Rayala ER (09.12.03)	Andhra	766.00
	Total			**13353.00**
XXX.	South Nilgiri	25. Nilambur ER (02.04.02)	Kerala	1419.00
XXXI.	(Kerala – Tamilnadu)	26. Coimbatore (19.09.03)	Tamilnadu	566.00
XXXII.	**Total**			**1985.00**

XXXIII. XXXIV. XXXV.	Western Ghat (Tamilnadu – Kerala)	27. Annamalai ER (19.09.03)	Tamilnadu	1457.00
		28. Anamudi ER (02.04.02)	Kerala	3728.00
	Total			**5185.00**
XXXVI. XXXVII. XXXVIII.	Periyar (Kerala – Tamilnadu)	29. Periyar (02.04.02)	Kerala	3742.00
		30. Srivilliputtur ER (19.09.03)	Tamilnadu	1249.00
	Total			**4991.00**
XXXIX. XL.	Northern India (Uttaranchal – U.P.)	31. Shivalik ER (28.10.02	Uttaranchal	5405.00
		32. Uttar Pradesh ER (09.09.09)	U.P.	744.00
	Total			**6149.00**
		Total		**69582.80**

Approved by Government of India, but not yet notified by the State government
[*Source:* MoEF, Annual Report, 2009-2010]

Table 4. Biosphere Reserves in India

S. No.	*Name of the Biosphere Reserve & total Geographical Area (km²)*	*Date of Notification*	*Location (State)*
1.	**Nilgiri** (5520)	01.08.1986	Parts of Wayanad, Nagerhole, Bandipur and Madumalai, Nilambur, Silent Valley and Siruvani hills (Tamil Nadu, Kerala and Karnataka)
2.	**Nanda Devi** (6497.03)	18.01.1988	Parts of Chamoli, Pithoragarh & Almora Districts and Valley of Flowers (Uttarakhand)
3.	Nokrek (820)	01.09.1988	Parts of Garo Hills (Meghalaya)
4.	Manas (2837)	14.03.1989	Parts of Kokrajhar, Bongaigaon, Barpeta, Nalbari, Kamprup and Darang Districts (Assam)
5.	**Sunderbans** (9630)	29.03.1989	Parts of delta of Ganges & Brahamaputra river system (West Bengal)
6.	**Gulf of Mannar** (10500)	18.02.1989	Indian part of Gulf of Mannar between India and Sri Lanka (Tamil Nadu)
7.	Great Nicobar (885)	06.01.1989	Southern most islands of Andaman and Nicobar (A&N Islands)
8.	Simlipal (4374)	21.06.1994	Parts of Mayurbhanj district (Orissa)
9.	Dibru-Saikhowa (765)	28.07.1997	Parts of Dibrugarh and Tinsukia districts (Assam)
10.	Dehang Debang (5111.5)	02.09.1998	Parts of Siang and Debang valley in Arunachal Pradesh
11.	Kanchanjanga (2619.92)	07.02.2000	Parts of North and West Sikkim.

12.	Panchmari (4926.28)	03.03.99	Parts of Betur, Hoshangabad and Chhindwara, Distt. of Madhya Pradesh.
13.	Agasthyamalai (3500.36)	12.11.2001 (Area expanded on 30.3.2005)	Parts of Tirunelveli and Kanya Kumari Districts in Tamil Nadu and Thiruvananthapuram, Kollam and Pathanamthitta.
14.	Achanakmar - Amarkantak (3835.51)	30.03.2005	Parts of Anuppur and Dindori districts of Madhya Pradesh and Parts of Bilaspur district of Chhattisgarh State
15.	Kachch (12454)	29.01.2008	Parts of Kachch, Rajkot, Surendranagar and Patan district of Gujarat

* *Sites with **bold** letters have been recognized by UNESCO on World Network of Biosphere Reserves.*

[*Source:* NBAP, MoEF, GoI, 2008]

Wetland and Lakes also form important components of aquatic ecosystem; 104 wetlands in 28 states (Table 5) have been brought under National Wetland Conservation Programme; more than 30 projects (Table 6) have been supported under Lake Conservation Programme. As such both species oriented and ecosystem based *in-situ* conservation programmes are providing support to the National Action Plan. Additionally, through the surveys, conducted by 'Botanical survey of India' and 'Zoological Survey of India', potential areas for future conservation are being identified.

Table 5. State wise Distribution of Wetlands under National Wetland Conservation & Management Programme

State	*Number of Wetlands*	*Area (ha)*
Andhra Pradesh	1	90100
Assam	2	4504
Bihar	3	11490
Chandigarh	1	148
Gujarat	8	1270875
Himachal Pradesh	5	15739
Haryana	2	288
Jammu and Kashmir	10	120450
Jharkhand	2	98965
Karnataka	7	4297
Kerala	5	213229
Madhya Pradesh	12	359814
Maharashtra	3	40298
Manipur	1	26600
Meghalaya	1	22150
Mizoram	2	185

Orissa	4	120407
Punjab	4	5965
Rajasthan	1	24000
Sikkim	6	164
Tamil Nadu	3	46283
Tripura	1	240
Uttar Pradesh	12	14080
Uttarakhand	2	1244
West Bengal	6	553099

[*Source:* NBAP, MoEF, GoI, 2008]

Table 6. List of 31 Projects for Conservation of 46 Lakes as on March 2007

S.No.	*Lake*	*State*
1.	Banjara	Andhra Pradesh
2.	Dal Lake, Srinagar,	J&K Jammu & Kashmir
3.	3 lakes of Bangalore namely Vengaiahnkere, Nagavara and Jarganahalli	Karnataka
4.	Bellandur lake, Bangalore	Karnataka
5.	Kotekere, Belgaum	Karnataka
6.	Bhishma, Gadag	Karnataka
7.	Lal Bagh, Bangalore	Karnataka
8.	Sharanabasveshwara	Karnataka
9.	Akkamahadevi, Haveri	Karnataka
10.	Chanapatna, Hasan	Karnataka
11.	Veli Akkulum, Thiruvananthapuram	Kerala
12.	Powai	Maharashtra
13.	9 lakes in Thane	Maharashtra
14.	Mahalaxmi lake, Vadagaon	Maharashtra
15.	Mansagar, Jaipur	Rajasthan
16.	Ooty	Tamil Nadu
17.	Kodaikanal	Tamil Nadu
18.	3 lakes of Agartala	Tripura
19.	4 lakes of Nainital District	Uttaranchal
20.	Nainital lake	Uttaranchal
21.	Rabindra Sarovar, Kolkata	West Bengal
22.	Mirik	West Bengal
23.	Bindusagar, Bhubaneswar	Orissa
24.	Kundwad lake, Davengere, Karnataka	Karnataka
25.	Rani Talab, Riva	Madhya Pradesh
26.	Kotetavarekere, Chickmaglore	Karnataka
27.	Tripuranthekeshwar, Bidar	Karnataka
28.	Rankala, Kolhapur	Maharashtra
29.	Varhala Devi, Bhiwandi	Maharashtra
30.	Sagar, Sagar	Madhya Pradesh
31.	Mansi Ganga	Uttar Pradesh

[*Source:* NBAP, MoEF, GoI, 2008]

In terms of *ex-situ* conservation, the 'Central Zoo Authority' set up in 1992 is mandated in oversee functioning of Indian Zoological Park/Gardens and to impose minimum standards and norms. For *ex-situ* conservation of flora, India has more than a century old practice of setting up Botanical Gardens. A central 'Botanic Garden of Indian region' (BGIR) at Noida, near New Delhi, has been set up 'to facilitate *ex-situ* conservation and propagation of rare and threatened indigenous plants of the country' (MoEF, Annual Report, 2007-2008) and serve as a 'Centre of Excellence'. The total numbers of zoological gardens and parks in India, since 1800 is reported to be 355. Not all zoological or botanical gardens are actively involved in research but even then most of them help in propagating the number of population to select species under captivity.

Ex-situ conservation of wild biodiversity apart, Indian Council of Agricultural Research has set up during post independence period, four major Bureaus for conservation and propagation of germplasm of Plants of agricultural importance (NBPGR), breeds of domesticated animals (NBAGR) & of fishes (NBFGR) and microbial organisms (NBMGR). These institutional set ups provide major support towards conservation of agrobiodiversity in India.

A green accounting programme has been initiated for India under 'Green Accounting for Indian States' project (www.gistindia.org). Out of eight monographs published so far, the monograph no. #4 of the programme is on 'The Value of Biodiversity in Indian Forest' (Gundimeda et al. 2006); such work will undoubtedly contribute towards Integration of Biodiversity Concern in Economic and Social Development (Action Point 5 & 10). Similarly, a project on Biodiversity a metadata base has been initiated by Government of India (Action point 7).

India has taken up a massive programme of environmental education in school, college and university level; the Environmental Science curricula in all Masters Degree level studies include 'Biodiversity' as a major component (Action point 9).

Being a part of CBD, India has established link at International level with UNDP, UNEP and other UN based organizations, as also with The World Bank and International NGOs like IUCN and WWF; India is also an active partner in SAARC and SACEP programmes on Biodiversity and provided valuable leadership support in the early years on assessment of faunal and floral biodiversity in South Asian countries (Action point 11).

During 2010 – the 'International Year of Biodiversity' – the 'National Environment Awareness Campaign' (NEAC), a programme for mass awareness on environment issues, promoted by Ministry of Environment & Forests, Government of India, made 'Biodiversity' as the focal theme. All major funding agencies including University Grant Commission, ICAR and Department of Science & Technology, besides Ministry of Environment and Forests supported a series of programmes, across the country. The initiative of India in conserving Biodiversity perhaps prompted CBD members to make India the venue for next conference of the Parties (COP 11) meet in India.

THE BIOLOGICAL DIVERSITY ACT, 2002 AND RULES, 2004

The seminal act passed by the Parliament and notified on February 5, 2003, by the Ministry of Law & Justice, provided the basis of all future programme and passed challenging agenda for conservation with People's participation. In a multilingual and multiethnic country, the new law obviously is not easy to implement but the democratic process of functioning of the set up proposed in the law is perhaps unique amongst 15 countries that have so far enacted separate law or passed ordinance.

The act stipulated a 3 tier system for governance of country's bioresource. Following CBD principle, it has strongly advocated a principle of benefit sharing in case bioresource is accessed with prior informed consent for commercial uses. As per law, National Biodiversity Authority (NBA <www.nbaindia.org>) has become functional since 2003 and 25 of the 28 States of India have set up State Biodiversity Board. The law envisages a proactive role at community level and recommends 07 members for 'Biodiversity Management Committee' (BMC) at the local level; all members are to be nominated by the local community with reservation for 02 women members and 18% from scheduled caste and tribal group.

To promote conservation and sustainable use, the preparation of 'People's Biodiversity Register' (PBR) is a novel concept. Such register in local language will act as the basis before a decision of access is taken up. Setting up Biodiversity Fund at National, State, Local level to augment conservation and provide incentive are other salient features of the act.

Such massive exercises need time. Civil Society Organizations also play a vital role and can act to review the appropriateness of action of NBA. 'BioWatch' – a website launched by NGO 'Kalpavriksh', in Pune, provides one such portal.

NBA has set up several expert committees on such subject as ABS, Database, format of PBR, Biodiversity Heritage Site, IPR, National

Repository, Agrobiodiversity etc. Wide institutional representation makes such effort more democratic and functional.

India has successfully mobilised external support to implement the provision of the Act. An UNDP Project has been launched in forest based biodiversity-rich states of Jharkhand and Madhya Pradesh in 2009. Similarly, an UNEP-GEF Project has been finalized with the objective to set up examples of funding 40 ABS instruments in five states of India viz., Andhra Pradesh, Gujarat, Himachal Pradesh, Sikkim and West Bengal, taking Agriculture, Forest and Wetland ecosystem as the major area of operation (December 2010 – December 2013); each programme is to have a component of co-financing at near 1:2 ratio (total outlay $9789000 at 36.37% and 63.63% ratio of GEF Trust fund and co-financing contribution). A separate project is being run by institute of Forest Biodiversity, Ranchi on 'Biodiversity Conservation through Community based natural Resource Management in the State of Jharkhand' in eastern India. The World Bank has lent support to WWF-I Project on preparation of 'State of Art Report on Biodiversity of Indian Sundarbans. These are just but a few examples of content programmes on Biodiversity Conservation. Obviously, lessons learnt from such diverse project outcome will help initiate effective action.

THE ROLE OF CIVIL SOCIETY & ACADEMIA

The Conservation of Biological Diversity and Traditional Knowledge connected therewith can never be achieved only with policy instrument and laws. The Society at large has to understand and appreciate the fact that even in 21st Century. Human Society depends heavily on living natural resources for its sustenance. That sustainable use can provide long term benefit can be focused only with field based academic research and participation of the Civil Society Organizations.

India has for long supported this idea. The benefit of use of indigenous seeds, non-traditional crops like small millets, chemical free agriculture have been appreciated through the untiring efforts of Community Based Organisations (CBOs) with technical support from liberal academicians. Today, one of the largest projects in conservation, cultivation and sustainable uses of medicinal plants is run by a CBO, 'Foundation for Revitalisation of Local Health Traditions' (FRLHT), in Karnataka. Similarly, largest numbers of indigenous rice varieties are being conserved on-farm by a West Bengal based NGO called 'Basudha' set up by 'Centre for Interdisciplinary Studies'. These are only possible with participation of the communities. NGOs like

'Deccan Development Society' in Andhra Pradesh, 'Development, Research, Communication & Service Centre' (DRCSC) in West Bengal, 'Research Foundation for Science, Technology & Ecology' in New Delhi & Dehradun, 'ENDEV-Society for Environment & Development' in West Bengal have all been active in propagating the message of conservation of Agrobiodiversity and associated Traditional Knowledge.

In case of Wild Biodiversity, Indian Civil Society has a long tradition of significant contribution from the days of Asiatic Society of Bengal in 18th Century, to Bombay Natural History Society in 19th Century, on to numerous wildlife related NGOs in the 20th Century. Many CBOs have appropriately shown the benefit of 'Ecotourism' with natural capital and local knowledge as the main resource.

The Role of Academic Institutions has largely been in the area of basic researches on Biodiversity of different ecosystem, Ecosystem services, Species richness and diversity, impact of Climate Change on Biodiversity. Basic research apart, Centre for Ecological Studies at the Indian Academy of Science, Bangaluru, contributed significantly towards developing a concept on People's Biodiversity Registers (PBRs) and completing 50 PBRs from South India long before the Biological Diversity Act, 2002, was passed.

INSTITUTIONAL COOPERATION

India has not only received support from UNDP, UNEP-GEF, The World Bank, DFID, JAICA, WWF, IUCN and other Institutional Agencies, for Conservation & Sustainable Utilisation of Biological Diversity, it has rendered valuable support to the South Asian Programme on Assessment of Biodiversity. India provided training to selected country expert on assessment of faunal and later floral diversity, with support from 'South Asian Cooperative Environment Programme' (SACEP), located in Sri Lanka, aided by NORAD. India actively participated in BioNet International's efforts to build up regional network and loop.

India continues to be an active participant in all the meetings of Conference of the Parties (COP) and keep in close linkage with CBD Secretariat in Montreal, Canada.

The International Cooperation in Biodiversity Conservation, especially in the area of exchange of information and material purely for academic research, continues through the initiatives of National level survey organisations, CSIR and ICAR and collaborative projects of Universities.

TASKS AHEAD

The Government of India and all the State Governments and Union Territories need to take urgent action programme and provide support towards formation of 'Biodiversity Management Committee' (BMC) at least at the level of Community Development Blocks in the Rural areas and at the level of Local Self Government in the Urban areas. The preparation of 'People's Biodiversity Register' (PBR) can only become useful, once the BMCs are in place. Additionally, setting up Local Biodiversity Fund is most vital to provide incentives to the conservers and providers of biodiversity and traditional knowledge related to bioresource utilisation. It is also becoming imperative that a system of payment for loss of biodiversity due to development projects is to be made mandatory; this may be applicable for forest ecosystem, agricultural ecosystem and wetland ecosystem. Currently, there is a system to charge 'Net Present Value' (NPV) for loss of goods and services from the forests (in case of conversion of forest land for non-forestry purposes), as also for compensatory afforestation and wildlife conservation. None of this revenue is diverted to the Local or State Biodiversity Fund. A review of the current system may help to provide support to the Local Biodiversity Fund, once the BMCs start functioning.

It is essential to understand that the future of Biodiversity Conservation depends on slowing down habitat destruction, combating invasive species and increasing pollution, controlling human over population and over harvesting of wild species; E O Wilson suggested an acronym for all these factors: HIPPO. One would like to add to the list, the issue of 'Combating Climate Change' through adaptation and mitigation process.

REFERENCES

MoEF, (2008). *National Biodiversity Action Plan*. Ministry of Environment and Forests, Government of India, New Delhi.

MoEF, (2008). *Annual Report, 2007–2008*. Ministry of Environment and Forests, Government of India, New Delhi.

MoEF, (2010). *Annual Report, 2009–2010*. Ministry of Environment and Forests, Government of India, New Delhi.

NBA, (2004). *Biological Diversity Act, 2002 and Biological Diversity Rules, 2004*. National Biodiversity Authority, India, Chennai.

20

Strategies for the Conservation of Crops Wild Relatives – Indian Context

S.K. Sharma and J.C. Rana

INTRODUCTION

The Crops Wild Relatives (CWR) are the species closely related to crop plants including crop progenitors, and may contribute beneficial traits such as resistance to biotic and abiotic stresses, yield improvement and other quality traits (Hoyt, 1998; Maxted et al. 2008). CWR are, therefore have been identified as critical resources for food security and long-term sustainability of agricultural production systems (Hoyt, 1998, Prescott-Allen and Prescott Allen, 1986, Swaminathan, 2010). Historically, conservation of plant genetic resources (PGR) has focused almost explicitly on cultivated plants (Maxted et al. 2008), however, in the last about one decade the importance of CWR to breed varieties for higher yield and to tolerate variety of stresses likely to be emerged from changing climate have been recognized and given priority for their conservation and use (CBD, 2002). As we see the number of hungry and malnutrition people has increased from 854 million in 2006 to 1.02 billion in 2009 the world over, and more particularly in Asia and the Pacific and Sub-Saharan Africa (FAO, 2009), despite the world leaders' pledge in November 1996, reaffirmed in June 2002 to reduce the number of hungry and malnourished people to 400 million by 2015. Further, the world population is expected to surpass 9 billion by 2050 demanding agriculture output to increase by 70% between now and then (WSFS, 2009; Anon. , 2010). The situation is expected to be more complicated under the current and predicted future climate change, spread and evolution of new pests, diseases and invasive species and requirement to reduce energy inputs to agriculture. Future crop species will need to be able to thrive in a drier, warmer, and more variable climate than at present, under agricultural regimes with lower requirements for energy-rich inputs, and in an environment increasingly populated by new pathogenic organisms (Moore et al. 2008). To meet out these challenges we need broader crop gene pools and CWR has greater role to play in it (WSFS, 2009; FAO 2009).

CROP WILD RELATIVES IN INDIA

India's diverse climatic conditions houses immense richness of agricultural biodiversity including diversity in crop plants, wild plants, livestock, aquatic species, below ground biota, and microbes (Rao, 1994; Nayar, 1997; Katwal et al. 2003). Indian gene centre is one of the 12 mega diversity centres of the world and it has Eastern Himalaya and Western Ghats as two "biodiversity hot spots" among the 34 identified the world over (Myers, 2000) housing about 49500 species of flowering and non-flowering plants out of the 260000 described the world over (Khoshoo, 1995; Nayar, 2008; Sharma et al. 2009). Flowering plants constitute 17500 species of which 5725 are endemic and distributed in the Himalaya and adjoining regions (3471), peninsular India (2015) and in the Andaman and Nicobar Islands (239) (Nayar, 1997; Singh et al. 2000). It also harbour 9,500 economic plant species, 1,256 grass species and over 2,000 species having ornamental value and 45 species of mangroves (Roy et al. 1998; Rao, 1994, 1997; MoEF, 2008).

Indian gene centre is also rich in domesticated crop diversity; nearly 45 species were indentified for Indian region by de Candole (1986); increased to 117 species by Vavailov (1926, 1951) and to 168 species by Zeven and de Wet (1982) out of 2489 species distributed in 12 regions of diversity of cultivated plants. Based on the number of species available in the region, their characters, distribution patterns and areas/habitats of build-up of diversity, the Indian region considered as a primary centre of diversity of several crops such as rice, Job's tear, chickpea, pigeon pea, rice bean, cowpea, velvet bean, brinjal, cucumber, radish, taro, yam, mango, banana, breadfruit, pummelo, tamarind, citron, sugarcane, sesame, coconut palm, safflower, jute, crotalaria, kenaf, black pepper, gum Arabic, cinnamon, bamboo etc.

Among CWR, there are 326 documented species that have originated and/or developed diversity in different phyto-geographical regions (Table 1) of India (Arora, 1991, 1993, 1996; Pandey and Arora, 2004). The documentation of CWR done by various workers (Dhillon et al. 2005, Hajra and Modgal, 1997, Pullaiah et al. 2000, Ganeshaiah, 2003, Pandey et al. 2008) suggests that the number of CWR occurring in India may be more, thus need to be updated particularly in the context of recent CWRs definition suggested by Maxted et al. (2006), which states that 'CWR is any taxon belonging to the same genus as a crop species. The gap analysis in this regard has been initiated by Botanical Survey of India and also NBPGR.

Table 1. Distribution of wild relative of crop plants in different phyto-geographical regions of India

Phyto-geographical zone	*No. of CWR*	*Phyto-geographical zone*	*No. of CWR*
Western Himalaya	125	Indus Plains	45
Eastern Himalaya	82	Malabar/Western Peninsular Region	145
North-Eastern region	132	Deccan/Eastern Peninsular Region	91
Gangetic Plains	66		

The number of CWR occurring in India has been further categorized into crops-group-wise: it represent cereals and millets (51), legumes (31), fruits (109), vegetables (54), oilseeds (12), fiber crops(24), spices and condiments (27) and others (26), as per Arora and Nayar, 1984; and Arora, 1991, 2000. The crops-group-wise enumeration of some important CWR occurring in India is as given below:

Cereals

India has about nine wild species *Oryza* viz. *O. nivara*, *O. sativa* var. *spontanea*, *O. rufipogon*, *O. malamphuzhensis*, *O. granulaat*, *O. malabarensis*, *O. periennis*, *O. jeyporensis*, *O. meyeriana*, out of 25 species reported the world over (Parthasarathy, 1972; Rana and Chatterjee, 2000; Sidique et al. 2005). Besides, there are three other species such as *Porteresia coarctata*, *Leersia hackelii* and *Zizania latifolia* which are related to genus *Oryza*. *O. nivara* is an annual weed found in ditches, water holes and edges of ponds while *O. rufipogon* is perennial and rhizomatous found in the deep water swamps. The North-West part of the Indian sub-continent adjoining the Hindukush Mountains is regarded as secondary centre of diversity of *Trirticum aestivum*. Archeological evidences revealed that *T. sphaerococuccum* and *T. aestivum* subsp. *compactum* are supposed to have originated in this region (Pal, 1968). The genetic diversity in its distant relatives such as *Aegilops*, *Agropyron*, and *Elymus* is extremely high in the mountainous regions. The genera *Zea* has two wild relatives- *Zea perennis* and *Z. diploperennis* both having unusual disease resistance properties but not occurring in India. Nevertheless, Himalayan region particularly the North-eastern region has rich racial diversity which showed maize with primitive characteristics like popcorn characteristics and prolificacy (Singh, 1977; Dhillon et al. 2006), and named as Sikkim Primitives 1 and 2 (Dhawan, 1964). The genetic diversity is high in *Horedum vulgare* for both hulled and hulless types but limited in terms of wild relatives. The wild form such as *H. vulgare* ssp. *spontaneum* and *H. vulgare* ssp. *vulgare* are found in the cold arid region of Himalaya.

Millets

Wild Sorghums that occur in India are only *Sorghum bicolor* ssp. *drummondii* and *S. halepense* but it has distinct racial diversity such as *S. bicolor, guinea, durra-bicolcor, S. bicolor* subsp *bicolor, durra-caudatum* and *caudatum-bicolor, Pennisetum glaucum* has rich genetic diversity in its cultivated form and limited for wild relatives except that of *P. purpureum*. Indian gene centre exhibits diversity in specific minor millet species like *Coix lacryma -jobi* (soft shelled forms), *Digitaria cruciata* var. *esculenta, D. sanguinalis, Echinochloa colonum, Eleusine coracana, Panicum sumatrense* and *Paspalum scrobiculatum* (Padulosi et al. 2009). Some of the millet species indigenous to India and have many wild species are *Paspalum scrobiculatum, Eleusine coracana, Ehinochloa oryzoides* and *Digitaria cruciata* (Rana and Chatterjee 2000).

Pseudocereals

The three major crops, namely amaranth, buckwheat and chenopod constitute the group pseudocereals. The genus *Amaranthus* has three cultivated species viz. *A. hyphochondriacus, A. caudatus,* and *A. cruentus;* while *A. dubius, A. blitoides, A. hybridus, A. lividus, A. retroflexus, A. spinosus, A. tricolor,* and *A. v*iridis are considered wild (Rana et al. 2005). The genus *Fagopyrum* has 15 known species such as *F. homotropicum, F. caudabum, F. esculentum, F. dibotrys, F. megaearpum, F. gracilipes, F. urophyllum, F. leptopodum, F. lineare, F. callianthum, F. pleioramosum, F. capillatum, F. statice*, occurring mainly in the high lands of Euro-Asia region (Farooq and Thair 1987; Anon 1988; Baniya 1994; Onishi 1995; Rana 2004). Further, *F. tataricum* ssp. *annum* occurs in the Eastern Himalayas, *F. tataricum* ssp. *potanini* in Tibet, Kashmir Himalayas and northern Pakistan (Ohnishi 1989, 1991, 1992), while *F. tataricum* ssp. *himalianum* and *F. esculentum* ssp. *emerginatum* are occurring in the cold arid region of Western Himalaya (Rana, 2004). Among chenopods, *Chenopodium album* is the cultivated species while *C. amranticolor, C. ambrosioides, C. botrys, C. foliosum, C. glaucum, C. hybridum, C. murale* and *C. opulifolium* are occurring wild (Chowdhery and Wadhwa, 1984).

Legumes and oilseeds: Legumes are second only to the cereals as a source of human food. India holds rich diversity in mung bean, urd bean, chick pea, pigeon pea, cowpea and horse gram. The important genera for which wide range of CWR exist are *Vigna, Cajanus, Canvalia, Cicer, Atylosia* and *Phaseolus.* Of the 34 wild *Cajanus* species, 32 are found in South, South-east and East Asia while 17 species viz. *Cajanus albicans, C. cajanifolius, C. candollei, C. kalnesis, C. mollis, C. rostrata, C. barbata, C. platycarpus, C. villosus, C. rugosus, C. scrabaeoides, C. sericeus C. grandiflorus, C. lineatus, C. niveus, C. elongatus,* and *C. rostrata* are occurring in India. Wild relatives of *Vigna* complex represent *V. glabrescens, V. dalzelliana, V. hainiana, V. khandalensis, V. vexillata, V. capensis, V. pilosa, V. vexillata* var. *wightii, V. bourneae, V. khandalensis, V. vexillata, V. trilobata, V. aconitifolia* var. *sylvestris, V. mungo* var. *sylvestris, V. radiata* var. *sublobata, V. unguiculata* var. *sesquipedalis.* Besides, there are other wild relatives enumerated in Indian Floras and Botanical accounts include *Phaseolus coccineus, P. lunatus, Cicer microphyllum, C. microphyllum, Macrotyloma sargarwalensis, Canavalia cathartica, C. maritima, Vicia* sp. , *Lathyrus* sp., *Mucuna* sp. , *Moghania* sp. and *Rhynchosia* sp.

Among oilseeds *Sesamum* and *Brassica* are important. The wild species are *Sesamum lacinatum, S. indicum, S. prostatum, S. malabaricum, S. alatum, S. capense* and *S. radiatum* in *Sesamum*; *Brassica tournefortii* in *Brassica, Linum perenne, L. mysorense* in *Linum,* and *Carthamus lanatus, C. oxyacantha* in *Carthamus* and many others like *Aesandra butyracea, Juglans regia, Prinsepia utilis* and *Prunus armeniaca* occurring in wild state and used as edible oils.

Fruit and Plantation Crops

India is the centre of diversity for *Mangifera indica* having largest number of cultivated varieties as well as wild species, namely *M. andamanica, M. khasiana, M. sylvatica* and *M. camptosperma* (Mukherjee, 1985; Yadav and Rajan, 1993). *M. khasiana* is endemic to Khasi hills of Meghalaya while *M. andamanica* is widespread in tropical, wet evergreen forests of Andaman Islands. Rich genetic diversity of wild *Musa* spp. in the North-eastern region and of cultivated types especially those belonging to the genomic groups *AA, AB, AAB* and *ABB* exist in the southern region. Maximum genetic variability of *M. acuminata* and *M. balbisiana* and several wild species such as *M. acuminata* ssp. *burmannica, M. nagensium, M. sikkimensis, M. cheesmanii, M. ornata, M. laterita, M. velutina, M. flaviflora* and *M. sanguinea* have been recorded here. In Citrus about 30 species occur in India (Bhattacharya and Dutta, 1956). Among these, *C. reticulata, C. sinensis, C. aurantifolia* and *C. limon* are the cultivated species; whereas *C. latifolia, C. grandis, C. paradisi* and *C. maderaspatana* occur in semi-wild state. Species like *C. limettioides, C. paratangerina, C. rugulosa* including *Glycosmis pentaphylla* are endemic to sub-Himalayan tract of North-west India. There are about 18 *Artocarpus* species including *A. heterophyllus* and *A. lakoocha* indigenous to Indian Western Ghats. Genetic diversity in terms of wild species does exist among *Carrica, Diospyros, Ficus, Morus, Phoenix, Syzygium, Vitis, Ziziyphus* etc.

Among temperate fruits, the genus Malus has 30 species worldwide but only two *M. baccata* and *M. sikkimensis* occur wild in India. Two varieties of *M. baccata* have also been identified as *M. baccata var. himalaica* in North-Western region and Meghalaya and *M. baccata var. dirangensis* in Arunachal Pradesh (Joshi and Rana, 1994). *Pyrus* and *Prunus* have 14 and 22 species respectively and constitute an important component of the temperate fruit diversity. One of the species *Prunus mira* grows in extremely cold areas of N-W Himalaya. *Juglans regia, Prunus dulcis* and *Corylus colurna* are important nuts for which intra-specific diversity and several wild forms are occurring in the Himalayan region. In soft fruits, *Rubus* is the most variable genus widely distributed in the Himalayan region. About 57 species of *Rubus* are occurring in India and 33 species and four taxonomic varieties are reported from Western Himalaya (Sharma and Chandel, 1996). The two species, *R. ellipticus* and *R. niveus*, have wide range of variability for foliar and fruit characters. Wild species of *Fragaria* such as *F. nubicola, F. daltoniana, F. nilgerrensis, F. vesca* and *Duchesnea indica* and *Actinidia callosa* of kiwi fruit also found here. There are many other fruit species called minor/wild fruits viz. *Ribes, Sorbus, Cotoneaster, Viburnum, Myrica, Docynia, Elaeagnus, Hippophae, Crataegus, Pyracantha, Pinus, Diospyros, Cydonia,* and *Cornus* occur in the temperate region.

Plantation crops constitute high value commercial crops viz. *Cocos nucifera, Areca catechu, Anacardium occidentale, Camellia sinensis* etc. The genetic diversity of *Camellia* consists of cultivated species, wild genera and species, weedy relatives, old seedling populations, polyclonal seed stocks, and clones. Besides tea producing taxa, several other species such as *C. caudata, C. irrawadiensis, C. taliensis, C. kissi* and *C. drupifera,* and other related genera, such as *Eurya, Pyrenaria, Schima* and *Gordonia* are also found in the forests of North-east India (Singh, 1999). Among, *Saccharum* complex, *S. officinarum, S. edule, S. barberi, S. sinense* are cultivated species; while *S. robustum* and *S. spontaneum* are occurring wild. It also has distantly related genera, such as *Erianthus, Narenga, Sclerostachya, Miscanthus*, which offer great scope for improvement of cultivated types.

Vegetables and Tuber Crops

Among vegetables, brinjal and some cucurbits are of Indian origin, whereas for okra and chilli, India considered as secondary centre of origin. *Solanum melongena* complex has three species, namely, *S. melongena, S. incanum* and *S. melongena* var. *insanum*. Wild relatives of *Solanum* viz. *Solanum torvum, S. indicum, S. insanum, S. surattense, S. pubescens, S. gilo,* and *S. khasianum* are widely distributed in sub-tropical and tropical regions. There are about 100 species of cucurbitaceous vegetables reported to occur in India including 34 endemic species viz. *Cucumis sativus* var. *hardwickii, C. trigonus, C. prophetarum, C. setosus, C. hystrix, Luffa graveolens, L. acutangula* var. *amara, L. cylindrica, L. echinata, L. umbellata, Momordica cymbalaria, Trichosanthes cucumeriana* var. *anguina, T. dioica, T. dicaeleosperma, T. khasiana, T. ovata, T. truncata, T. multiloba, T. anamalaeiensis, T. tricuspidata, T. cuspidata, T. nervifolia, T. perotteliana, T. ovigera, Momordica cochinchinensis, M. macrophylla, M. subangulata, M. cymbalaria, M. dioica, M. cymbalaria, M. denticulata, M. balasamina, Neoluffa sikkimensis* and *Citrullus colocynthis* (Chakravarty, 1982; Arora and Nayar, 1984). In *Abelmoschus*, nine species and two varieties namely, *A. angulosus, A. tuberculatus, A. manihot, A. moschatus, A. ficulneus, A. esculentus, A. crinitus, A. caillei, A. manihot* ssp *tetraphyllus* var. *tetraphyllus* and *A. manihot* ssp. *tetraphyllus* var. *pungens* are occurring. Intra-specific diversity is also high in *Capsicum annuum, Coccinia cordifolia* and *Moringa oleifera*.

Rich genetic diversity in terms of species and landraces in root and tuber crops, namely *Amorphophallus paeoniifolius, Manihot esculenta, Ipomoea batatas, Dioscorea alata, Dioscorea rotundata, Psophocarpus tetragonolobus, Dioscorea esculenta, Dioscorea bulbifera* var. *sativa, Colocasia esculenta, Alocasia macrorrhiza, Xanthosoma sagittifolium* occurs here. Besides, there are several tubers of minor importance such as *Maranta arundinacea, Solenostemon rotundifolius, Moghania vestita, Canna edulis, Psophocarpus tetragonolobus* and *Pachyrrhizus erosus* are found wild and used as food in western peninsular region. Genus *Allium* represents a major group and about 30 species are found in the Indian region (Babu, 1977). The cultivated Alliums are *Allium cepa* var. *cepa, A. cepa* var. *aggregatum, A. xproliferum, A. fistulosum, A. tuberosum, A. sativum, A. porrum, A. schoenoprasum,* whereas *A. carolinianum, A. chinense, A. consanguineum, A. humile,*

A. przewalskianum, A. stolczkii, A. stracheyi, A. victorialis and *A. wallichii* are important wild species mostly occurring in the Himalayan region.

Fibre Crops

Corchorus is represented by about 40 species, of which eight species are found in India. *C. capsularis* and *C. olitorius* are cultivated for fibre, while *C. aestuans, C. depressus, C. fascicularis, C. tridens, C. trilocularis* and *C. urticaefolia* are found in wild state. There are about 90 species of *Crotolaria* occurring in India which include some important species like *C. juncea, C. pallida, C. burhia, C. retusa, C. spectabilis, C. verrucosa. C. maduraensis, C. sandoorensis, C. grahamiana, C. retusa, C. spectabilis, C. albida, C. mysorensis, C. medicaginea, C. striata, C. breviflora* var. *pohliana, C. triquetra, C. sandoorensis, C. peduncularis, C. lutescens, C. grahamiana,* and *C. lutescens.* The genus *Gossypium* has around 40 species, of which four *G. arboreum* (native to India), *G. barbadense, G. herbaceum* and *G. hirsutum* including races such as *bengalense, burmanicum, cemum, arborium* are cultivated types. Truly wild perennial types are found scattered throughout the range of species while no truly wild annual forms are found here.

Spices and Condiments

India occupies an important position with respect to genetic diversity of spices and condiments viz. *Piper, Elettaria, Cardamomum, Zingiber, Curcuma, Coriandrum, Trigonella, Foeniculum, Trachyspermum, Cinnamomum, Myristica, Syzygium, Tamarindus, Garcinia* and *Vanilla.* The species diversity of *Piper* in South India, represent *P. argyrophyllum, P. atttenuatum, P. barberi, P. betle, P. galeatum, P. hapnium, P. hookeri, P. hymenophyllum, P. longum, P. psuedonigrum, P. schmidtii, P. silentvalleyensis, P. sugandhi, P. sugandhi* var. *brevipilis, P. trichostachyon* and *P. wightii.* For *Zingiber* and *Curcuma* the South Asian region is believed to be the centre of origin; possibly India, in view of the fact that maximum diversity for these crops occurs. While Baker (1886) described 24 species from the Indo-Malayan region, Sabu (1991) described eighteen species of *Zingiber Z. officinale, Z. ligulatum, Z. zerumbet, Z. spectabilis, Z. purpureum, Z. clarkia, Z. roseum, Z. marginatum, Z. wightianum, Z. intermedium, Z. macrostachyum, Z. chrysanthum, Z. cernuum, Z. rubens, Z. capitatum, Z. squarrosum, Z. cylindricum,* and *Z. elatum* from Western Ghats and adjacent areas. Velayudhan et al. (1999) have described 32 species of *Curcuma,* of which nine are endemic to India.

The genus *Murraya* is represented by three species in India, namely *M. exotica, M. elongata* and *M. koenigii*. In tree spices, *Myristica, Cinnamon* and *Garcinia* are important. In *Myristica, M. fragrans* is cultivated type while *M. amygdalina, M. andamanica, M. attenuata, M. dactyloides, M. beddomeii, M. gibbosa, M. glabra, M. glaucescens, M. irya, M. kingii, M. longifolia* and *M. Magnifica* occur in wild form. The species of *Cinnamon* occurring in India are: *Cinnamomum tamala, C. citriodorum, C. filipedicellatum, C. goaense, C. macrocarpum, C. malabathrum, C. nicolsonianum, C. perottettii, C. riparium, C. keralense, C. travancoricum, C. verum, C. wightii, C. sulphuratum, C. heynianum, C. gracile, C. chemungianum, C. walaiwarense, C. dubium, C. ovalifolium, C. litseafolium, C. citriodorum, C. rivulorum,. C. sinharajense* and *C. capparu-corende* (Kostermans, 1983). In *Garcinia*, 36 species are native to India, out of which 15 are endemic to Western Ghats (7), Andaman and Nicobar Islands (7) and one in North-east India (Singh, 1993; Srivastava, 1994; Abraham et al. 1999). The centre of diversity for the genus *Elettaria* is the Malaysian-Indonesian region from where eight species have been reported recently (Sakai and Nagamasu, 2000). Intra-specific variability is also high for seed spices and important among these are *Coriandrum sativum, Foeniculum vulgare, Trigonella foenum-graecum, T. corniculata, Apium graveolens, Trachyspermum ammi, Ammi visnaga, Carum carvi, Bunium persicum* and *Nigella sativa*.

Ornamental Crops

The important ornamentals that are native to India and are under cultivation are *Rosa moschata, Lilium bulbiferum, L. longiflorum, Eremurus himalaicus, Primula denticulata, P. rosea, Michelia champaka, Nelumbo nucifera, Nymphaea* spp., *Tulipa clusiana* forma stellata and *Tulipa clusiana*. The genera where inter and intra-specific diversity is high in India represent *Rosa, Lilium, Jasminum,* and *Primula* etc. Eleven species of *Rosa* found growing in wild are *Rosa brunnonii, R. foetida, R. gigantea, R. involucrata, R. longicuspis, R. macrophylla, R. moschata, R. rubiginosa, R. sericea, R. webbiana,* and *R. wichuriana*. It is believed that four of the 11 wild species have contributed to the development of modern garden roses. The species endemic to Himalayas are *R. brunnoni, R. gigantea, R. involucrata, R. longicuspis, R. sericea* and *R. webbiana*. Some species of *Lilium* occurring in India include *L. nepalense, L. polyphyllum,* and *L. Wallichianum,* and *L. nilgiriensis*. The species of *Chrysanthemum* native to India are- *C. atkinsoni, C. indicum, C. leucanthemum, C. retsria* and *C. stiliszkai*. Indian orchids also form a major group of ornamental crops and many with proven floricultural and/or herbal traits, namely, *Arachnis, Ascocentrum, Calanthe, Cymbidium, Dendrobium, Habenaria, Paphiopedilum, Phalaenopsis, Phaius, Renanthera, Rhynchostylis, Spathoglottis, Vanilla,* and *Vanda*. Amongst ornamental trees, *Bauhinia variegata, Cassia fistula, Cassia nodosa, Butea monosperma, Erythrina indica, Lagerstroemia speciosa, Pongamia glabra, Thespesia populnea, Saraca indica* and *Anthocephalus cadamba* are important; while *Jasminum* spp., *Bauhinia acuminate, Hiptage benghalensis, Mussaenda frondosa, Clerodendron inerme, Plumbago zeylanica* and *Passiflora leschenaultia* are native shrubs and climber. Sharma and Rana (2000) have also described 134 naturally occurring plants of ornamental value from N-W Himalaya.

IMPORTANCE AND UTILIZATION OF CROP WILD RELATIVES

The increasing genetic uniformity of crop varieties combined with climate change effects makes crops more vulnerable to various biotic and abiotic stresses. There had been examples of large scale devastations of crops due to genetic uniformity, for instance potato famine of the 1840s due to late potato blight epidemic across Ireland, Europe and North America; 1970s southern corn blight outbreak in the US maize and rice losses due to blast in the Philippines, Indonesia and India leading to the great Bengal famine. Intensive modern breeding efforts have contributed to a narrowing of the gene pool by further concentrating on favourable alleles already present in early domesticates (Simmonds, 1976; Ladizinsky, 1985; Debouck, 1991). CWR are therefore important for maintaining genetic diversity and preventing such losses, which may have serious consequences for food security. Importance of wild species of crop taxa, in the wake of technological advances, in study and utilization of species as a source of germplasm for pre-breeding and genetic enhancement of crops for specific traits, has made these a priority in PGR management. At the same time, post-CBD (Convention on Biological Diversity), concerns of habitat loss and genetic erosion have resulted in emphasis on *in situ* conservation; problems of access and benefit sharing and sovereign rights and other policy issues too have come to the forefront. As we know CWR survive in nature by different rules from crops – the survival of the fittest and by this rule they have developed resistance to survive drought and floods, extreme heat and cold, and adapted to cope with many natural hazards. They often developed resistance to pest and disease that caused so much damage to the related crops. This is why they are so valuable to agriculture. Scientists have estimated the commercial value of CWR estimating their contribution crops improvement (Box 1).

Box 1. Some astonishing example of contribution of wild relative

- The potato was brought from Peru to Europe in the 15th century and eventually it was cultivated as a crop plant all over Europe. It became a staple in the entire Ireland and in the 19th century a disease late blight of potato caused a major disaster and millions of people starved to death. A wild species *S. demissum* of potato was found and when it was hybridized with the standard crop plant a disease-resistant variety was obtained.
- A single Ethiopian barley plant has a gene that now protects US$ 160 million annual barley crop from lethal yellow dwarf virus, which is fatal to barley plants.
- A wild hop have "better bitterness" to English beer and in 1981 brought US$15 million to the British brewing industry.
- The desirable traits of wild sunflowers are estimated US$267 million to US$384 million annually in the United States
- One wild tomato variety has contributed to a 2.4 per cent increase in solids contents worth US$250 million
- Three wild peanuts have provided resistance to the root knot nematode, which cost peanut growers around the world US$100 million each year
- In both India and Africa yields of cassava (tapioca) – one of the most important root crops throughout the tropics – were increased 18 fold because of disease resistance bought in from wild Brazilian cassava.
- A single variant of a related wild species of rice *Oryza nivara* collected from a water logged field in UP saved the entire rice crop in South and South East Asia from grassy stunt virus.
- The sugarcane industry in the US was saved form collapse by disease resistance genes brought in from wild Asian species.
- Wild relatives have given strawberries their extraordinary high yield, colour and sweetness.

Despite having valuable genes with immense value for crops improvement for variety of traits and adaptation to changing environmental conditions, utilization of CWR for crops improvement has been enjoyed a great success only in few crops, while disappointing for numerous others. Many genes are still lies untapped in these genetic resources (Wang et al. 1994), presumably due to the genetic bottlenecks. Tanksely and McCouch (1997) argued that breeders were not fully exploiting the potential of CWR; historically they relied on searching genes for beneficial traits associated with certain CWR rather than searching more generally for beneficial genes. Hajjar and Hodgkin (2007) while reviewing the utilization of CWR in the past 20 years found that over 60 wild species have been used in 13 major crops, with over 100 beneficial traits having been derived from them. They found that most common use of CWR continues to be as a source of pest and disease resistance, although there have been many examples of gene introductions from CWR with other important traits such as drought and salt tolerance, improved quality and cytoplasmic male sterility etc. Gene flow between wild and domesticated populations has been important following domestication, and would have partially restored the genetic diversity of domesticated populations (Badr et al. 2000). In many traditional production systems, landraces are still cultivated and gene flow from wild to cultivated remains an important source of evolution (Jarvis and Hodgkin, 1999).

There are many examples where genes have been derived/identified from CWR against various stresses. Some of the examples include: use of *Oryza nivara* genes provided long-lasting resistance to grassy stunt virus (Brar and Kush 1997, Khush et al. 1977; Plucknett et al. 1987 and the use of *O. spontanea* as the source of wild abortive cytoplasmic male sterility, which has provided the cornerstone for today's hybrid rice (Li and Zhu 1988). *Oryza rufipogon* for tolerance of soils with high acidic-sulfate content (Nguyen et al. 2003), and *O. longistaminata* genes for drought tolerance (Brar 2005); *Solanum demissum* providing resistance to potato late blight (National Potato Council 2003); stem and leaf rust resistance from *Thinopyrum elongatum* and *Aegilops umbellulata* in wheat (Prescott-Allen and Prescott-Allen 1986). Tomatoes have provided many classic examples of improved quality traits such as increased soluble solid content, fruit colour, and adaptation to harvesting, over 40 resistance genes have been derived from *Solanum peruviannum*, *L. cheesmanii*, *Solanum pennellii* Correll var. *pennellii*, *L. pimpinellifolium* (Rick and Chetelat 1995). Broccoli varieties producing high levels of anti-cancer compounds have been developed using genes obtained from wild Italian *Brassica oleracea*.

In the Indian context, attempts have been made to identify and transfer genes from CWR; for instance: *O. rufipogon*, *O. longistaminata*, *O. punctata*, *O. brachyantha*, *O. minuta*, *O. offcinalis* have high level of resistance yellow stem borer, *Porteresia coarctata* a distant relative of *Oryza* have genes for tolerance/resistance to saline and marshy habitats; *Ochthochloa compressa* to drought tolerance; *Vigna tribolata*, *V. mungo* var. *sylvestris*, *V. radiata* var. *sublobata* to yellow mosaic virus, *V. vexillata* has high protein and resistance to cowpea pod sucking bug and buchids and is crossable with *V. unguiculata* and *V. radaita*; *Cajanus platycarpus* occurring in Shivalik hills is closely related to *C. cajan* and is photo-insensitive, highly resistant to *Phytophthora* bight and reniform nematode and pod borer. *Macrotyloma sar-garhwalensis* have crude protein of 38.0%, which is about two times higher than the cultivated horse gram. There are many other species such as *Cicer microphyllum* to cold hardiness, drought tolerance and seeds/pod; and *Sesamum laciniatum* to leaf phyllode, *S. malabaricum, S. mulyanum* and *S. alatum* for powdery mildew and leaf rolle; *Linum perenne* for drought and cold hardiness, *L. grandiflorum* for linseed bud fly and alternaria blight, *L. tenuifolium* for high linolenic acid; *Brassica oxyrhina, Moricandia arvensis, Trachystoma balli, Diplotaxis catholica* as sources of CMS, *B. tournefortii, Diplotaxis acris, D. harra, Eruca sativa* for drought tolerance in *Brassica; Corchorus aestuans, C. tridens, C. fascicularis* resistant to Jute semi-looper and *Macrophomina* sp., *C. olitorius* for yellow mite, *C. capsularis* for stem rot and anthracnose, *C. depressus* for drought tolerance and *Erianthus* sp. have gene for cold and drought tolerance, high yield for *Saccharum* improvement.

The species of *Solanum* complex such as *S. incanum, S. viarum, S. melongena* var. *insanum* have gene for resistance to *Fusarium* wilt, bacterial wilt and frost tolerance; *S. torvum S. sisymbrifolium* for *Verticillium* wilt and *Meloidogyne incognita* nematode; *S. gilo, S. integrifolium* for *M. incognita* race 1 and 2 and *S. khasianum* for shoot and fruit borer resistance in brinjal. The wild species of okra *Abelmoschus tuberculatus* to Yellow common bean, *A. crinitus* occurring in the Shivalik hills with a marked dry season, and *A. angulosus* grows at high altitudes, being cold tolerant and have shown complete resistance to Yellow Mosaic Virus. *Cucumis hardwickii* and *C. callosus* for downy mildew and fruit fly, *Cucumis melo* var. *chito* for *Fusarium* wilt resistance in cucumber; *Allium ampeloprasum* for downy mildew in onion and *Momordica dioica* for drought tolerance in bitter gourd. In cassava, *Manihot esculenta* subsp *flabellifolia* for protein enrichment, *M. caerulescence* for CMD resistance, *M. pseudoglaziovii* and *M. glaziovii* for drought tolerance, *Manihot pruinosa* for high tuber production and *M. anomala* for high photosynthetic efficiency have been identified. Among *Dioscorea,* wild species were indentified for having genes for bulbils production in *D. bulbifera,* higher yield and bulbils production in *D. pentaphylla* and to anthracnose resistance in *D. oppositifolia.* In black pepper, *Piper barberi, P. chaba, P. hymenophyllum, P. longum, P. nigrum* (wild form) have resistance to Pollu beetle *(Longitarsus nigeipennis*).

Malus baccata in apple; *Pyrus pashia* and *P. pyrifolia* in pear; *Prunus cerasoides* in cherry and *P. mira* in peach are used as rootstocks with multiple disease and insect resistance and drought tolerance. In Citrus, *Citrus jambhiri, C. limonia* and *C. karna* are used as rootstocks for cultivated species and are tolerant to citrus tristeza virus and most promising rootstocks for mandarin, orange and Kinnow in the lower hills. *Musa acuminata* ssp.

burmannica found resistant to leaf spot and Foc wilt and Rhodochlamys such as *M. laterita, M. velutina, M. ornate* and *M . aurantiaca* have resistance to leaf spot, wilt and tolerance to nematodes. *Phylanthus acidus* has tolerance for rust and frost. *Vitis parviflora* showed multiple disease resistance while *V. himalayana* is cold hardy, drought tolerant and late ripener, hence escape fruit cracking in rainy season in grapes. *Malus baccata* has shown resistance to collar rot, root rot and woolly aphis and tolerance to drought in apple while *Pyrus pyrifolia* and *P. pashia* showed resistance to powdery mildew and root rot in pear.

CONSERVATION STRATEGIES TO CROPS WILD RELATIVES

The natural populations of many species of CWR are increasingly at risk primarily due to habitat loss, degradation and fragmentation, the impact of invasive species, over exploitation and global warming, thus resulting into reducing in numbers, narrowing of distributional range and genetic erosion. Many cereal and pulses crop wild relatives, including wild wheat such as *Aegilops*, *Agropyron*, *Elymus*, millet species, *Vigna*, occur in arid or semi-arid lands are severely affected by over-grazing and desertification. Wild and minor fruits, medicinal plants, spices and condiments occurring in deep forests are vulnerable to over exploitation and forest fires. Mountain areas, which have diverse range of wild relatives, are particularly vulnerable to the loss due to high ecosystem fragility. CWR traditionally found as natural inhabitants of agro-ecosystems, in and around farms; the increasing industrialization of agriculture is reducing their occurrence. There are large numbers of CWR species of *Amaranthus, Chenopodium*, *Solanum*, *Vigna, Trichosanthes* and many other economic herbs, which are occurring in community and waste lands in and around villages, road sides, are severely affected by invasive species. Climate change has also emerged a greatest threat to genetic diversity worldwide. The changes in temperature and precipitation regimes is likely to alter species distribution, phenology, and regeneration (Rana and Sharma, 2009), thus creating strong selection pressure for more adaptive genotypes (Maxted, et al. 2008). Scientists the world over are dipping into wild and cultivated gene pools to develop crop varieties that will have to replace existing varieties in disturbed agro-climatic zones, should the climate change significantly. Wild species which are highly endemic, narrowly distributed and lack migration corridors, for instance *Cicer microphyllum, Linum perenne, Allium carolinianum* in cold arid region are more vulnerable (Rana and Sharma, 2009).

The need to conserve genetic diversity has long been recognized in National and International deliberations including CBD, GPA, and IUCN (IUCN, 2006; Heywood et al. 2008). This would need a variety of conservation approaches depending on specific biological factors. Such approaches should include conservation of ecosystems/agro-ecosystems, specific habitats, naturally occurring gene pools, and special genetic stocks. There is need to do gap analysis on *have* and *have not,* documenting the existing species along with their status, prioritizing species and areas for conservation to develop an integrated *in situ* and *ex situ* conservation and use strategies ensuring role of CWR for the benefit of mankind.

CWR Cataloguing and Prioritization

As a first step in the conservation and use of CWR, we need to know how many taxa there are, what they are, where they are and how much they are (Kell et al. 2005, 2008a). Inventories are needed to establish what is already conserved and what is left out; where the gaps are in their conservation and to provide the data needed for integrating CWRs into existing conservation initiatives (Ford-Lloyd et al. 2008). Taxon inventories provides an insight on the baseline data critical for biodiversity assessment and monitoring to policy makers, conservation practitioners, plant breeders and other user groups (Kell et al. 2008b). They provide the essential foundations for the formulation of strategies for *in situ* and *ex situ* conservation and on the species current and potential uses as novel crops (Maxted et al. 2007, Scholten et al. 2008). The Catalogue also provides the information needed for plant breeders to source new material and for conservationists to collect material from as wide a range of a taxon's distribution as possible (Kell et al. 2008).

The national programme efforts have been initiated to revise the status of CWR in India. The Botanical Survey of India and NBPGR has taken the lead in this direction to document the range of species diversity in primary, secondary and tertiary gene pools of CWR. Once we have a complete inventory of CWR available, conservation priorities can be decided. The species which require the most urgent attention and conservation planning can be dealt on priority. Prioritization of species has been done by using the criteria laid down by various workers. Maxted et al. (1997) considered factors such as current conservation status, potential economic use, threat of genetic erosion, genetic distinctiveness, eco-geographical distribution, biological and cultural importance as important for prioritizing the PGR for conservation. They envisaged that these prioritizing factors could effectively be used to compare a limited number of species competing for limited conservation resources. Rana and Chaterjee, (2000) explained that those wild relatives which have been identified to be closest to their domesticated counterparts morphologically and genetically; have limited distributional range, rare, endemic, threatened due to overexploitation and those have useful traits of high socio-economic significance needs to prioritize. The help of Red data book (Nayar and Sastry, 1987–90) to decide the priority species has also been envisaged as one of the most important criteria. Having set priorities, there is need to look for various conservation approaches which can be one or a combination of more than one depending upon the nature of species to be conserved.

In Situ Conservation

In situ conservation refers to the maintenance and use of species in the habitats where they naturally occur and have evolved without human aids. It is suitable for wild species occurring in forests and protected areas mainly forest and horticultural crops genetic resources including wide range of forages. Here wild populations regenerate naturally, and are dispersed naturally by various external influences thus maintain an intricate relationship, often interdependence, between the species and components of the environment in which they occur. This dynamic co-evolution leads to greater diversity and better adapted germplasm to diverse range of environments. It also facilitate studies on CWR linked to autecology, biology, community ecology, interspecific association and their interdependence with other components of ecosystems, with overall emphasis on continuous monitoring of ongoing ecological changes and appropriate management of habitats (Gadgil et al. 1996, 2003; Singh and Gadgil, 1996).

The Ministry of Environment and Forests facilitated conservation of natural flora, through its field organizations. The entire spectrum of this component of plant biodiversity is distributed over 10 bio-geographical zones and is being conserved *in situ* in 15 Biosphere Reserves, 99 National Parks, 523 wild life sanctuaries, 46 conservation reserves, 668 protected areas, and 39 wetland and mangroves, 19,000 documented sacred groves and 309 forest preservation plots provide refuge to large number of agriculturally important plants that disappeared out of the surrounding landscapes. The establishment of sanctuaries in Tura range in Garo Hills of Meghalaya for conservation of rich native diversity of wild *Citrus* and *Musa* species, and for *Rhododendron* and orchids in Sikkim is also helping *in situ* conservation of economically important species. The identification of National Agro-biodiversity Heritage/ Hotspots Sites would be playing an important role in safeguarding not only CWR but the entire spectrum of agro-biodiversity.

On-farm Conservation

Often *in situ* and on-farm conservation are used in same context, we have discussed it separately with the intention that *in situ* allows only conservation while on-farm encourage conservation though use under human supervision. The effectiveness of strategies to maintain and use crop diversity on-farms depends on the extent to which local varieties continue to meet the needs of farmers and communities. On-farm conservation in the present agricultural context has been seen as most effective approach particularly for PGR related to Food and Agriculture. Majority of the species of CWR managed as part of agricultural production systems because of their occurrence in and around human settlements sometimes as escapes and or in waste community lands. Therefore, traditional agricultural productions systems allow more genetic diversity of CWR to perpetuate than modern and more intensive production systems. Conservation of traditional agricultural production systems or on-farm conservation of genetic diversity has been advocated as one of the approaches to achieve ever green revolution (Sawminathan, 2006). On-farm conservation has been recognized as participatory and dynamic approach involving farmers and their long established skills and knowledge on genetic resources. But, at the same time increasing commercialization, industrialization and intensification of agriculture coupled with large scale migration leaving agricultural field abandoned and couple of other factors making it even far more difficult to operate not only in cash crops dominating regions but also in remote and marginal hill areas which by and large practice traditional farming systems (Maikhuri et al. 1996; Rana et al. 2001, 2010).

Nonetheless, being dynamic it maintains the continuum in the evolutionary processes; thereby create useful genetic variability within traditional agricultural systems (Brush, 1999, Bisht et al. 2006). It also support large number of rare alleles and genotypes evolved spatially and temporally than the accessions conserved in the gene banks (Jarvis et al. 2008). These rare combinations of new alleles are capable to cope with wide variety of stresses and have been reviewed in detail (Oldfield and Alcorn 1987; Iwanaga 1995; Maxted et al. 1997a, b; Brush and Meng, 1998 and Sthapit et al. 2008). Germplasm specialists must therefore, explore genetic variations to identify germplasm with traits for developing crops cultivars adapted to theses stresses from materials conserved and cultivated on-farm as they have withstood the competition with their improved allies over the years and is testimony to their resilience to frequently changing environments.

The initiatives for on-farm conservation and/or conserving traditional agricultural production systems have been undertaken in Western Himalayan region, Bastar area of Chhattisgarh for rice landraces, small millets, rajmash and pseudocereals by several civil society organizations. M.S. Swaminathan Research Foundation for rice landraces such as navara, kalazeera and small millets in Kolli hills in Tamil Nadu and Jeypore, Koraput in Orissa (MSSRF, 2010), and other like Society for Research and Initiatives for Sustainable Technologies and Institutions, Navdanya, Gene Campaign, Beej Bachao Andolan, Foundation for Revitalization of Local Health Tradition and Deccan Development Society are involved in the conservation traditional production systems. Some states in the north-eastern region have been identified for survey of crops and their wild relatives of local importance for immediate conservation efforts. They are: Arunachal Pradesh for *Citrus jambhiri*, Assam for species of bamboos, Manipur (Loktak lake area) for wild rice, *Zizania latifolia*, Tripura for various species of *Ziziphus* and Mizoram for wild *Musa* and *Ensete* and Zingiberaceae (SPGRFAI, 2007). Community-based management of wild relatives in non protected areas, with the help of governmental and nongovernmental departments are involved in undertaking on-farm conservation activities. Significant progress has been made for conserving some species such as *Acorus calamus, Picrorhiza kurroa, Costus speciosus, Rauvolfia serpentina, Pyrus pyrifolia, P. pashia, Juglans regia, Docynia indica*, and *Punica granatum* (wild form).

Ex Situ Conservation

Ex situ conservation of germplasm takes place outside the natural habitat or outside the production system, in facilities specifically created for this purpose. Gene banks conserve genetic resources in a way that its viability is prolonged in order to minimize the regeneration cycles to the maximum extent possible. However, they monitor viability periodically, regenerate and replace the seed samples with new if viability goes down the standards and also conduct basic studies to improve the seed health and longevity. The information systems used by genebanks are becoming increasingly important tools for researchers and breeders seeking data on the distribution of crops and their wild relatives. The importance of genebanks has increased significantly over the last several decades (Frankel and Bennett, 1970;

Frankel and Hawkes, 1975, Hawkes and Ford-Lloyed, 1992). As commercial agriculture has expanded; many farming systems that preserved local agricultural diversity have been transformed and local varieties have been lost. Controlling powerful social and economic forces so that they do not result in genetic erosion is often not possible, certainly not in the short term (http://www.bioversityinternational.org). As a result, genebanks often represent the only option for conserving biodiversity. Depending on the type of species to be conserved, different *ex situ* conservation methods may be used.

Seed Genebanks

The most widely used technique for conserving plant genetic resources is seed banking and collection maintained in genebanks are referred to as base collections. According to FAO, this technique accounts for 90% of the accessions conserved *ex situ* globally. The CWRs are under-represented in most of the gene banks including CGIAR *ex situ* collections. The base collections are maintained at –20°C in vacuum packed standard aluminum foil pouches at 3–7% seed moisture content, depending on the crop species and with initial seed viability above 85% as per International Genebank standards. Base collections are stored under conditions (–20°C) ensuring long-term viability of material (more than 50 years) as a security to the active collection while active collections are stored under conditions (6–8°C and 25–35% RH) that ensure medium-term viability (10–20 years). Ideally, these are maintained in sufficient quantity to be available on request. Hamilton et al. (2003) have described considerations for improved conservation and utilization concepts and strategies. In India, we are conserving around 0.4 million accessions of 1,134 species in seed gene bank, of these around 30,000 accessions represent CWRs belonging to more than 400 species majoring *Piper* (18 species), *Dioscorea* and *Vigna* (16 species each), *Curcuma* (14 species), *Solanum* (12 species), *Citrus, Syzygium* and *Zingiber* (11 species each), *Cinnamomum* (10 species), *Allium* (9 species), *Momordica, Oryza* and related genera, *Trichosanthes* and *Sesamum* (6–7 species each). Keeping in view the poor representation of CWR in genebanks collections, fresh impetus has been laid by NBPGR to enrich the collection with CWR.

In Vitro Storage and Cryopreservation

Conserving germplasm in seed gene banks is only possible for species with seeds that can tolerate desiccation and low temperatures; however, there are many species whose seeds cannot survive under such conditions. For species with so-called 'recalcitrant' seeds or species that are vegetatively propagated, such as roots tubers and aroids different conservation techniques are used. *In vitro* storage and cryopreservation are two technologies used in genebanks to conserve such species. *In vitro* storage involves keeping plant tissues under strict sterile conditions in glass tubes and vessels. Cryopreservation involves storing the living tissues at ultra-low temperature, usually at –196°C in liquid nitrogen, to guarantee long-term preservation of germplasm in genetically unaltered state. Research work on cryopreservation have led to the development of protocols for cryopreservation of no less than over 150 different plant species. A wide range of species can now be routinely cryopreserved: *Musa* spp. , *Manihot esculenta, Rubus, Pyrus, Coffea arabica, Elaeis guineensis, Camellia sinensis* and many tree species generally produce recalcitrant seeds. *In vitro* storage is not always suitable for long-term conservation and requires advanced infrastructure - adding to the cost. Cryopreservation is a more promising technique for the long-term conservation of 'difficult' species. Presently 1,969 accessions of 158 vegetative propagated crops under *in vitro* and 8,493 accession of 720 species have been *cryo* preserved at NBPGR. New protocols have been developed for micro-propagation and *in-vitro* conservation of vegetative propagated species and also for cryopreservation of vegetative propagated and non-orthodox seed species like tea, black pepper, almond, neem, etc. Pollen grains of 279 accessions of *Mangifera indica* and *Garcinia* spp and dormant buds of 197 accessions of *Morus* spp. were also cryopreserved. Besides, there are many crop based institutes in India who are doing *in vitro* conservation of their respective crops, for example Central Potato Research Institute for potato; Directorate of onion and garlic for *Allium spp*; Indian Institute of Spices Research for spices and many others (www.icar.org.in) .

Field Gene Banks

The conservation of germplasm in field gene banks involves the collecting of materials and planting in the orchard or field in another location. Field gene banks have traditionally been used for perennial plants, including species producing recalcitrant seeds, little or no seeds, stored as clonal material and species have long juvenility. Field gene banks are commonly used for species as cocoa, rubber, coconut, coffee, sugarcane, banana, tuber crops, tropical and temperate fruits, vegetatively propagated crops, such as wild onion and garlic, and forage grasses. We are conserving large number of wild and weedy species, especially of threatened and rare in nature, in the field gene banks being maintained at 10 regional stations/base centres of NBPGR, 58

National Active Germplasm Sites and 245 botanical gardens, arboreta, herbal gardens and other field repositories located in different agro-climatic and phyto-geographical regions of the country.

DNA banking

The rapidly expanding study of DNA in different areas of science has created an odd surplus: the DNA itself (de Vicente and Andersson, 2006). Reasonably easy and inexpensive to store, with established techniques for almost infinite multiplication, the samples of DNA created in laboratories around the world have become an important resource for future research. The demands from breeders are likely to emerge for particular segment of DNA or a cloned GENE rather than asking for entire range of genetic diversity or seed samples stored in genebanks and needs further evaluation for traits most sought. DNA resources can be maintained at 20°C for short- and mid-term storage (up to 2 years), and at 70°C or in liquid nitrogen for longer periods. Although, DNA banking is not at present widespread, hence should not be viewed as a substitute for existing techniques for the conservation of genetic resources. However, DNA banks can complement conservation strategies that make use of *ex situ* and *in situ* conservation. For many species that are difficult to conserve by conventional means and are highly threatened in the wild, DNA storage may provide a way to conserve the genetic diversity of these species and their populations in the short-term, until effective methods are developed (Ehsan Dulloo et al. 2003). Keeping in view the enhanced use and value of genomic resources instead of seed material as such and also to screen the germplasm under the expected climate change scenarios, the ICAR has established genomic resources centre in NBPGR. This venture is expected to enhance the utilization of genomic resources and its availability to researchers across the country. There is need to have integrated conservation of genomic resources in genome research centre to collect, validate and facilitate the use of useful genes and gene constructs. The indigenous genetic resources possessing gene and combination of genes for desirable traits provide a buffer output in times of drought, heat tolerance, floods and physiological responses like photo-insensitivity, high photosynthetic rates, low respirations, flood and many other biotic stresses. Current economic decisions are largely based on direct use values, although the other uses may be of equal or greater importance and in the context of genetic resources are indeed likely to be positive. Genomic resources will comprise of genomic DNA including DNA of rare and endangered species, genomic & c DNA libraries, BAC libraries, gene constructs and promoters etc. and will be conserved for long-term. It will also help to protect our indigenous genetic resources because gene and gene constructs are patentable.

Complementary Conservation

It has been recognized widely that adoption of any single approach will not work to conserve the entire spectrum of genetic diversity including CWRs. Thus, a complementary conservation strategy can be defined as the combination of different conservation actions, which together lead to an optimum sustainable use of a gene pool. We should not forget that the main objective in any PGR conservation programme is to maintain the highest possible level of genetic variability present across the gene pool of a given species or crop both in its natural range and in a germplasm collection. The various conservation approaches discussed above have distinct advantages and disadvantages, but the most effective conservation system should incorporate elements of both. They need to be used in a complementary manner. A complementary conservation strategy involves striking the right balance between different methods employed. It depends on the species being conserved, the local infrastructure and human resources, the number of accessions in a given collection, its geographic site and intended use of the conserved germplasm (http://www.bioversityinternational.org). A good complementary conservation strategy does not categorize crops or species into definitive classes. It is dynamic, and lends itself to meet the challenges of changes that are occurring in the field of genetic resources as it is open to new technologies and new needs.

EPILOGUE

The focus in wild relatives of crop plants as a genetic resource requires consolidation and analysis of information on taxonomy, distribution, reproductive behaviour vis-à-vis their population structure, strategies for conservation of seeds and populations, the genetics of resistance to abiotic and biotic stresses and the potential for use in crop breeding programmes. Herbarium-based and experimental studies are therefore needed on this group to identify areas of diversity, alternative conservation strategies, including *in-situ* conservation. Species/species complexes that have been studied are scanty considering the large number of native economically important taxa in India. Species introduced for their potential importance are another large resource which needs to be conserved on a priority basis as future introductions are likely to be few especially since these are now exchanged only with prior consent and under material transfer and benefit-sharing agreements i.e. new IPR regimes. *Ex situ* conservation is not just about conserving biodiversity for its own sake. The main purpose of the collections is to ensure agricultural growth and keep our options open for innovation. To make the genetic resources useful to farmers, breeders and researchers, curators carefully document the collected materials, make the information available and establish a transparent and safe system for its distribution so that breeders, farmers and researcher can use it in their work. To cope with problems arising from intensive modern agriculture, it is essential to maintain genetic diversity within crop gene pools; so the breeders are well equipped with the genes to keep themselves one step ahead of the rapidly evolving pests and diseases and even new climate regimes. The value of diversity in crops is similar to the value of a diversified portfolio to an investor. If one stock fails, the investor does not lose everything. Nature does not produce individual plants resistant to all diseases, pests and environmental stresses, and neither can the breeder. It is natural genetic diversity, in which many individuals, each somewhat different genetically, thrive together, allows a population to withstand challenges to its survival and produce traits of needs.

Many a times, when crossing a CWR to a crop to transfer a specific characteristic, the breeder comes across progenies having unwanted traits. A further problem is that desired genes are sometimes linked with unwanted genes and a breeder may spend years trying to separate them. Therefore, to make CWR more accessible and easier to use, germplasm enhancement programmes needs to be taken up to remove undesirable traits. This is a way of "packaging" genes from wild relatives. The rapid development of genetic engineering through recombinant DNA technology offers new opportunities for using genes from wild relatives not only from primary, secondary and tertiary gene pools but far beyond this. However, conventional plant breeding still hold promise in transferring genes from primary gene pools, there is ample scope of transferring genes from distant relatives through genetic engineering. Therefore, plant breeders have to work closely with genetic engineers to develop new cultivars. There is need to design altogether new crops, plant types and varieties, which have the ability to perform well under adversities and developing for such needs is only possible if wide range of genes is available. The amalgamation of genes from diverse sources including CWR to broaden the existing gene pool could be a probable answer to address the very pertinent and much warranted issue of food and environment security both in the present and future context. To have a long lasting impact, conservation must be done in a holistic and dynamic manner, based on the principle of sharing the benefits derived from conserved gene pools with the communities who conserve them.

References

Abraham, Z. Mithal, S.K. and Gautam, P.L. (1999). Genetic resources of *Garcinia* species at NBPGR Regional Station, Thrissur, Kerala, India. In: *Biodiversity, Conservation and Utilization of Spices, Medicinal and Aromatic Plants.* Sasikumar, B., Krishnamoorthy, B., Rema, J., Ravindran, P.N. and Peter, K.V. (Eds.). Indian Institute of Spices Research, Calicut, Kerala, India, pp. 131–133.

Ahmedullah, M. and Nayar, M.P. (1999). *Red Data Book of Indian Plants.* Vol. 4. Botanical Survey of India.

Altieri, M.A. (1990). Why study traditional agriculture? In: *Agroecology.* Carroll, C.R. Vandermeer, J.H. and Rosset, P. (Eds.). 1992. McGraw-Hill Publishing Company, New York.

Anonymous, (2010). What it will take to feed the world. *Nature,* 464: 969, doi: 10. 1038/464969a.

Anonymous, (1988). *Hill Crop Monitoring Tour* (Sept. 28–Oct. 12, 1988) *Report.* NHCRP-NARC, ARPP / USAID and IDRC, Canada, Nepal.

Arora, R.K. (2000). Wild relatives of cultivated plants. In: *Flora of India. Introductory Volume.* Part 2, Botanical Survey of India, Calcutta, India, pp. 218–234.

Arora, R.K. and Nayar, E.R. (1984). Wild relatives of crop plants in India. *NBPGR Monograph* No. 7, National Bureau of Plant Genetic Resources, New Delhi, India, p. 90.

Arora, R.K. (2000). Wild relatives of cultivated plants. In: *Flora of India, Introductory.* Vol. 2, Botanical Survey of India, Calcutta, West Bengal, India, pp. 218–234.

Arora, R.K. Roshini, Nayar, E. and Pandey, Anjula, (2006). Indian Centre of Floristic and Economic Plant Diversity: A Review. In: *Hundred Years of Plant Genetic Resources Management in India.* Singh, K., Anurudh, Srinivasan, K., Saxena, S. and Dhillon, B.S. (Eds.). NBPGR (ICAR), Pusa Campus, New Delhi, India.

Arora, R.K. (1991). Plant diversity in the Indian gene centre. In: *Plant Genetic Resources Conservation and Management*. Paroda, R.S. and Arora, R.K. (Eds.). International Board for Plant Genetic Resources, New Delhi, India, pp. 25–54.

________. (1993). Biodiversity in crop plants and their wild relatives. *Proceedings of the First National Agricultural Science Congress, National Academy of Agricultural Sciences,* New Delhi, India, pp. 160–180.

________. (1996). Indian region provides treasure house of wild plant genetic resources. *Diversity,* 12(3): 22–23.

Arora, R.K. and Nayar, E.R. (1984). Wild relatives of crop plants in India. *Scientific Monograph,* 7. National Bureau of Plant Genetic Resources, New Delhi, India.

________. (1999). Wild relatives of crop plants: Diversity, distribution and conservation in the Indian gene centre. In: *Biodiversity, Taxonomy and Ecology*. Tandon, R.K. and Singh, P. (Eds.). Scientific Publishers (India), Jodhpur, India.

Babu, C.R. (1977). *Herbaceous Flora of Dehradun*. Publication and Information Department, Council of Scientific and Industrial Research, New Delhi, India, p. 721.

Baker, J.G. (1886). Scitamineae. In: *Flora of British India*. Hooker, J.D. (Ed.). 6: 198–264.

Baniya, B.K. (1994). *Buckwheat Genetic Resources in Nepal: A Status Report* (submitted to IPGRI-APO, Singapore) NARC, Nepal, p. 51.

Badr, A.K. Muller, Schafer-Pregl R., Rabey H. El, Effgen S, Ibrahim, H. H., Pozzi, C., Rohde, W., Salamini, W. (2000). On the origin and domestication history of barley (*Hordeum vulgare*). *Molecular Biology and Evolution*, 17: 499–510.

Bhattacharya, S.C. and Dutta, S. (1956). Classification of *Citrus* fruits of Assam. *Sc. Monogr.,* 20. p. 110. ICAR, New Delhi.

Bisht, I.S., Rao, K.S., Bhandari, D.C., Nautiyal Sunil, Maikhuri, R.K. and Dhillon, B.S. (2006). A suitable site for in-situ (on-farm) management of plant diversity in traditional agro-ecosystems of western Himalaya in Uttaranchal state: A case study. *Genetic Resources and Crop Evolution* (2006), 53: 1333–1350.

Brar, D. (2005). Broadening the gene pool and exploiting heterosis in cultivated rice, In: *Rice is Life: Scientific Perspectives for the 21st Century.* Toriyama, K., Heong, K.L., Hardy, B. (Eds.). *Proceedings of the World Rice Research Conference, Tokyo and Tsukuba, Japan*, 4–7 November, 2004.

Brar, D. and Kush, G. (1997). Alien introgression in rice. *Plant Mol. Biol.*, 35: 35–47

Brown, A.H.D., Marshal, D.R.l., Frankel, O.H. and Williams, J.T. (1989). *The use of Plant Genetic Resources.* Cambridge University Press, Cambridge.

Brush, S.B. (1999). *Genes in the Field: On-farm Conservation of Crop Diversity.* Lewis Publishers, Boca Raton, Florida, USA.

Brush, S.B. and Meng, E. (1998). Farmers' valuation and conservation of crop genetic resources. *Genet. Resour. Crop Evol.*, 45: 139–150

CBD (1992). *Convention on Biological Diversity: Text and Annexes.* Secretariat of the Convention on Biological Diversity, Montreal, Canada. Available at: http://www. biodiv. org/convention.

CBD (2002). *Biodiversity Target (*2010). Secretariat of the Convention on Biological Diversity, Montreal. Available at: http://www. biodiv. org/2010-target/default. aspx (accessed 27 March 2007).

CBD (2002). *Global Strategy for Plant Conservation.* Secretariat of the Convention on Biological Diversity, Montreal, Canada. Available at: http://www. biodiv. org/decisions.

Chakravarty, H.L. (1982). *Fascicles of Flora of India, Fascicle II, Cucurbitaceae.* Botanical Survey of India, Howrah, India.

Chatterjee, S. (1995). Global hotspots of biodiversity. *Current Science,* 68(12): 1178–1180.

Chowdhery, H.J. and Wadhwa, B.M. (1984). *Flora of Himachal Pradesh, Vol. 3.* Botanical Survey of India, Deptt. of Environment. New Delhi.

de Candole, A. (1986). *Origin of Cultivated Plants.* (Reprint of 2nd edition) Hafner Publ. Co. New York and London, p. 468.

de Vicente, M.C. and Anderson, M.S. (Eds.). (2006). DNA banks—providing novel options for genebanks? *Topical Reviews in Agricultural Biodiversity.* International Plant Genetic Resources Institute, Rome, Italy.

Debouck, D.G. (1991). Genetic variation in crop species and their wild relatives: A viewpoint for their conservation, In: *Genetic Diversity and Crop Strategies for Roots and Tubers.* Becker, B. (Ed.). Bonn, Germany, Arbeitsgemeinschaft Tropische und Subtropische Agrarforschung e. V. and International Board for Plant Genetic Resources. pp. 41–51.

Dhawan, N.L. (1964). Primitive Maize in Sikkim. *Maize Genetics Co-op Newsletter,* 38: 69–70.

Dhillon, B.S., Sharma, A.K. Kumar, Dinesh, Malhi, N.S. and Singh N.N. (2006). Maize. In: *Plant Genetic Resources: Foodgrain Crops.* Dhillon, B.S., Saxena, S., Agrawal, A., Tyagi, R.K. (Eds.). Narosa Publishing House, New Delhi, pp. 90–136.

Dhillon, B.S., Rana, J.C. and Singh, A.K. (2005). Collection, Conservation and Utilization of Indigenous Horticultural Crops Germplasm. *Proc. First Indian Horticulture Congress.* Chaddha, K.L. and Singh, S.P., (Eds.). IARI, New Delhi, India, pp. 11–35.

Dulloo, Ehsan, Nagamura Yoshiaki and Ryder Oliver. (2003). DNA storage as a complementary conservation strategy. In: *DNA Banks—Providing Novel Options for Genebanks? Topical Reviews in Agricultural Biodiversity.* de Vicente, M.C. and Andersson, M.S. (Eds.). International Plant Genetic Resources Institute, Rome, Italy, pp. 11–24

FAO, (2009). The state of food insecurity in the world: *Economic Crisis – Impacts and Lesson Learned.*

Farooq, S., Tahir, S. (1987). Comparative study of some growth attributes in buckwheat. *Fagopyrum,* 7, 9–12

Ford-Lloyd, B.V., Kell, S.P. and Maxted, N. (2008). Establishing conservation priorities for crop wild relatives. In: *Crop Wild Relative Conservation and Use.* Maxted, N., Ford-Lloyd, B.V., Kell, S.P., Iriondo, J., Dulloo, E. and Turok, J. (Eds.). CABI Publishing, Wallingford, pp. 110–119.

Frankel, O.H. and Bennett, E. (1970). *Genetic Resources in Plants – Their Exploitation and Conservation.* Blackwell Scientific Publications, Oxford, UK.

Frankel, O.H., Hawkes, J.G. (1975). *Crop Genetic Resources for Today and Tomorrow.* Cambridge University Press, Cambridge.

Gadgil, M., Singh, S.N., Nagendra, H.C. and Subhash, M.D. (1996). In: *Situ Conservation of Wild Relatives of Cultivated Plants: Guiding Principles and a Case Study.* Food and Agricultural Organization of the United Nations and Indian Institute of Science. Bangalore, India, p. 21.

Gadgil, M., Olsson, P., Berkes, F. and Folke, C. (2003). Exploring the role of local ecological knowledge for ecosystem management: Three case studies. pp. 189–209. In: *Navigating Social–Ecological Systems: Building Resilience for Complexity and Change.* Berkes, F. Colding, J. and Folke, C. (Eds.). Cambridge University Press, Cambridge, UK.

Ganeshaiah, K.N. (2003). Sasya Sahyadri CD ROM 2003. *Distribution, Taxonomy and Diversity of Plants of Western Ghats.* UAS Dharwad, Bangalore, India.

Guarino, L., Ramanatha Rao, V. and Reid, R. (1995). *Collecting and Plant Genetic Diversity: Technical Guidelines.* CAB International, Wallingford, UK.

Hajjar, R. and Hodgkin, Toby. (2007). The use of wild relatives in crop improvement: A survey of developments over the last 20 years. *Euphytica,* 156, (1–2): 1–13.

Hajra, P.K. and Mudgal, V. (Eds.). (1997). *Plant Diversity Hotspots in India. An Overview,* Botanical Survey of India, Calcutta, India, p. 197.

Hamilton, S.H., Engles, J.M.M. and Hintum, T. van. (2003). Considerations for improved conservation and utilization concepts and strategies. pp. 43–59 In: *A Guide to Effective Management of Germplasm Collections.* Engles, J.M.M. and Visser, L. (Eds.). IPGRI Handbooks for Genebanks No. 6. Rome, Italy: International Plant Genetic Resources Institute.

Hawkes, J.G., Maxted, N. and Ford-Lloyd, B.V. (Eds.). (1992). *The Ex-situ Conservation of Plant Genetic Resources.* Kluwer Acad. Pub. Dordrecht, pp. 32–39.

Heywood, V.H., Kell, S.P. and Maxted, N. (2008). Towards a global strategy for the conservation and use of crop wild relatives. In: *Crop Wild Relative Conservation and Use.* Maxted, N., Ford-Lloyd, B.V., Kell, S.P., Iriondo, J., Dulloo, E. and Turok, J. (Eds.). CABI Publishing, Wallingford, pp. 657–666. http://cwris. ecpgr. org.

Hoyt, E. (1988). *Conserving the Wild Relatives of Crops.* IPGRI/IUCN/WWF, Rome, Italy.

IUCN, (2006). *The IUCN Species Survival Commission 2006 Red List of Threatened Species,* Gland, Switzerland. Available at: http://www. redlist. org.

Iwanaga, M. (1995). IPGRI strategy for in-situ conservation of agricultural biodiversity. In: *In-situ Conservation and Sustainable use of Plant Genetic Resources for Food and Agriculture in Developing Countries.* Engels, J.M.M. (Ed.). IPGRI/DSE, Rome, pp. 13–26.

Jarvis, D.I. and Hodgkin, T. (1999). Wild relatives and crop cultivars: Detecting natural introgression and farmer selection of new genetic combinations in agroecosystems. *Molecular Ecology,* 8, S159–S173.

Jarvis, A., Upadhyaya Hari, Gowda, C.L.L., Aggerwal, P.K. and Fujisaka Sam. (2008a). Climate change and its effect on conservation and use of plant genetic resources for food and agriculture and associated biodiversity for food security. *Report to ICRISAT/FAO* November, 2008.

Joshi, B.D. and Rana, J.C. (1994). Collecting temperate fruits and their wild relatives in North-West Himalayas. *IBPGR News Letter*, p. 15.

Katwal, R.P.S., Srivastva, R.K., Kumar, S. and Jeeva, V. (2003). State of forest genetic resources conservation and management in India. Indian Council of Forestry Research and Education, *Forest Genetic Resources Working Papers,* FAO, Rome, p. 36.

Kell, S.P., Knüpffer, H., Jury, S.L., Maxted, N. and Ford-Lloyd, B.V. (2005). *Catalogue of Crop Wild Relatives for Europe and the Mediterranean.* University of Birmingham, Birmingham, UK.

Kell, S.P., Knüpffer, H., Jury, S.L., Ford-Lloyd, B.V. and Maxted, N. (2008). Crop wild relatives of the Euro-Mediterranean region: Making and using a conservation catalogue. *In: Crop Wild Relative Conservation and Use.* Maxted, N., Ford-Lloyd, B.V., Kell, S.P., Iriondo, J., Dulloo, E. and Turok, J. (Eds.). CABI Publishing, Wallingford, pp. 69–109.

Kell, S.P., Moore, J.D., Iriondo, J.M., Scholten, M.A. et al. (2008a), CWRIS: An information management system to aid crop wild relative conservation and sustainable use. *Crop Wild Relative Conservation and Use.* Maxted, N. et al. (Ed.). CABI Publishing, Wallingford, pp. 471–491.

Khoshoo, T.N. (1995). Census of India's biodiversity: Tasks ahead. *Curr. Sci.*, 69: 14–17.

Khush, G.S., Ling, K.C., Aquino, R.C. and Aguiero, V.M. (1977). Breeding for resistance to grassy stunt in rice. *Proceedings of the 3rd International Congress of the Society for the Advancement of Breeding Researchers in Asia and Oceania* (SABRAO). Plant Breeding Paper, 1(4b): 3–9.

Kostermans, A.J.G.H. (1983). The South Indian species of *Cinnamomum* Schaeffer (Lauraceae). *Bull. Bot. Survey.* India, 25: 90–133.

Ladizinsky, G. (1985). Founder effect in crop-plant evolution. *Econ. Bot.*, 39: 191–199.

Li, Z. and Zhu, Y. (1988). Rice male sterile cytoplasm and fertility restoration, pp. 85–102. In: *Hybrid Rice.* IRRI, Manila, Philippines.

Maikhuri, R.K., Rao, K.S. and Saxena, K.G. (1996). Traditional crop diversity for sustainable development of Central Himalayan agroecosystems. *Int. J. Sust. World Ecol.*, 3: 8–31.

Maxted, N., Ford-Lloyd, B.V. and Hawkes, J.G. (1997a). Complementary conservation strategies. In: *Plant Genetic Conservation: The In-situ Approach*. Maxted, N., Ford-Lloyd, B.V. and Hawkes, J.G. (Eds.). Chapman and Hall, London, pp. 15–40.

Maxted, N., Hawkes, J.G., Ford-Lloyd, B.V. and Williams, J.T. (1997b). A practical model for in-situ genetic conservation. In: *Plant Genetic Conservation: The In-situ approach.* Maxted, N., Ford-Lloyd, B.V. and Hawkes, J.G. (Eds.). Chapman and Hall, London, pp. 339–367.

Maxted, N. and Kell, S. (2008). Establishment of a network for the in-situ conservation of crop wild relatives: Status and needs. *Commission on Genetic Resources for Food and Agriculture.* Food and Agriculture Organization of the United Nations.

Maxted, N., Ford-Lloyd, B.V. and Kell, S.P. (2008). Crop wild relatives: Establishing the context. In: *Crop Wild Relative Conservation and Use.* Maxted, N., Ford-Lloyd, B.V., Kell, S.P., Iriondo, J., Dulloo, E. and Turok, J. (Eds.). CABI Publishing, Wallingford, pp. 3–30.

Maxted, N., Ford-Lloyd, B.V., Jury, S.L., Kell, S.P. and Scholten, M.A. (2006). Towards a definition of a crop wild relative. *Biodiversity and Conservation,* 15(8), 2673–2685.

Maxted, N., Scholten, M.A., Codd, R. and Ford-Lloyd, B.V. (2007). Creation and use of a national inventory of crop wild relatives. *Biological Conservation*, 140: 142–159.

MoEF, (2008). *Annual Report*, Government of India, Ministry of Environment & Forests Paryavaran Bhavan, CGO Complex, Lodhi Road, New Delhi – 110 003.

Moore Jonathan Shelagh, Kell, D.P., Jose, M. Iriondo, Brian, V. Ford-Lloyd and Nigel Maxted. (2008). CWRML: Representing crop wild relative conservation and use data in XML. *BMC Bioinformatics*, 9:116 doi: 10.1186/1471-2105-9-116.

MSSRF, (2010). Integrated management of biodiversity resources in partnership with communities. MSSRF, Chennai.

Mukherjee, S.K. (1985). Systematic and ecogeographic studies on crop genepools 1. *Mangifera* L. IBPGR, Rome.

Myers, N., Mittermeier, R.A., Mittermeier, C.G. da Fonseca, G.A.B. and Kent, J. (2000). Biodiversity hotspots for conservation priorities. *Nature,* 403: 853–858.

National Potato Council (2003). Potato statistics: Seed certification, approved acreage, varieties planted and production.

Nayar, M.P. (1977). Changing pattern of Indian flora. *Bull. Bot. Surv.* India, 19: 145–154.

________. (1980). Endemism and patterns of distribution of endemic genera (Angiosperms). *J. Econ. Tax. Bot.,* 1: 99–110.

________. (1997). 'Hotspots" of plant diversity in India – strategies. In: *Conservation and Economic Evaluation of Biodiversity.* Pushpangadam, P., Ravi, K. and Santosh, V. (Eds.). Oxford and IBH Pulb. House, New Delhi, India, p. 316.

________. (2008). Hotspots and warmspots of biodiversity. In: *Biodiversity Conservation.* Binojkr, M.S. and Gopalakrishnanam P.K. (Eds.). Scientific Publishers, Jodhpur, India.

Nayar, M.P. and Sastry, A.R.K. (1987–1990). *Red Data Book of Indian Plants.* 3 Vols, Botanical Survey of India, Calcutta, West Bengal, India. Vol. 1. p. 367. Vol. 2. p. 268. Vol. 3. p. 290.

Nguyen, B., Brar, D., Bui, B., Nguyen, T., Pham, L., Nguyen, H. (2003). Identification and mapping of the QTL for aluminum tolerance introgressed from the new source, *Oryza rufipogon* Griff, into indica rice (*Oryza sativa* L.). *Theor. Appl. Genet.,* 106: 583–593.

Ohnishi, O. (1989). Cultivated buckwheat and their wild relatives in the Himalayas and Southern China. *Proceedings of 4th International Symposium on Buckwheat,* 562–571.

________. (1991). Discovery of wild ancestor of common buckwheat. *Fagopyrum,* 11: 5–10.

________. (1992). Buckwheat in Bhutan. *Fagopyrum,* 12: 5–13.

Oldfield, M.L. and Alcorn, J.B. (1987). Conservation of traditional agro-ecosystems. *Bioscience,* 37: 199–209.

Onishi, O. (1995). Discovery of new *Fagopyrum* species and its implication for the studies of evolution of *Fagopyrum* and of the origin of cultivated buckwheat. *Proceedings of 6th International Symposium on Buckwheat,* 26–29.

Padulosi, S., Mal, Bhag, Ravi, S Bala, Gowda, J., Gowda, K.T.K., Shanthakumar , G., Yenagi, N., Dutta, M. (2009). Food security and climate change: Role of plant genetic resources of minor millets. *Indian Journal of Plant Genetic Resources,* Vol. 22(1): 1–17.

Pal, B.P. (1968). *Wheat.* Indian Council of Agricultural Research, New Delhi, India.

Pandey, A. and Arora, R.K. (2004). Potential for domestication and use of wild species in the Indian gene centre. In: *Plant Genetic Resources Management*. Dhillon, B.S., Tyagi, R.K., Lal, A. and Saxena, S. (Eds.). Narosa Publishing House, New Delhi, India.

Pandey, A., Bhandari, D.C., Bhatt, K.C., Pareek, S.K., Tomer, A.K. and Dhillon, B.S. (2005). *Wild Relatives of Crop Plants in India: Collection and Conservation*. National Bureau of Plant Genetic Resources, New Delhi, India.

Pandey, A., Tomer, A.K., Bhandari, D.C. and Pareek, S.K. (2008). Towards collection of wild relatives of crop plants in India. *Genetic Resources and Crop Evolution*, 55(2): 187–202.

Parthasarathy, N. (1972). Rice breeding in tropical Asia up to 1960 In: *Rice Breeding.* International Rice Research Institute, Los. Baños, Philippines, pp. 5–29.

Plucknett, D.L., Smith, N.J.H., Williams, J.T. and Anisships, N.M. (1987). A case study in rice germplasm: IR36, pp. 171–185. In: *Genebanks and The World's Food*. Plucknett, D.L., Smith, N.J.H, Williams, J.T. and Anishetty, N.M., (Eds.). Princeton University Press, Princeton, NJ.

Prescott-Allen, C., Prescott-Allen, R. (1986). *The First Resource: Wild Species in the North American Economy*. Yale University, New Haven.

Pullaiah, T., Ramakrishniah, V., Sandhya, Rana, S. and Rao, P.N. (2000). Flora of Guntur Dt. AP, India. Regency Publication, 32–36.

Rana, J.C. (2004). Buckwheat Genetic Resources Management in India. *Proceedings of 9th International Symposium on Buckwheat,* 271–282.

Rana, J.C. and Sharma, S.K. (2009). Plant genetic resource management under emerging climate change. *Indian J. Genet.*, 69(4): 1–17.

Rana, J.C., Yadav, S.K., Mandal, S. and Yadav, Sangeeta, (2005). Genetic divergence and interrelationship analysis in grain amaranth germplasm. *Indian J. Genet.,* 65(2): 99–102.

Rana, J.C., Sharma, B.D., Gautam, P.L. (2001). Agri-diversity erosion in the north-west Indian Himalayas–some case studies. *Indian Journal of Genetics and Plant Breeding,* 13: 252–258.

Rana, J.C., Singh, Archna, Sharma, Yogender, Pradheep, K. and Mendiratta, Nisha. (2008). Dynamics of plant bioresources in western Himalayan region of India-watershed based case study. *Current Sci.*, 98(2): 192–203.

Rana, R.S. and Chatterjee, S. (2000). Prioritization of Wild Relatives and Crop Plants of India and Domesticated Animals. In: *Setting Biodiversity Conservation Priorities for India: Summary of the Findings and Conclusions of the Biodiversity Conservation Prioritisation Project.* Singh, S., Sastry, A.R.K., Mehta, R. and Uppal, V. (Eds.). WWF-India, New Delhi, India.

Rao, K.S. and Saxena, K.G. (1994). *Sustainable Development and Rehabilitation of Degraded Village Lands in Himalaya.* Himavikas Publication. No. 8. Bishen Singh, Mahendra Pal Singh. Dehradun, pp. xiv+287.

Rao, R.R. (1997). Diversity of Indian flora. *Proc. Indian Natn. Sci. Acad.,* B63 3: 127–138.

________. (1994). *Biodiversity in India.* Bishen Singh, Mahendra Pal Singh. Dehradun, India.

Rathore, D.S. (2001). Networking for genetic resources management of horticultural crops. *Paper Presented at Symposium on Plant Genetic Resources Management Advances and Challenges,* National Bureau of Plant Genetic Resources, New Delhi.

Rick, C.R. and Chetelat, (1995). Utilization of related wild species for tomato improvement. First international symposium on Solanacea for fresh market. *Acta. Hort.,* 412: 21–38.

Roy, B., Hardev, R.C. and Pal, D.C. (1998). *Plants for Human Consumption in India.* Botanical Survey of India, Calcutta, West Bengal, India, p. 188.

Sabu, M. (1991). A taxonomic and phylogenetic study of South Indian Zingiberaceae. Ph.D Thesis, University Calicut, Kerala, India.

Sakai, S. and Nagamasu, H. (2000). Systematic studies of Bornean Zingiberaceae: II. *Elettaria* of Sarawak. *Edinb. J. Bot.,* 57: 227–243.

Saxena, S., Singh, Anurudh, K., Srinivasan, K. and Dhillon, B.S. (2006). Plant genetic resources management: Future perspective. In: *Hundred Years of Plant Genetic Resources Management in India.* Singh, A.K., Srinivasan, K., Saxena, S. and Dhillon, B.S. (Eds.). NBPGR, New Delhi, pp. 227–243.

Scholten, M., Maxted, N., Kell, S.P. and Ford-Lloyd, B.V. (2008). Creation of a national crop wild relative strategy: A case study for the United Kingdom. In: *Crop Wild Relative Conservation and Use.* Maxted, N., Ford-Lloyd, B.V., Kell, S.P., Iriondo, J., Dulloo, E. and Turok, J. (Eds.). CABI Publishing, Wallingford, pp. 120–142.

Sharma, B.D. and Rana, J.C. (2000). Survey, collection and scope of floriculture germplasm. In: *Horticulture Technology.* Sharma, V.K. and Azad, K.C. (Eds.). Vol. 1. (Production). Deep and Deep Publication New Delhi, pp. 381–398.

Sharma, S.K., Joshi, B.K., Lakra, W.S. and Arora, D.K. (2009). Biodiversity. In: *State of Indian Agriculture:* A NAAS Publication. Rai, Mangala, Acharya, S.S., Virmani, S.M., Aggarwal, P.K. (Eds.).

Sharma, B.D. and Chandel, K.P.S. (1996). Occurrence, distribution and diversity of soft fruits in N-W and W-Himalayas and prospects of their conservation and utilization. *Indian J. Pl. Genet. Resources,* 9(2): 237–246.

Siddiq, E.A., Saxena, S. and Malik, S.S. (2005). Rice. In: *Plant Genetic Resources: Food Grain Crops.* Dhillon, B.S., Saxena, S., Tyagi, R.K. and Agarwal, A. (Eds.). Narosa Publishing House, New Delhi, 27–57.

Simmonds, N.W. (1976). *Evolution of Crop Plants.* Longman, London, New York.

Singh, A.K., Srinivasan, K., Jain, S.K. and Saxena, S. (2001). Germplasm conservation: Seed genebank. In: *National Bureau of Plant Genetic Resources: A Compendium of Achievement.* Dhillon, B.S., Varaprasad, K.S., Srinivasan, K., Singh, M., Archak, S., Srivastava, U. and Sharma, G.D. (Eds.). NBPGR, New Delhi, pp. 183–208.

Singh, S.N. and Gadgil, M. (1996). Some suggestions for *in-situ* conservation of wild rice. *International Rice Research Notes,* 21(2–3), 6–7.

Singh, I.D. (1999). Plant improvement. In: *Global Advances in Tea Science.* Jain, N.K. (Ed.). Aravalli Books International Pvt. Ltd., New Delhi, India, pp. 427–448.

Singh, N.P. (1993). Clusiaceae. In: *Flora of India.* Sharma, B.D. and Sanjappa, M. (Eds.). Portulacaceae–Ixonanthaceae, Vol. 3. Botanical Survey of India, Calcutta. pp. 86–151.

Singh, N.P., Singh, D.K., Hajra, P.K. and Sharma, B.D. (Eds.). (2000). *Flora of India Introductory Volume (Part II).* Botanical Survey of India, Calcutta.

Singh, B. (1977). *Races of Maize in India.* Indian Council of Agricultural Research, New Delhi, India.

SPGRFAI, (2007). State of plant genetic resources for food and agriculture in India (1996–2006): *A Country Report.* National Bureau of Plant Genetic Resources (Indian Council of Agricultural Research), New Delhi, p. 70.

Srivastava, S.K. (1994). *Garcinia dhanikhariensis* (Clusiaceae), a new species from the Andaman Islands, India. *Nordic J. Bot.*, 14: 51–53.

Sthapit, B.R., Gauchan, D., Subedi, A. and Jarvis, D. (2008). On-farm management of agricultural biodiversity in Nepal: Lessons learnt. *Proceedings of National Symposium*, 18–19 July 2006. Kathmandu, Nepal.

Swaminathan, M.S. (2010). Science and sustainable food security: Selected papers of (IISC Centenary Lecture Series). *World Scientific*, p. 436.

________. (2006). An evergreen revolution. *Crop Sci.*, 46: 2293–2303.

Tanksley, S.D., McCouch, S.R. (1997). Seed banks and molecular maps: Unlocking genetic potential from the wild. *Science,* 277: 1063–1066.

Thuiller, W., Lavorel, S., Araujo, M.N., Sykes, M.T. and Prentice, I.C. (2005). Climate change threats to plant diversity in Europe. *Proceeding of the National Academy of Sciences of the United States of America,* 102(23): 8245–8250.

Vavilov, N.I. (1926). Studies on the origin of cultivated plants. *Bull. Appl. Bot.*, 10: 262–268 (Origin and Geography of Cultivated Plants. English ed., Translated from Russian by Love, D. 1992. Cambridge University Press, Cambridge, UK, p. 498).

Vavilov, N.I. (1951). The origin, variation, immunity and breeding of cultivated crops. *Chronica Bot.*, 13: 364.

Velayudhan, K.C., Asha, K.I., Mithal, S.K. and Gautam, P.L. (1999). Genetic resource of turmeric and its relatives in India. In: *Biodiversity, Conservation and Utilization of Spices, Medicinal and Aromatic Plants.* Sasikumar, B., Krishnamoorty, B., Rema, J., Ravindran, P.N. and Peter, K.V. (Eds.). Indian Institute of Spices Research, Kerala, India, pp. 101–109.

Wang, G., Mackill, D.J., Bonman, J.M., McCouch, S.R. and Champoux, M.C. (1994). RFLP mapping of genes conferring complete and partial resistance to blast in a durably resistant rice cultivar. *Genetics,* 136: 1421–1434.

WSFS, (2009). Declaration of the World Summit on Food Security. Rome, 16–18 November, 2009.

Yadav, I.S. and Rajan, S. (1993). Genetic resources of *Mangifera.* In: *Advances in Horticulture.* Chadha, K.L. and Pareek, O.P. (Eds.). Vol. 1 Part 1. Malhotra Publishing House, New Delhi, pp. 77–93.

Zeven, A.C. and de Wet, J.M.J. (1982). *Dictionary of Cultivated Plants and their Regions of Diversity.* Centre of Agricultural Publishing and Documentation, Wageningen, the Netherlands, p. 219.

Zuhukovsky, P.M. (1968). New centres of origin and new gene centres of cultivated plants including specifically endemic microcenters of species closely allied to cultivated species. *Bot. Zh.,* 53: 430–460.

Index